Iterative Detection

*Adaptivity, Complexity Reduction,
and Applications*

THE KLUWER INTERNATIONAL SERIES
IN ENGINEERING AND COMPUTER SCIENCE

ITERATIVE DETECTION
Adaptivity, Complexity Reduction, and Applications

KEITH M. CHUGG
University of Southern California
TrellisWare Technologies, Inc.

ACHILLEAS ANASTASOPOULOS
University of Michigan

XIAOPENG CHEN
Marvell Semiconductor, Inc.

Kluwer Academic Publishers
Boston/Dordrecht/London

Distributors for North, Central and South America:
Kluwer Academic Publishers
101 Philip Drive
Assinippi Park
Norwell, Massachusetts 02061 USA
Telephone (781) 871-6600
Fax (781) 681-9045
E-Mail <kluwer@wkap.com>

Distributors for all other countries:
Kluwer Academic Publishers Group
Distribution Centre
Post Office Box 322
3300 AH Dordrecht, THE NETHERLANDS
Telephone 31 78 6392 392
Fax 31 78 6546 474
E-Mail <services@wkap.nl>

 Electronic Services <http://www.wkap.nl>

Library of Congress Cataloging-in-Publication Data

A C.I.P. Catalogue record for this book is available
from the Library of Congress.

Copyright © 2001 by Kluwer Academic Publishers

Printed on acid-free paper.

Printed in the United States of America.

To our parents

Contents

Preface

Along with many other researchers and practicing communication engineers, we were excited to learn of the existence of turbo codes in the mid 1990's. Initially, however, we were not working in the area of coding and found it difficult to educate ourselves on turbo codes. We came to understand and utilize iterative detection methods in an effort to solve other data detection problems. In particular, we were working in two distinct areas where we eventually found the methods of iterative detection to be extremely useful. One was two-dimensional (2D) data detection with applications to page access storage (the focus of the third author). The other was data detection for trellis coded modulation (TCM) over interleaved, intersymbol interference (ISI) channels (the focus of the second author). With a vague understanding of the turbo decoding algorithm, we basically rediscovered the principles of iterative detection in the context of these applications. As a result of this (sometimes tiring) experience, we developed a different perspective on iterative detection than those involved solely with turbo decoding and a good appreciation for the pitfalls of executing the algorithm. This latter appreciation was enhanced by the first author's opportunity to teach iterative detection as part of *EE666 Data Communications* at the University of Southern California (USC) during the Spring 1999 term.

For the 2D data detection problem, we were able to decompose one detection problem (a 2D problem) into coupled subproblems (two 1D problems). Based on this, we began to think of iterative detection as a potential tool for performing reduced complexity, near-optimal detection. That is, just as one can consider a turbo decoder as a good approximation to the optimal decoder for a code constructed as a concatenated network of constituent codes, we began to think of decomposing a system into subsystems and applying iterative detection. The page-access memory application also provided us with the impetus to develop algorithms with parallel, or locally-connected, architectures. After developing such algorithms, we found the 2D application an intuitive example to understand the view of iterative detection based on message-passing and graphical models. We began to discover that the way that one formed a system decomposition, or formulated a graphical model with cycles, was not unique and we sought to select such models to minimize the computational complexity while maintaining near-optimal performance.

The TCM-ISI application, or "joint equalization/decoding," provided the impetus to investigate methods of parameter estimation in concert with iterative detection. We had been working extensively on adaptive/blind data detection for isolated systems with memory – *e.g.*, hard-decision adaptive sequence detection algorithms for ISI channels. While trying to identify the proper soft-decision information for systems with parametric uncertainty, we developed what we call adaptive iterative detection (AID). AID is a powerful method for joint parameter and data estimation for the same reason that iterative detection is so useful: it allows one to exploit global system structure by locally exploiting subsystem structure and exchanging marginal beliefs on digital data symbols. Thus, in a TCM-Interleaver-ISI system, one can estimate and track the ISI channel using estimators that exploit the TCM structure. Intuitively, it is possible to use a decision-directed channel estimator that exploits the coding gain.

This book is a summary of the work done from 1997 to 1999 while we were all at USC. Our goals are to provide the following (hence the title):

- An introduction to the principles of iterative detection

- The notion of iterative detection as a tool for complexity reduction

- The theory and practice of adaptive iterative detection

- Detailed examples of the above through a variety of applications

This book is intended for readers who have a good background in digital detection and estimation theory (*e.g.*, [Va68]) and would like to apply iterative detection or conduct research in the area. Our goal is to present a general view of iterative detection that is built upon the fundamentals of Bayesian decision theory. If a reader is solely interested in turbo codes and their decoding, other references may be more suitable (*e.g.*, [BeMo96, HeWi98, VuYu00]). Similarly, a reader without the background in detection and estimation may find it simplest to learn the (equivalent) message passing on graphical models directly (*e.g.*, [Pe88, Je96, Fr98]). Even the reader with the proper background will benefit greatly by keeping a pad and pencil (and computer and compiler!) nearby so the numerous examples and numerical results can be verified. Each chapter concludes with a list of exercises and open issues. This book could be used as a text for a focused class in the second year of graduate study (*e.g.*, EE666 at USC).

The first bullet above is addressed in Chapters 1 and 2. This includes a general view of iterative detection based only on detection theory and block diagrams. This "block diagram" view relies heavily on the work of Benedetto, et. al. [BeDiMoPo98]. Despite the appearance that [BeDiMoPo98] has been relatively unappreciated, we have found it to be the most accessible reference for those getting started, who do not have an extensive background in coding or graph theory. Part of this overview is a detailed development of the forward-backward algorithm and its variants. Chapter 2 also includes a summary of belief (probability) propagation and graphical models. This development was heavily influenced by Wiberg's thesis [Wi96] and the well-known papers by McEliece, et. al. [McMaCh98] and Kschischang and Frey [KsFr98]. Excellent talks by Guido Montorsi and David Forney at the 1997 IEEE Communication Theory Workshop [Mo97, Fo97] provided our early exposure to [BeDiMoPo98] and [Wi96], repectively, and great motivation for the work presented herein. In Chapters 1 and 2, we include many detailed examples to familiarize the reader not only with our notation and terminology, but with the details of the computational methods. Specific examples include parallel concatenated convolutional codes (PCCCs), serially-concatenated codes (SCCCs), and the TCM-ISI application.

The use of iterative detection for complexity reduction is covered in Chapter 3. This includes the development of a reduced-state forward-backward algorithm and the concept of "self-iteration." Basically, this allows one to trade additional iterations for fewer states in the trellis. We demonstrate that such an approach can be useful for performing data detection for ISI channels with large delay spread – *i.e.*, for an isolated system without any concatenation. Specifically, we show that this method can be less complex and more robust that traditional reduced-state sequence detection algorithms. We also consider long but sparse ISI channels. It is demonstrated that one can decompose such a system into a conceptual model comprising the parallel concatenation of several ISI channels, each with small delay spread. Applying iterative detection to this conceptual model yields near-optimal performance with substantial complexity reduction. We also develop several useful generic tools for complexity reduction in iterative detection including filtering of soft-information to slow convergence and soft-information thresholding (*i.e.*, decision feedback).

Adaptive iterative detection is covered in Chapter 4. This is essentially an extension of the material in Chapters 1 and 2 to the case where some parameters (typically associated with the channel) are not known at the receiver. A family of algorithms that are the conceptual equivalent of the forward-backward algorithm is developed. From this framework, we motivate a family of adaptive forward-backward algorithms that can be viewed as the marriage of the forward-backward algorithm and forward-only adaptive hard decision data detection algorithms (*e.g.*, Per-Survivor Processing). We demonstrate these techniques by applying AID to phase or fading channel tracking for PCCCs and SCCCs and to channel tracking for TCM over interleaved, frequency-selective fading channels.

Two-dimensional applications are considered in Chapter 5. The two applications considered in detail are the mitigation of 2D ISI and half-toning of gray-scale images. The latter application is an example of how the principles of iterative detection can be applied to encoding or data fitting problems as well as data detection. For the 2D ISI application, we also consider the optimal detector. Specifically, we develop upper and lower bounds on the associated performance. This provides an opportunity to compare the performance of the optimal (impractical) detector with that of the practical iterative detector which illustrates that the iterative algorithm performs near optimally. In both cases we demonstrate how the complexity reduction tools can be applied. We also illustrate the nonuniqueness of the graphical formulation or system decomposition and show that the resulting algorithms may have significantly different characteristics.

In Chapter 6, which is co-authored by Prof. Peter Beerel of USC, we describe implementation issues through a baseline design of a turbo decoder (*i.e.*, the decoder for a rate 1/2 PCCC). This is based on the class *EE577b VLSI System Design* taught by Peter at USC during the Spring 2000 term. In particular, Pornchai Pawawongsak's class project serves as a baseline design. Through this baseline design, we discuss quantization effects, metric normalization methods, internal bit-width choices, and more advanced architectures. Thus, this provides a case study which may serve as a starting point for those interested in developing hardware for iterative detection.

It was mentioned that the authors performed the bulk of this work from 1997-1999 while at USC. In fact, a significant amount of recent work performed by several Ph.D. students at USC has been included. In particular, we thank significant contributions from Kyuhyuk Chung (Examples 2.9, 2.21, 2.22), Robert Golshan (Section 2.4.4), Jun Heo (Fig-2.31, Section 4.2.3, Fig-6.2, Fig-6.3), Phunsak Thiennviboon (Examples 2.15, 2.19, 2.20, 3.5-3.6, Fig-5.19, Fig-5.21, and Sections 2.6.1.2, 5.3.2.2, 5.5, 5.4.3). In addition to these students, who also provided excellent feedback on the material, we thank the following for helpful comments and discussions: Gianluigi Ferrari (visiting USC from U. of Parma), Anders Hansson (visiting USC from Chalmers U. Tech.), Idin Motedayen-Aval (U. Michigan), and Prof. Vijay Kumar of USC.

We thank our research collaborators and the students from EE666 and EE577b. Specifically, thanks to Prof. Mark Neifeld at the U. of Arizona for his contributions to the 2D work (including the motivation!) and Prof. Antonio Ortega at USC for his contributions to the halftoning work. We gratefully acknowledge our financial supporters which include the National Science Foundation (NCR-9616663 and CCR-9726391), ViaSat, Inc. and the Intelligence and Information Warfare Division (IIWD) of the U.S. Army. Special thanks to Dr. Tom Carter at TrellisWare, Dr. Chul Oh at IIWD, and Prof. Andreas Polydoros from U. of Athens and TrellisWare for their encouragement and support.

KEITH M. CHUGG

ACHILLEAS ANASTASOPOULOS

XIAOPENG CHEN

Introduction

Nearly every textbook in the area of digital communications begins with a system block diagram similar to that shown in Fig-I.1(a)-(b). In fact, the receiver block diagram in Fig-I.1(b) mirrors the processing performed in most practical receiver implementations. This *segregated* design paradigm allows each component of the receiver to be designed and "optimized" without much regard to the inner workings of the other blocks of the receiver. As long as the each block does the job it is intended for, the overall receiver should perform the desired task: extracting the input bits.

Despite the ubiquity of the diagram in Fig-I.1(b) and the associated design paradigm, it is clearly not optimal from the standpoint of performance. More specifically, the probability of error for the bit estimates or bit-sequence estimate is not minimized by this structure. This segregated processing is adapted for tractability – both conceptual tractability and tractability of hardware implementation. The optimal receiver for virtually any system is conceptually simple, yet typically prohibitively complex to implement. For example, consider the transmission of 1000 bits through a system of the form in Fig-I.1(a). These bits may undergo forward error correction coding (FEC), interleaving, training insertion (pilots, synchronization fields, training sequences, etc.), before modulation and transmission. The channel may corrupt the modulated signal through random distortions (possibly time-varying and non-linear), like-signal interference (co-channel, multiple access, cross-talk, etc.), and additive noise. The point is, regardless of the complexity of the transmitter and/or channel, the optimal receiver would compute 2^{1000} likelihoods and select the data sequence[1] that most closely matches the assumed

---

[1] As described in detail in Chapter 1, one could average these likelihoods to obtain bit likelihoods if optimal bit decisions are desired instead of the optimal sequence decision.

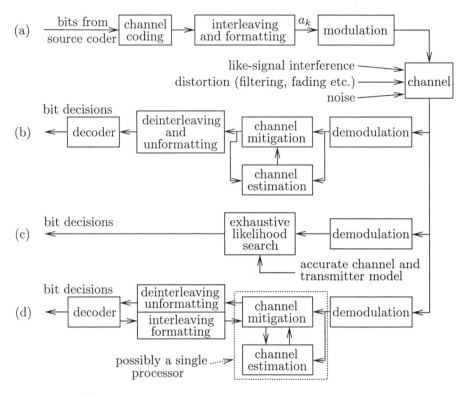

Figure I.1. (a) A typical communication block diagram, (b) a traditional segmented receiver design, (c) the optimal receiver processing, and (d) a receiver based on iterative detection principles. The formatting includes insertion of framing and training information (overhead). Channel mitigation includes tasks such as equalization, diversity combining, array combining, etc.

model. This is shown in Fig-I.1(c). Ignoring the obvious complexity problems, this requires a good model of the transmitter formatting and the channel effects. For example, the likelihood computation mentioned above may include averaging over the statistics of a fading channel model or the possible data values of like-signal interferers.

The approaches described in this book are exciting because they enable receiver processing that can closely approximate the above "holy grail" with feasible conceptual and implementation complexity. Specifically, data detection and parameter estimation are done using the *entire global system structure*. Unlike the direct approach in Fig-I.1(c), the iterative receiver in Fig-I.1(d) exploits this structure indirectly. The key concept in this approach is the exchange and updating of "soft information" on digital quantities in the system (*e.g.*, the coded modulation symbols). This concept is shown in Fig-I.1(d). The iterative detection receiver is similar to the conventional segregated design in that, for each subsystem block in the model, there is a corresponding processing block. In fact, each of these corresponding processing blocks in the receiver of Fig-I.1(c) exploits only *local* system structure – *e.g.*, the FEC decoder does not use any explicit knowledge of the channel structure. As a consequence, the complexity of the receiver in Fig-I.1(d) is comparable to the traditional segregated design in Fig-I.1(b) (*i.e.*, the increase in complexity is usually linear as opposed to the exponential increase in complexity associated with the optimal processing in Fig-I.1(c)).

The distinction between the conventional segregated design and that in Fig-I.1(d), however, is that the processing for each sub-block in Fig-I.1(d) is *biased* by some beliefs on its inputs and outputs. These beliefs (also referred to as reliabilities, soft decision, soft information) are provided to each local processing unit by other processing units connected to it. These beliefs represent *marginal* soft information in that they are beliefs on the individual symbols as opposed to the entire sequence. The task of the processing unit is to *update* the beliefs on the input and output variables of the corresponding system sub-block in Fig-I.1(a). Each sub-block processing unit will be activated several times, each time biased by a different (updated) set of beliefs.

For example, suppose that a system using convolutional coding and interleaving experiences severe like-signal interference and distortion over the channel. In this case, the channel mitigation block in Fig-I.1(b) will output hard decisions on the coded/interleaved bit sequence a_k. Suppose that, given the severity of the channel, the error probability associated with these coded-bit decisions will be approximately 0.4. Deinterleaving these decisions and performing *hard-in* (Hamming distance) decoding of the convolutional code will provide a very high bit error rate (BER) – *i.e.*, nearly 0.5.

For the receiver in Fig-I.1(d), however, the channel mitigation block produces soft-decision information on the coded/interleaved bit sequence a_k. For example, this may be thought of as two numbers $P[a_k = 1]$ and $P[a_k = 0]$ that represent a measure of current probability or belief that the k-th coded bit a_k takes on the value 1 or 0, respectively. Clearly, soft decisions contain more information than the corresponding hard decisions. In this example, it is possible that even though the hard decisions on a_k associated with the receiver of Fig-I.1(b) are hopelessly inaccurate, the soft decision information contains enough information to jump-start a decoding procedure. For example, two different possible sequences of soft decision information are shown in Table I.1. Note that each of these correspond to exactly the same hard decision information – *i.e.*, the hard decisions obtained by *thresholding* the soft information

k:	0	1	2	3	4 ...
true data:	0	0	1	0	1
case A:	(0.99, 0.01)	(0.97, 0.03)	(0.51, 0.49)	(0.48, 0.52)	(0.03, 0.97)
case B:	(0.51, 0.49)	(0.55, 0.45)	(0.51, 0.49)	(0.48, 0.52)	(0.48, 0.52)
decisions:	0	0	0	1	1

Table I.1. Example of two sequences of soft information implying the same hard decisions, but containing very different soft information. The soft information is given as $(P[a_0 = 0], P[a_0 = 1])$.

is the same.[2] However, the soft information in case B is much worse than that in case A. Specifically, for case A, there is a high degree of confidence for correct decisions and very low confidence for incorrect decisions. For case B, there is little confidence in any of the decisions.

A receiver of the form in Fig-I.1(d) would pass the soft information through a deinterleaver to a soft-in decoder for the convolutional code. This decoder is a modified version that produces beliefs on the coded bits as well as the uncoded bits. Thus, after activation of this decoder, one could make a decision on the uncoded bits. Alternatively, the updated beliefs on the coded bits could be interleaved and used in the role of a-priori probabilities to bias another activation of the channel mitigation processing unit in Fig-I.1(d). In fact, this process could be repeated with the channel mitigation and FEC decoder exchanging and updating beliefs on the coded bits through the interleaver/deinterleaver pair. After several iterations, final decisions can be made on the uncoded bits by thresholding the corresponding beliefs generated by the code processing unit. This is what is meant by iterative detection.

Note that, in this example, even though the hard decision information on the coded bits after activating the channel mitigation processing unit is very unreliable (*e.g.*, an error rate of 0.4), the soft information may allow the FEC decoder to draw some reasonable inference. For example, if the soft information is that of case A in Table I.1, then the FEC decoder may update the beliefs as to over-turn the unreliable (incorrect) decisions and re-enforce the reliable decisions (*i.e.*, the correct ones in this example). Note that this updating takes into account only these marginal beliefs on the coded bits and the local code structure.

In summary, the receiver processing in Fig-I.1(d) closely approximates the performance of the optimal processing in Fig-I.1(c) with complexity roughly comparable to that of the traditional segmented design in Fig-

---

[2]In general, hard decisions are obtained from a soft information measure by selecting the conditional value (*e.g.*, $a_k = 1$ or $a_k = 0$) associated with the largest belief.

I.1(b). It does so by performing locally optimal processing which exploits the local system structure and updating and exchanging marginal soft information on the subsystem inputs and outputs.

Complexity Reduction   Since the processing of Fig-I.1(d) approximates the performance of the optimal processing with much lower complexity, iterative detection may be viewed as a complexity reduction tool. This complexity reduction is based on the partitioning or decomposition of the system into subsystems. Specifically, the complexity of the receiver in Fig-I.1(d) is roughly the sum of the complexities of the individual processing units, multiplied by the number of times each is activated (*i.e.*, the number of iterations). Since each of these processors uses only local structure, the associated complexity is roughly the same as the optimal decoder for that subsystem *in isolation*. For example, in the soft-decoder described in the above example has roughly the same complexity as the well-known Viterbi algorithm. A key conceptual point is that the partitioning described is entirely arbitrary. One can choose to decompose the system into any number of equivalent block diagrams, each of which faithfully represents the system structure. This modeling choice, however, impacts the complexity of the associated processors in Fig-I.1(b) and Fig-I.1(d). For example, the receiver in Fig-I.1(c) corresponds to a model without decomposition and is prohibitively complex. Furthermore, the notion that exchanging and updating soft information can be a replacement for hypothesis searching effort can be exploited aggressively to provide complexity reduction for systems which may not even have been modeled as a concatenation of subsystems. This provides motivation for the material in Chapter 3.

Adaptivity   Acquisition and tracking of channel parameters (*e.g.*, carrier phase and frequency, symbol synchronization, channel impulse response, etc.) can also be included into the iterative processing of Fig-I.1(d). In the above example, the interference and distortion associated with the channel may be unknown and/or time-varying. Therefore, the channel mitigation processor should estimate and track these parameters (possibly aided by some training signals). In an iterative detector, it is possible to re-estimate these parameters at each activation of the channel mitigation processing unit. The estimates are different for each of these activations because the reliability information biasing this estimation changes. Specifically, it is refined by the code processor using the structure of the code. Thus, through the passing of marginal soft information and iteration, the channel estimator indirectly uses the global system structure. Intuitively, a decision directed estimation processor can operate with the coding gain even when isolating interleavers are present. For an isolated system, this may be possible using joint parameter estimation and data detection (*e.g.*, Per-Survivor Processing). However, these adaptive iterative detection approaches, which are discussed in Chapter 4, are applicable to systems comprising concatenated subsystem (*e.g.*, concatenated codes with interleaving).

Applicability and Impact   Notice that in the above discussion, we have not mentioned "turbo codes." While the decoding of these powerful concatenated codes is the most well-known and celebrated special case of iterative detection, it represents a very special case of a broadly applicable concept. Iterative detection is applicable to virtually every practical digital communication system and can provide significant gains in performance and/or complexity reduction. Most significantly, we assert that this can lead to a true paradigm shift in digital communication systems design. Specifically, this shift is from segmented design paradigm to the joint-design paradigm which is enabled by the methods described above. With this new paradigm, systems can operate reliably in much more severe regimes. Examples of these gains will be given throughout this book. With the new paradigm, practical codes virtually achieve the Shannon bound. Like-signal interference that cripples the most powerful multiuser detectors in the segmented design paradigm, is bearable when the global system structure (*e.g.*, FEC) is used to aid interference mitigation. Phase dynamics that break phase-lock loops at moderate signal to noise ratios (SNRs) function well at extremely low SNR by exploiting coding gain in the estimation process. Furthermore, these significant gains can be achieved with hardware complexity that is either well within today's technology or feasible in the near term.

Abbreviations

A number of abbreviations used throughout the book are listed in Table I.2. In addition, for those chapters where a number of additional abbreviations are used, a similar table of those chapter-specific abbreviations is included at the end of the chapter.

Notational Conventions

We try to follow general conventions for notation throughout this book. In some cases more explicit notation is desirable (*i.e.*, in introducing concepts) which later becomes cumbersome and is abbreviated for compactness. A summary of the notation and conventions used is given in Table I.3, which is divided roughly into typefaces, standard signal and systems notation, quantities related to probability, quantities

8-PSK	8-ary Phase Shift Keying
A-SISO	Adaptive Soft Input Soft Output
A-SODEM	Adaptive Soft-output Demodulator
ACS	Add Compare Select
AL	Average Likelihood
APP	A-Posteriori Probability
AR	Auto Regressive
ARMA	Auto Regressive Moving Average
AWGN	Additive White Gaussian Noise
BCJR	Bahl Cocke Jelineck and Raviv
BER	Bit Error Rate
BPA	Belief Propagation Algorithm
BPSK	Binary Phase Shift Keying
CC	Convolutional Code
CDMA	Code Division Multiple Access
CM	Combining and Marginalization (problem)
CSI	Channel State Information
DD-PLL	Decision Directed Phase Locked Loop
DDFSE	Delayed Decision Feedback Sequence Estimate
DFE	Decision Feedback Equalization
DLM	Digital Least Metric (problem)
EC	Estimator Correlator
FBT	Forward/Backward-Tree
FI	Fixed-Interval
FIR	Finite Impulse Response
FL	Fixed-Lag
FL-SISO	Fixed Lag Soft Input Soft Output (module)
FSM	Finite State Machine
GAP	Generalized A-posteriori Probability
GDL	Generalized Distributive Law
GL	Generalized Likelihood
GM	Gauss Markov
HID	Hard-In Decoding
ID	Iterative Detection
ISI	Intersymbol Interference
iid	Independent, Identically Distributed
KF	Kalman Filter
L^2VS	Lee and Li Vucetic Sato (algorithm)
LDPC	Low Density Parity Check (code)
LMS	Least Mean Squares
LS	Least Squares

Table I.2.   Table of abbreviations (continued on next page)

related to soft information. Further notation is introduced as needed throughout the book.

Our emphasis is on post-correlator receiver processing, so we deal with discrete index sets almost exclusively (*i.e.*, referencing the literature for

M-PSK	M-ary Phase Shift Keying
MAI	Multiple Access Interference
MAP	Maximum A-posteriori Probability
MAP-PgD	MAP Page Detection
MAP-SqD	Maximum A-posteriori Probability Sequence Detection
MAP-SyD	Maximum A-posteriori Probability Symbol Detection
ML	Maximum Likelihood
ML-SqD	Maximum Likelihood Sequence Detection
ML-SyD	Maximum Likelihood Symbol Detection
MMSE	Minimum Mean Square Error
MPF	Marginalized Product Function
MSE	Mean Square Error
MSM	Minimum Sequence (or Sum) Metric
M*SM	Min* Sum Metric
PAM	Pulse Amplitude Modulation
PCCC	Parallel Concatenated Convolutional Code
pdf	Probability Density Function
PLL	Phase Locked Loop
pmf	Probability Mass Function
POM	Page-oriented Optical Memory
PSP	Per-Survivor Processing
QPSK	Quadrature Phase Shift Keying
RA	Repeat-Accumulate (code)
RLS	Recursive Least Squares
RSC	Recursive Systematic Code
SCCC	Serial Concatenated Convolutional Code
SEP	Symbol Error Probability
SID	Soft-In Decoding
SISO	Soft Input Soft Output (module)
SNR	Signal-to-Noise Ratio
SOA	Soft Output Algorithm
SOBC	Soft Output BroadCaster
SODEM	Soft Output Demodulator
SOMAP	Soft Output MAPper
SOVA	Soft Output Viterbi Algorithm
SW	Sliding-Window
TCM	Trellis Coded Modulation
TDMA	Time Division Multiple Access
TH	THreshold detector
USI	Uniform Side Information
VA	Viterbi Algorithm

Table I.2.   (Continuation) Table of abbreviations

the models assumed). Unless specified otherwise, we use the implicit convention that a digital quantity v_n takes on values in $\mathcal{V}$ with index range $n = 0, 1, \ldots (N-1)$. Furthermore, the cardinality of the alphabet $\mathcal{V}$ is denoted by $|\mathcal{V}|$ and, unless otherwise specified, $\mathcal{V} = \{0, 1 \ldots (|\mathcal{V}| -$

1)}. A digital sequence denoted by b is implicitly binary. Using the same time index variable implies that the two sequences are defined on the same time-scale (*e.g.*, a_k and x_k). This is often not the case in systems considered (*e.g.*, multiple bits per symbol), so we use different index variables to emphasize this when appropriate (*e.g.*, a_m and x_n). Digital symbols corrupted unintentionally are typically denoted by letters near the end of the alphabet (*i.e.*, , x, y). Continuous valued variables are limited for the most part to channel parameters, noise, and channel observations. Channel parameters are typically denoted by f, g, h, or θ. Noise is denoted by n or w. The channel observation is denoted by z or possibly r.

A random quantity is indicated by explicitly denoting the sample space variable ζ. For example, $v(\zeta)$ is a random variable. By implication, the same variable without the sample space argument is a realization or conditional value of that random quantity – e.g., v is a realization $v(\zeta)$. We will often work with both probability density functions (for continuous random variables/vectors), probability mass functions (for discrete random quantities), and mixtures thereof. In our notation, these different measures are to be distinguished implicitly since, for simplicity, we denote all by $p(\cdot)$.

We consider baseband equivalent models throughout this book. The associated normalization is based on the convention that $\tilde{v}(t) = \Re\left\{v(t)\sqrt{2}\exp(-j2\pi f_c t)\right\}$ where $\tilde{v}(t)$ is a passband signal and $v(t)$ is the complex baseband equivalent, f_c is the reference (carrier) frequency, and $j = \sqrt{-1}$. Under this convention, after correlation to a normalized version of $v(t)$, a signal with energy E_v will be obtained. If additive white Gaussian noise (AWGN) with two-sided spectral level $N_0/2$ is present, the corresponding post-correlator noise will have variance $N_0/2$ in each dimension. Thus, when we consider a real-valued scalar model of the form $z_k = a_k + w_k$, then $\mathbb{E}\{a_k^2\} = E_a$, $\mathbb{E}\{w_k\} = 0$, and $\mathbb{E}\{w_k^2\} = N_0/2$. If the same baseband model is complex, then $\mathbb{E}\{|a_k|^2\} = E_a$, $\mathbb{E}\{w_k\} = 0$, and $\mathbb{E}\{|w_k|^2\} = N_0$ with the noise being a circular complex process so that $\mathbb{E}\{\Re\{w_k\}^2\} = \mathbb{E}\{\Im\{w_k\}^2\} = N_0/2$. The shorthand used for Gaussian and complex Gaussian densities is

$$\mathcal{N}_n(\mathbf{z}; \mathbf{m}; \mathbf{K}) = \frac{1}{\sqrt{2\pi|\mathbf{K}|}} \exp\left[-\frac{1}{2}(\mathbf{z} - \mathbf{m})^{\mathrm{T}}\mathbf{K}^{-1}(\mathbf{z} - \mathbf{m})\right]$$

$$\mathcal{N}_n^{cc}(\mathbf{z}; \mathbf{m}; \mathbf{K}) = \frac{1}{\pi|\mathbf{K}|} \exp\left[-(\mathbf{z} - \mathbf{m})^{\mathrm{H}}\mathbf{K}^{-1}(\mathbf{z} - \mathbf{m})\right]$$

Soft information on various random digital quantities can be expressed in a number of ways. While these variations are described in detail in the following chapters, we use the general notation S[v] (*i.e.*, square brackets) to denote some soft information on a digital random variable $v(\zeta)$. Much like in a pmf, v represents a conditional value for $v(\zeta)$, so that $S[v]$ corresponds to $|\mathcal{V}|$ numbers. To emphasize a soft information on a specific conditional value (*e.g.*, 0), we will use $S[v = 0]$.

Typefaces

bold lower case	column vector (*e.g.*, $\mathbf{v}$)
bold upper case	matrix (*e.g.*, $\mathbf{V}$)
calligraphy	set (*e.g.*, $\mathcal{A}$)
san serif	system (*e.g.*, M)

Signals, Systems, Algebra

$|\mathcal{A}|$	cardinality of set $\mathcal{A}$
$(\cdot)^*$	complex conjugate
$(\cdot)^{\mathrm{T}}$	vector or matrix transpose
$(\cdot)^{\mathrm{H}}$	vector or matrix complex conjugate and transpose
$\Re\{\cdot\}$	real part
$\Im\{\cdot\}$	imaginary part
$\mathbf{K}^{\mathrm{I}}$	pseudo-inverse of matrix $\mathbf{K}$
$\mathbf{v}_m^n$	column vector $[v_n, v_{n-1}, \ldots, v_m]^{\mathrm{T}}$
	($\mathbf{v}$ implies the entire index range)
$|\mathbf{K}|$	determinant of matrix $\mathbf{K}$
δ_k	Kronecker delta function with argument k
$\mathrm{I}_{\{q\}}$	Indicator function (1 iff q is true)
$\triangleq$	equality by definition
$\cong$	approximately equal
$\mathcal{R}$	set of real numbers
$\mathcal{C}$	set of complex numbers
$\mathcal{O}(N)$	order N (big-0)
$N_0/2$	Noise spectral level
E_b (E_s)	energy per bit (symbol)
$\circledast$	convolution
s_k	trellis (or finite state machine) state
t_k	trellis (or finite state machine) state transition
L	memory of a simple FSM

Table I.3. Table of notational conventions (continued on next page)

Probability

$z(\zeta)$	random variable
$p_{z(\zeta)}(z)$	pdf (pmf) of a continuous (discrete) random variable
$p_{z(\zeta),a(\zeta)}(z,a)$	joint mixed pdf/pmf for a continuous $(z(\zeta))$ and
	a discrete $(a(\zeta))$ random variable
$P(\mathcal{A})$	probability of the event $\mathcal{A}$
$\Pr\{v(\zeta)\in\mathcal{D}\}$	probability of the event $\{\zeta:v(\zeta)\in\mathcal{D}\}$
$\mathbb{E}\{\cdot\}$	ensemble averaging (expectation)
$\mathbb{E}_{z(\zeta)}\{\cdot\}$	expectation with respect to $z(\zeta)$
$\mathrm{var}\,[z(\zeta)]$	variance of the random variable $z(\zeta)$
$\mathcal{N}_n(\mathbf{z};\mathbf{m};\mathbf{K})$	n-dimensional Gaussian pdf with mean $\mathbf{m}$,
	(nonsingular) covariance matrix $\mathbf{K}$ and argument $\mathbf{z}$
$\mathcal{N}_n^{cc}(\mathbf{z};\mathbf{m};\mathbf{K})$	same as above, for complex circular Gaussian pdf

Soft Information

$\equiv$	equivalence
$\triangleq$	defined as an equivalent quantity
$\sim$	related by isomorphism
©	combining operator
©$^{-1}$	inverse combining operator
ⓜ	marginalization operator
$\min^*$	min-star operator
$\mathrm{S}[a]$	generic soft information on a
$\mathrm{SI}[a]$	generic (extrinsic) soft-in information on a
$\mathrm{SO}[a]$	generic (extrinsic) soft-out information on a
$\mathrm{P}[a],\mathrm{PI}[a],\mathrm{PO}[a]$	same as above, for soft information in the probability domain
$\mathrm{M}[a],\mathrm{MI}[a],\mathrm{MO}[a]$	same as above, for soft information in the metric domain
M^{-s}	soft inverse of system M
$x:u$	all possible x consistent with u

Table I.3. (Continuation) Table of notational conventions

Chapter 1

OVERVIEW OF NON-ITERATIVE DETECTION

1.1 Decision Theory Framework

A general decision problem can be formulated for our purposes as involving three components

- *The source*: producing a hypothesis, modeled as a random variable $H(\zeta)$ that takes on a finite number of values $\{H_0, H_1, \ldots H_{|\mathcal{H}|-1}\}$. A statistical model of the source is the *a-priori* probability $p_{H(\zeta)}(H_m)$ for each hypothesis.
- *The channel*: providing an observation $\mathbf{z}(\zeta)$. The statistical description of $\mathbf{z}(\zeta)$, conditioned on each of the possible hypotheses, is assumed available at the channel.
- *The decision rule*: mapping a realization of $\mathbf{z}(\zeta)$ to an action (*e.g.*, declaring a hypothesis to be true) in the action space based on knowledge of the source and channel models.

The objective is to obtain a decision rule which is optimal with respect to some criterion given a specific model of the source and channel. This framework is illustrated in Fig-1.1. There are several excellent books that treat decision theory for engineers [He60, Mi60, We68, Va68, Po94]. Our goal is to introduce these concepts to the point where the relationship between different data detection methods will be clear. Thus, our focus is on a small subset of the material covered in these references.

We focus on the case where the a-priori statistics of the source are known to the receiver and thus can be used in the decision rule. Furthermore, we assume that there are a finite number of allowable actions. In most cases of interest there is a one-to-one correspondence between the actions and the hypotheses such that action A_m corresponds to deciding or declaring that H_m is true. The decision rule is defined by a

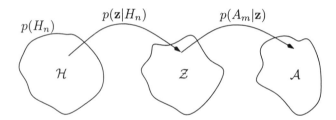

Figure 1.1.   The set-up for general decision problems considered.

probability mass function (pmf) $d(A_m|\mathbf{z}) = p_{A(\zeta)|\mathbf{z}(\zeta)}(A_m|\mathbf{z})$. Admissible decision rules are those for which the action $A(\zeta)$ is independent of the hypothesis when conditioned on the observation. A *deterministic* decision rule is one with all probability mass located at one action (*i.e.*, $d(A_m|\mathbf{z})$ is one for some value of m and zero for others). A deterministic rule yields, conceptually at least, a partitioning of the observation space into the decision regions $\mathcal{Z}_m$ – *i.e.*, $\mathcal{Z} = \cup_m \mathcal{Z}_m$ and $\{\mathcal{Z}_i\}$ are disjoint – as illustrated in Fig-1.2. It is desirable to reduce the decision rule to this form since it represents a simple receiver implementation.

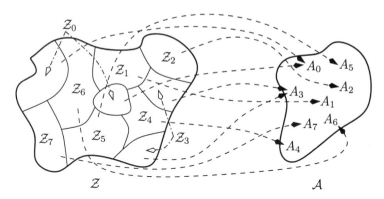

Figure 1.2.   Decision rule implemented as a partition of the observation space.

In some cases the statistical description of the channel (*e.g.*, $p(\mathbf{z}|H_n)$ for $n = 0, 1, \ldots |\mathcal{H}| - 1$) will be most conveniently expressed with an additional condition on a finite set of parameters Θ. One can associate a statistical model for these unknown parameters provided by $\{p(\Theta|H_m)\}$ or simply model the parameter as an unknown deterministic parameter. In the latter case, one may still impose some structure on the parameters. For example, the energy of a signal may be modeled as a deterministic constant, but it is known to be non-negative. It is important to distinguish the model used for the purposes of designing a decision rule and other possible models. For example, one may select a deterministic model for a random parameter if there is little confidence in the available probabilistic models. As another example, an accepted model for a fading channel may be a random process with nonrational power spectral density, but for the purposes of designing the decision rule one may select an approximation to this model with rational power spectral density.

1.1.1   The Bayes Decision Rule

A Bayes decision rule minimizes the average Bayes risk $R(d)$ over all admissible decision functions, where

$$R(d) = \int_{\mathcal{Z}} p_{\mathbf{z}(\varsigma)}(\mathbf{z}) \left[\sum_m d(A_m|\mathbf{z}) C(A_m|\mathbf{z}) \right] d\mathbf{z} \qquad (1.1)$$

where $\mathcal{Z}$ is the observation space and $C(A_m|\mathbf{z})$ is the *cost* (or risk) associated with taking action A_m, given $\mathbf{z}(\varsigma) = \mathbf{z}$, averaged over the source statistics

$$C(A_m|\mathbf{z}) = \sum_i C(A_m, H_i) p_{H(\varsigma)|\mathbf{z}(\varsigma)}(H_i|\mathbf{z}) \qquad (1.2)$$

The finite set of coefficients $\{C(A_m, H_i)\}_{m,i}$ specifies the cost of taking action A_m with H_i occurring. These coefficients relate the Bayes risk to some more tangible optimization criterion as described in the subsequent development.

It follows from (1.1) that any rule that takes action A_m with the property that $C(A_m|\mathbf{z}) \leq C(A_i|\mathbf{z})$ for all i is a Bayes rule. The Bayes rule is generally not unique because of "tie" conditions where two or more actions have the same cost. In this case the total probability mass may be spread among these best actions in any way. One can always obtain a deterministic Bayes rule by breaking these tie conditions in a deterministic manner (*e.g.*, select the action with the smallest index). Thus, we will use the notation

$$\text{Bayes action} = \arg \min_m C(A_m|\mathbf{z}) \qquad (1.3)$$

Note that the *A-Posteriori Probability (APP)* of $H(\varsigma) = H_m$ given a realization of $\mathbf{z}(\varsigma)$ can be written as

$$p_{H(\varsigma)|\mathbf{z}(\varsigma)}(H_m|\mathbf{z}) = \frac{p_{\mathbf{z}(\varsigma)|H(\varsigma)}(\mathbf{z}|H_m) p_{H(\varsigma)}(H_m)}{p_{\mathbf{z}(\varsigma)}(\mathbf{z})} \qquad (1.4\text{a})$$

$$\equiv p_{\mathbf{z}(\varsigma)|H(\varsigma)}(\mathbf{z}|H_m) p_{H(\varsigma)}(H_m) \qquad (1.4\text{b})$$

where $\equiv$ is used to denote an *equivalence* between quantities in terms of information on the hypothesis $H(\varsigma)$. Specifically, the term $p_{\mathbf{z}(\varsigma)}(\mathbf{z})$ may be dropped since it is not a function of the hypothesis.

The *Maximum A-Posteriori Probability (MAP)* decision rule is the special case of the Bayes rule when A_m corresponds to deciding that H_m is true and $C(A_m, H_i) = 1 - \delta_{m-i}$. This may be seen by substituting these cost coefficients into (1.2) and noting that

$$C(A_m|\mathbf{z}) = \sum_{i \neq m} p_{H(\zeta)|\mathbf{z}(\zeta)}(H_i|\mathbf{z}) = 1 - p_{H(\zeta)|\mathbf{z}(\zeta)}(H_m|\mathbf{z}) \qquad (1.5)$$

so that minimization of the Bayes risk is equivalent to maximization of $p_{H(\zeta)|\mathbf{z}(\zeta)}(H_m|\mathbf{z})$. Furthermore, substituting the coefficients $C(A_m, H_i) = 1 - \delta_{m-i}$ into (1.1)-(1.2), it is straightforward to show that the average risk is the *probability of error*. Thus, the MAP decision rule minimizes the probability of error for the $|\mathcal{H}|$-ary decision problem. For the special case of uniform *a-priori* probability on the hypotheses, the APP $p(H_m|\mathbf{z})$ is equivalent to the *likelihood* $p(\mathbf{z}|H_m)$. Thus, the term Maximum Likelihood (ML) is often used to describe the MAP detector for this special case. While one could use an ML rule in a case when the a-priori probabilities are not uniform (*i.e.*, not MAP detection), when we use the term ML detection, it is implicitly assumed that the a-priori probabilities are uniform.

1.1.2   Composite Hypothesis Testing

When the conditional statistical description of the observation also depends on a parameter Θ, the decision problem is often called a composite hypothesis test. The term "composite" refers to the fact each hypothesis H_m represents many possibilities of the form (H_m, Θ).

Specifically, suppose that $p(\mathbf{z}|H_m, \Theta)$ is known for each hypothesis and each allowable value of the *nuisance parameter (set)* $\Theta(\zeta)$. A random model has been assumed for the parameter and $p(\Theta|H_m)$ is also assumed to be known for the purposes of inferring on $H(\zeta)$. Defining the decision rule in this case is no more difficult, conceptually at least, since

$$p_{\mathbf{z}(\zeta)|H(\zeta)}(\mathbf{z}|H_m) = \int p_{\mathbf{z}(\zeta)|H(\zeta),\Theta(\zeta)}(\mathbf{z}|H_m, \Theta)p_{\Theta(\zeta)|H(\zeta)}(\Theta|H_m)d\Theta$$

$$= \mathbb{E}_{\Theta(\zeta)|H(\zeta)}\{p_{\mathbf{z}(\zeta)|H(\zeta),\Theta(\zeta)}(\mathbf{z}|H_m, \Theta(\zeta))|H_m\} \qquad (1.6)$$

Thus, from the composite problem, one can obtain the likelihood in (1.6) and proceed as described in Section 1.1.1. To emphasize that the resulting Bayes rule incorporates the expectation in (1.6), it may be said that the Bayes rule minimizes the risk averaged over $\Theta(\zeta)$. Similarly, the likelihood $p_{\mathbf{z}(\zeta)|H(\zeta)}(\mathbf{z}|H_m)$ is often referred to as the *average likelihood*. In many cases of interest, $\Theta(\zeta)$ is statistically independent of $H(\zeta)$. Finally, the special case where the nuisance parameter is a sinusoidal carrier phase $\phi(\zeta)$, uniformly distributed over an interval of length 2π, is very common. In this case, the decision rule is often referred to as (phase) *noncoherent* detection.

In the case where Θ is modeled as unknown and deterministic, defining an optimality criterion for the decision rule is more nebulous. Specifically, the risk associated with a reasonable rule often depends on the actual value of Θ. One common approach is to use the most likely value of Θ, conditioned on $\mathbf{z}$ and H_m to provide a representative from the set of likelihoods $p_{\mathbf{z}(\zeta)|H(\zeta)}(\mathbf{z}|H_m; \Theta)$, where the notation $p(\cdot; \Theta)$ denotes parameterization by Θ. Specifically, the *generalized likelihood* is defined by [Va68]

$$g_{\mathbf{z}(\zeta)|H(\zeta)}(\mathbf{z}|H_m) \triangleq p_{\mathbf{z}(\zeta)|H(\zeta)}(\mathbf{z}|H_m; \Theta)\big|_{\Theta = \hat{\Theta}(\mathbf{z}, H_m)} \qquad (1.7a)$$

$$\hat{\Theta}(\mathbf{z}, H_m) = \arg\max_{\Theta} p_{\mathbf{z}(\zeta)|H(\zeta)}(\mathbf{z}|H_m; \Theta) \qquad (1.7b)$$

As mentioned earlier, the optimization in (1.7b) may be restricted to some set if such information is available about Θ. The generalized likelihood may then be used in place of $p_{\mathbf{z}(\zeta)|H(\zeta)}(\mathbf{z}|H_m)$ in designing a decision rule. We refer to the rule that uses the generalized likelihood in place of the (average) likelihood in (1.2) as the *generalized Bayes decision rule*, or, for the special case of (1.5), the *generalized MAP rule*. Use of generalized likelihood, however, is ad-hoc and, in general, does not optimize a well-defined cost criterion [Va68].

Notice that the average and generalized likelihoods differ only in the manner in which they *marginalize* out the parameter from the composite test. Specifically, this is done by maximizing in the generalized likelihood case and ensemble averaging in the average likelihood case. It may be reasonable to consider using maximization even when a probabilistic model is assumed for the nuisance variable. Specifically, we will also consider "likelihoods" of the form

$$g_{\mathbf{z}(\zeta)|H(\zeta)}(\mathbf{z}|H_m) = \max_{\Theta} p(\mathbf{z}|H_m, \Theta)p(\Theta|H_m) \qquad (1.8)$$

which is simply (1.6) with the expectation replaced by maximization. We also refer to this as generalized likelihood. Thus, generalized likelihood refers to (1.7) and (1.8) in the contexts of deterministic and random nuisance parameter models, respectively. When we wish to emphasize the difference, we use the term *d-generalized* and *p-generalized* for (1.7) and (1.8), respectively, to remind the reader of the deterministic or probabilistic nuisance parameter model.

Note that (1.7) and (1.8) may be viewed as using the ML and MAP estimates of Θ in place of the true value of Θ in the probability density function (pdf) of $\mathbf{z}$ conditioned on $H(\zeta) = H_m$. Thus, when estimation of Θ may be expected to be reliable (*e.g.*, high SNR), one expects the generalized likelihood approach to work well. Problem 1.7 provides an example to compare average and generalized likelihoods.

1.1.3   Statistical Sufficiency

The Bayes decision rule can always be expressed in terms of the likelihoods $p_{\mathbf{z}(\zeta)|H(\zeta)}(\mathbf{z}|H_m)$. Thus, these likelihoods form a set of sufficient statistics for obtaining the Bayes rule from the observation $\mathbf{z}$. In many cases of interest, however, a simpler set of sufficient statistics exists. Informally, if one begins with the likelihood and simplifies via equivalence relationships (*i.e.*, dropping terms that do not depend on H_m), then the result is a set of sufficient statistics. In other words, if the equivalent of the likelihoods $p(\mathbf{z}|H_m)$ can be expressed in terms of $s(\mathbf{z}|H_m)$ for each possible hypothesis, then $\{s(\mathbf{z}|H_m)\}$ is a set of sufficient statistics for deciding on $H(\zeta)$ from $\mathbf{z}$.

Example 1.1. ——————————————————————
Consider the case of deciding between signals corrupted by AWGN

$$\mathbf{z}(\zeta) = \mathbf{x}(H(\zeta)) + \mathbf{w}(\zeta) \tag{1.9}$$

It is straightforward to verify that $s(\mathbf{z}|H_m) = \mathbf{z}^{\mathrm{T}}\mathbf{x}(H_m)$ for $m = 0, 1, \ldots$ $|\mathcal{H}| - 1$, is a set of sufficient statistics. Note that the dimension of observation $\mathbf{z}$ can be larger than $|\mathcal{H}|$ so that it is often the case that one cannot reconstruct $\mathbf{z}$ from $\mathcal{S} = \{s(\mathbf{z}|H_m) = \mathbf{z}^{\mathrm{T}}\mathbf{x}(H_m)\}$, but the associated likelihood can be reconstructed.

——————————————————————— *End Example*

More formally, $\mathcal{S}(\zeta)$ is a set of *sufficient statistics* for deciding on $H(\zeta)$ from $\mathbf{z}$ if

$$p_{\mathbf{z}(\zeta)|\mathcal{S}(\zeta),H(\zeta)}(\mathbf{z}|\mathcal{S}, H_m) = p_{\mathbf{z}(\zeta)|\mathcal{S}(\zeta)}(\mathbf{z}|\mathcal{S}) \quad \forall \, m \tag{1.10}$$

Notice that if (1.10) holds then

$$p_{\mathbf{z}(\zeta),\mathcal{S}(\zeta)|H(\zeta)}(\mathbf{z}, \mathcal{S}|H_m) = p_{\mathbf{z}(\zeta)|\mathcal{S}(\zeta)}(\mathbf{z}|\mathcal{S})p_{\mathcal{S}(\zeta)|H(\zeta)}(\mathcal{S}|H_m) \tag{1.11}$$

which means that the likelihood $p(\mathbf{z}|H_m)$ is equivalent to the likelihood $p(\mathcal{S}|H_m)$. In the engineering literature (1.11) is often summarized by the notion of *irrelevance* [WoJa65]. Specifically, if (1.11) holds, then $\mathbf{z}(\zeta)$ is said to be *irrelevant* in the presence of $\mathcal{S}(\zeta)$ for the purposes of deciding on $H(\zeta)$. This also often leads to reformulating the problem as an equivalent decision problem with $\mathcal{S}(\zeta)$ treated as the observation.

For example, in Example 1.1, one could obtain the joint pdf of $v_m(\zeta) = \mathbf{z}^T(\zeta)\mathbf{x}(H_m)$ for all m conditioned on $H(\zeta) = H_m$ and find the MAP rule via this equivalent problem.

When there is a nuisance parameter, the conditional likelihoods $p(\mathbf{z}|H_m, \Theta)$ and $p(\mathbf{z}|H_m; \Theta)$ always form a set of sufficient statistics for the stochastic and deterministic parameter models, respectively. For the average likelihood method, the notion of a sufficient statistic is clear since the marginalization over $\Theta(\zeta)$ in (1.6) yields the likelihood in (1.10). For the generalized likelihood cases in (1.7) and (1.8), we define the statistical sufficiency analogously with $p(\cdot)$ in (1.10) replaced by $g(\cdot)$ as appropriate (*i.e.*, with the averaging over Θ replaced by maximization). More precisely, $\mathcal{S}$ is a set of sufficient statistics for deciding on $H(\zeta)$ from $\mathbf{z}$ using a generalized rule for the nuisance parameter Θ if

$$g_{\mathbf{z}(\zeta)|\mathcal{S}(\zeta),H(\zeta)}(\mathbf{z}|\mathcal{S}, H_m) = g_{\mathbf{z}(\zeta)|\mathcal{S}(\zeta)}(\mathbf{z}|\mathcal{S}) \quad \forall\, m \qquad (1.12)$$

where $g(\cdot)$ represents maximization of the associated pdf over Θ according to (1.7) and (1.8) for the deterministic and random parameter models, respectively.

1.2   MAP Symbol and Sequence Detection

In order to begin constructing the building blocks of iterative detection receivers, we consider a typical example system which maps a sequence of independent symbols $\{a_m\}_{m=0}^{M-1}$ to a sequence of outputs $\{x_n\}_{n=0}^{N-1}$ by some manner that is known at the receiver. The sequence x_n is then corrupted by a memoryless probabilistic channel to produce the observation sequence z_n. The arbitrary system is illustrated in Fig-1.3 in three different block diagram conventions. With the *implicit index*

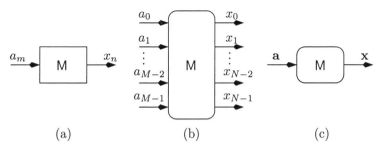

(a)   (b)   (c)

Figure 1.3.   The basic system block diagram in the (a) implicit index, (b) explicit index, and (c) vector mapping conventions.

block diagram convention, shown in Fig-1.3(a), a set of input and output variables are represented implicitly by a_m and x_n respectively, as is the

convention in much of the signal processing and communications literature. Each element of the input and output sequences is shown explicitly
in the *explicit index* convention of Fig-1.3(b). These explicit index block
diagrams are the graphical system models adopted in Chapter 2. In the
third convention, the system in viewed as a mapping from the input vec-
tor $\mathbf{a}_0^{M-1}$ to the output vector $\mathbf{x}_0^{N-1}$. The following provides a simple
system example which is expanded upon through much of this chapter.

Example 1.2. ─────────────────────────────────────
Consider bit-labeling of a non-binary signal set. In this context, the M
binary symbols a_m provide a bit label that uniquely corresponds to a
signal $x(\mathbf{a}_0^{M-1})$ in an 2^M-ary signal set. As a specific example, we consider the case of $M = 2$ and the simple (pulse amplitude modulation or
PAM) signal set. This convention is shown in Fig-1.4. Note that, since

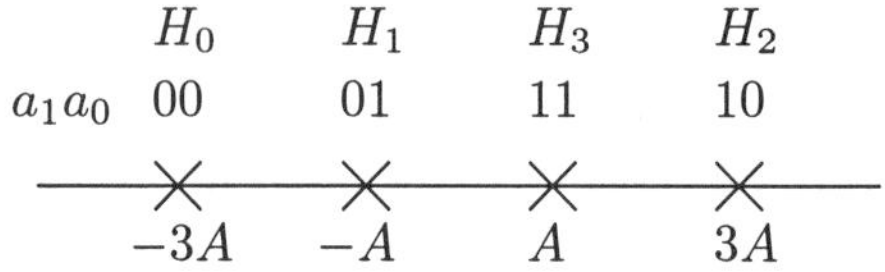

$$
\begin{array}{ccccc}
 & H_0 & H_1 & H_3 & H_2 \\
a_1 a_0 & 00 & 01 & 11 & 10
\end{array}
$$

$$
\begin{array}{cccc}
\times & \times & \times & \times \\
-3A & -A & A & 3A
\end{array}
$$

Figure 1.4. Bit labeling of a 4-PAM signal set used as a simple example of a system.

there is only one output signal, the form in Fig-1.3(b) is a natural way
to represent this system.
───────────────────────────────── *End Example*

With this fairly general system model, MAP detection of either $\mathbf{a}(\zeta)$
or $a_m(\zeta)$ for each value of m may be considered. The former, is an
$|\mathcal{A}|^M$-ary decision between sequences and is referred to as *MAP Se-
quence Detection (MAP-SqD)*. The latter is an $|\mathcal{A}|$-ary decision between
symbols and is referred to as *MAP Symbol Detection (MAP-SyD)*. It
is important to note that any sequence decision rule implies a set of
symbol decision rules (*i.e.*, one for each time index) and vice-versa since
there is a one-to-one correspondence between a set of symbol hypotheses
and sequence hypothesis. Thus, both the symbol and sequence decision
rules associated with MAP-SyD and MAP-SqD can be considered. In
the following, we develop these rules with an emphasis on their common
traits.

MAP Sequence Detection (MAP-SqD)   Denoting the decision
for $\mathbf{a}(\zeta)$ by $\hat{\mathbf{a}}$, for the MAP-SqD case, we obtain

$$\hat{\mathbf{a}} = \arg\max_{\mathbf{a}} p_{\mathbf{z}(\zeta)|\mathbf{a}(\zeta)}(\mathbf{z}|\mathbf{a}) p_{\mathbf{a}(\zeta)}(\mathbf{a}) \tag{1.13}$$

The MAP-SqD decision rule implies a decision rule for a_m defined by

$$\hat{a}_m = \arg\max_{a_m} P[a_m] \tag{1.14a}$$

$$P[a_m] = \max_{\mathbf{a}:a_m} p_{\mathbf{z}(\zeta)|\mathbf{a}(\zeta)}(\mathbf{z}|\mathbf{a})p_{\mathbf{a}(\zeta)}(\mathbf{a}) \tag{1.14b}$$

where the notation $u : v$ means all u consistent with v (*i.e.*, it may be read as "all u with v fixed"). While (1.14) simply says that the maximization of (1.13) can be performed in two stages, it is intended to emphasize that the symbol decision under the MAP-SqD criterion may be obtained by thresholding (maximizing) the *soft information* $P[a_m]$. The notation $P[\cdot]$ is used to denote some form of soft information such as probability or likelihood. In fact, this corresponding symbol decision rule is a p-generalized MAP decision rule with nuisance parameter $\{a_i(\zeta)\}_{i\neq m}$.

MAP Symbol Detection (MAP-SyD)   The MAP-SyD is obtained by the MAP rule for $a_m(\zeta)$ based on a realization of $\mathbf{z}(\zeta)$. Specifically, the MAP-SyD decision for $a_m(\zeta)$ is obtained by maximizing $p_{a_m(\zeta)|\mathbf{z}(\zeta)}(a_m|\mathbf{z})$ over all $a_m \in \mathcal{A}$. Since $p_{a_m(\zeta)|\mathbf{z}(\zeta)}(a_m|\mathbf{z}) \equiv p_{\mathbf{z}(\zeta),a_m(\zeta)}(\mathbf{z}, a_m)$ this may be expressed as

$$\hat{a}_m = \arg\max_{a_m} P[a_m] \tag{1.15a}$$

$$P[a_m] = \sum_{\mathbf{a}:a_m} p_{\mathbf{z}(\zeta)|\mathbf{a}(\zeta)}(\mathbf{z}|\mathbf{a})p_{\mathbf{a}(\zeta)}(\mathbf{a}) \tag{1.15b}$$

It should be noted that the symbol decision rules implied by the MAP symbol and the MAP sequence optimality criteria may be written in similar forms with the only difference being the manner in which the joint information $p(\mathbf{z}, \mathbf{a})$ is converted to marginal information on a_m. Specifically, this *marginalization* is accomplished by maximization and summation over all other symbols at locations other than m for the MAP-SqD and MAP-SyD cases, respectively. This may be viewed as an example of the generalized and average likelihood computation for (1.14b) and (1.15b), respectively, with nuisance parameter set $\Theta(\zeta) = \{a_i(\zeta)\}_{i\neq m}$.

The form of the sequence decision rule implied by the set of MAP-SyD rules for each symbol is less clear initially. It is straightforward, however, to show that the $|\mathcal{A}|^M$-ary sequence decision rule implied is that which minimizes the *average number of symbol errors* associated with the sequence decision. To show this, let $C(\tilde{\mathbf{a}}, \mathbf{a})$ denote the cost associated with deciding $\mathbf{a}(\zeta) = \tilde{\mathbf{a}}$ when the true sequence is $\mathbf{a}$ so that

(1.2) specializes to

$$C(\tilde{\mathbf{a}}|\mathbf{z}) \equiv \sum_{\mathbf{a}} C(\tilde{\mathbf{a}}, \mathbf{a}) p_{\mathbf{z}(\zeta)|\mathbf{a}(\zeta)}(\mathbf{z}|\mathbf{a}) p_{\mathbf{a}(\zeta)}(\mathbf{a}) \qquad (1.16)$$

By comparison, the sequence decision implied by the MAP-SyD criterion may be written as a sequence decision using the indicator function $I_{\{\cdot\}}$ as

$$\hat{\mathbf{a}} = [\arg\max_{\tilde{a}_0} \sum_{\mathbf{a}:\tilde{a}_0} p(\mathbf{z}, \mathbf{a}), \dots, \arg\max_{\tilde{a}_{M-1}} \sum_{\mathbf{a}:\tilde{a}_{M-1}} p(\mathbf{z}, \mathbf{a})]^{\mathrm{T}} \qquad (1.17\text{a})$$

$$= [\arg\max_{\tilde{a}_0} \sum_{\mathbf{a}} I_{\{a_0=\tilde{a}_0\}} p(\mathbf{z}, \mathbf{a}), \dots, \arg\max_{\tilde{a}_{M-1}} \sum_{\mathbf{a}} I_{\{a_{M-1}=\tilde{a}_{M-1}\}} p(\mathbf{z}, \mathbf{a})]^{\mathrm{T}}$$

$$= \arg\max_{\tilde{\mathbf{a}}} \sum_{\mathbf{a}} \left[\sum_{i=0}^{M-1} I_{\{a_i=\tilde{a}_i\}} \right] p(\mathbf{z}, \mathbf{a}) \qquad (1.17\text{b})$$

$$= \arg\min_{\tilde{\mathbf{a}}} \sum_{\mathbf{a}} d_H(\mathbf{a}, \tilde{\mathbf{a}}) p(\mathbf{z}, \mathbf{a}) \qquad (1.17\text{c})$$

Each term inside the brackets in (1.17a) corresponds to P$[a_m]$ in (1.15b) and the maximization can be decoupled and performed on each term independently yielding the MAP-SyD on each symbol as defined in (1.15a). For a given $\mathbf{a}$, the term $p(\mathbf{z}|\mathbf{a})p(\mathbf{a})$ will occur in a number of the summations in (1.17a). Specifically, using the indicator function for the condition that $a_m = \tilde{a}_m$, $I_{\{a_m=\tilde{a}_m\}}$ (1.17b) is obtained. Finally, with $d_H(\tilde{\mathbf{a}}, \mathbf{a})$ denoting the number of symbol disagreements in $\mathbf{a}$ and $\tilde{\mathbf{a}}$ (*i.e.*, the Hamming distance), (1.17c) follows from the fact that $\sum_{m=0}^{M-1} I_{\{a_m=\tilde{a}_m\}} = M - d_H(\mathbf{a}, \tilde{\mathbf{a}})$. Comparing (1.16) and (1.17), it is apparent that MAP-SyD yields the Bayes sequence decision associated with $C(\tilde{\mathbf{a}}, \mathbf{a}) = d_H(\tilde{\mathbf{a}}, \mathbf{a})$. More precisely, minimization of $C(\tilde{\mathbf{a}}|\mathbf{z})$ in (1.16) with $C(\tilde{\mathbf{a}}, \mathbf{a}) = d_H(\tilde{\mathbf{a}}, \mathbf{a})$ is equivalent to maximization shown in (1.17). Furthermore, it follows from (1.16) that this choice of cost coefficients yields a sequence decision rule that minimizes the average number of symbol errors. In summary, the sequence decision rule induced by MAP-SyD for each symbol is a (non-MAP) Bayes rule with costs proportional to number of symbol differences between two hypothesized and potential decision sequences.

The above development does not exploit the assumptions about the source and the channel. The independent symbol sequence assumption, implies the factorization $p_{\mathbf{a}(\zeta)}(\mathbf{a}) = \prod_m p_{a_m(\zeta)}(a_m)$. Also, the memoryless channel assumption implies that

$$p_{\mathbf{z}(\zeta)|\mathbf{a}(\zeta)}(\mathbf{z}|\mathbf{a}) = p_{\mathbf{z}(\zeta)|\mathbf{x}(\zeta)}(\mathbf{z}|\mathbf{x}(\mathbf{a})) = \prod_{n=0}^{N-1} p_{z_n(\zeta)|x_n(\zeta)}(z_n|x_n(\mathbf{a})) \qquad (1.18)$$

where the notation $x_n(\mathbf{a})$ denotes the value of $x_n(\zeta)$ arising when $\mathbf{a}(\zeta) = \mathbf{a}$. The only assumption on the structure of the system is that specifying the input $\mathbf{a}$ uniquely determines the output $\mathbf{x}$, denoted by $\mathbf{x}(\mathbf{a})$.

In summary, we have that MAP-SqD and MAP-SyD are both optimal under different criteria and both may be expressed as symbol decision rules in terms of the marginal channel likelihoods and the a-priori probabilities:

$$\hat{a}_m = \arg\max_{a_m} \left[\max_{\mathbf{a}:a_m} p(\mathbf{z}|\mathbf{x}(\mathbf{a}))p(\mathbf{a}) \right] \quad \text{(MAP-SqD)} \quad (1.19a)$$

$$\hat{a}_m = \arg\max_{a_m} \left[\sum_{\mathbf{a}:a_m} p(\mathbf{z}|\mathbf{x}(\mathbf{a}))p(\mathbf{a}) \right] \quad \text{(MAP-SyD)} \quad (1.19b)$$

$$p(\mathbf{z}|\mathbf{x}(\mathbf{a}))p(\mathbf{a}) = \prod_{n=0}^{N-1} p(z_n|x_n(\mathbf{a})) \times \prod_{m=0}^{M-1} p(a_m) \quad (1.19c)$$

Example 1.3. ―――――――――――――――――――――

Continuing with the Example 1.2, we consider MAP detection of a_0^1 (*i.e.*, MAP-SqD) and MAP detection of each a_m individually (*i.e.*, MAP-SyD). It may seem odd to view detection of a_m as "symbol detection" in this case since it may be most natural to refer to x as the symbol and $\{a_m\}$ as bits. However, this is a consequence of the simplicity of this system and the current context. Consider the case where $z(\zeta) = x(\mathbf{a}(\zeta)) + w(\zeta)$, and $w(\zeta)$ is a mean zero Gaussian random variable with variance $N_0/2$ (*i.e.*, $p_{w(\zeta)}(w) = \mathcal{N}(w; 0; N_0/2)$) which is independent of the input symbols. In this case

$$p_{z(\zeta)|x(\zeta)}(z|x(\mathbf{a}))p_{\mathbf{a}(\zeta)}(\mathbf{a}) = \mathcal{N}(z; x(\mathbf{a}); N_0/2))p(\mathbf{a}) \quad (1.20a)$$

$$= \frac{p(a_0)p(a_1)}{\sqrt{\pi N_0}} e^{\frac{-[z-x(a_1,a_0)]^2}{N_0}} \quad (1.20b)$$

Thus, MAP-SqD is achieved by a 4-ary maximization

$$\hat{\mathbf{a}} \Leftarrow \max \left[p_{a_1(\zeta)}(0)p_{a_0(\zeta)}(0)e^{\frac{-(z+3A)^2}{N_0}}, p_{a_1(\zeta)}(1)p_{a_0(\zeta)}(0)e^{\frac{-(z-3A)^2}{N_0}}, \right.$$

$$\left. p_{a_1(\zeta)}(1)p_{a_0(\zeta)}(1)e^{\frac{-(z-A)^2}{N_0}}, p_{a_1(\zeta)}(0)p_{a_0(\zeta)}(1)e^{\frac{-(z+A)^2}{N_0}} \right] \quad (1.21)$$

If the input symbols are uniformly distributed so that $p(a_0)p(a_1) = 1/4$, this reduces to the rule

$$\hat{\mathbf{a}} = \arg\max_{a_0, a_1} e^{\frac{-[z-x(a_1,a_0)]^2}{N_0}} = \arg\min_{a_0, a_1} \frac{1}{N_0}|z - x(a_1, a_0)|^2 \quad (1.22)$$

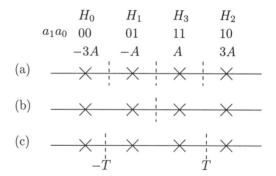

Figure 1.5. Decision regions for MAP-SqD in the 4-PAM example: (a) for $\mathbf{a}(\zeta)$, (b) for $a_1(\zeta)$, and (c) for $a_0(\zeta)$ with $T = 2A$.

the latter of which is the familiar *minimum distance* rule. The decision regions for this case are illustrated in Fig-1.5(a). It is apparent that the decision region for selecting $\hat{a}_1 = 0$ is $(-\infty, 0]$ with $\hat{a}_1 = 1$ selected otherwise. Similarly, $\hat{a}_0 = 0$ is selected iff $|z| > 2A$.

For the MAP-SyD case, first consider detection of $a_1(\zeta)$ via

$$\hat{a}_1 = \arg\max_{a_1} P[a_1] \tag{1.23}$$

$$P[a_1 = 1] = p_{a_1(\zeta)}(1)p_{a_0(\zeta)}(0)e^{\frac{-(z-3A)^2}{N_0}} + p_{a_1(\zeta)}(1)p_{a_0(\zeta)}(1)e^{\frac{-(z-A)^2}{N_0}}$$

$$P[a_1 = 0] = p_{a_1(\zeta)}(0)p_{a_0(\zeta)}(0)e^{\frac{-(z+3A)^2}{N_0}} + p_{a_1(\zeta)}(0)p_{a_0(\zeta)}(1)e^{\frac{-(z+A)^2}{N_0}}$$

For the special case of uniform a-priori probabilities on the inputs, this can be simplified to

$$e^{\frac{-(z-3A)^2}{N_0}} + e^{\frac{-(z-A)^2}{N_0}} \overset{\hat{a}_1 = 1}{\underset{\hat{a}_1 = 0}{\gtrless}} e^{\frac{-(z+3A)^2}{N_0}} + e^{\frac{-(z+A)^2}{N_0}} \tag{1.24}$$

Note that the left side of (1.24) is obtained by negating the z argument in the right side of (1.24). It follows that there is a tie condition at $z = 0$ and that the decision rule is a partition into positive and negative values of z. Clearly, for $z < 0$, $P[a_1 = 0] > P[a_1 = 1]$, so the rule is $\hat{a}_1 = 1$ iff $z > 0$. This is the same as that implied by the MAP-SqD rule.

The MAP-SyD rule for $a_0(\zeta)$ with uniform a-priori probabilities does not reduce to that implied by the MAP-SqD rule. Specifically, following similar steps as above and using the fact that $\cosh(z) = \cosh(-z)$, we obtain

$$\exp\left(\frac{-4A^2}{N_0}\right)\cosh\left(\frac{6A}{N_0}|z|\right) \overset{\hat{a}_0 = 0}{\underset{\hat{a}_0 = 1}{\gtrless}} \exp\left(\frac{4A^2}{N_0}\right)\cosh\left(\frac{2A}{N_0}|z|\right) \tag{1.25}$$

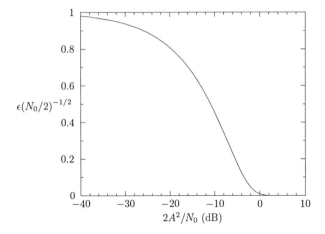

Figure 1.6.   The deviation of the threshold for MAP-SyD a_0 from that associated with the MAP-SqD in the 4-PAM example. Note that this deviation is approximately zero over the entire range of useful SNR.

The hyperbolic cosine is strictly monotonically increasing function (SMIF) for non-negative arguments, so this may be written as

$$|z| \overset{\hat{a}_0 = 0}{\underset{\hat{a}_0 = 1}{\gtrless}} T = \left(\frac{N_0}{2A}\right) f^{-1} \left(\exp\left[\frac{8A^2}{N_0}\right]\right) \tag{1.26}$$

where $f(x) = \cosh(3x)/\cosh(x)$ is invertible for $x > 0$ by the SMIF property of the $\cosh(\cdot)$. Notice that this test is of the same form as the MAP-SqD rule. However, the threshold T for the MAP-SqD rule is $T = 2A$, while for this MAP-SyD rule the threshold is a function of the parameters A and N_0. At moderate to high SNR (*i.e.*, proportional to A^2/N_0), we expect the rules to be approximately the same. In fact, note that $f(x) \cong e^{2x}$ for large x so that T tends toward $2A$ for large SNR. This motivates expressing T in (1.26) as $T = 2A + \epsilon$. The factor $\epsilon/\sqrt{N_0/2}$ is plotted against SNR in Fig-1.6. Notice that $\epsilon/\sqrt{N_0/2}$ tends toward a value of 1 as $N_0 \to \infty$ and tends toward zero for large SNR.

——————————————————————————————— *End Example*

In order to further develop the commonality of the MAP sequence and symbol detectors, it is instructive to consider both operating in the *"metric"* domain. Specifically, given a probability or likelihood on a quantity $\mathbf{v}$, we define the associated metric as $-\ln[p(\mathbf{v})]$. Note that the equivalence operation in the "probability" domain is multiplication by any positive constant which translates to addition of any finite-valued constant in the metric domain. Specifically, $-\ln p'(\mathbf{v}) \equiv -\ln p(\mathbf{v}) + h$ where h is any finite constant. Using these conventions, the MAP-SqD relations in (1.19a) can be expressed directly in the metric domain as

$$\hat{a}_m = \arg\min_{a_m} \left[\min_{\mathbf{a}:a_m} \left(-\ln p(\mathbf{z}|\mathbf{x}(\mathbf{a})) - \ln p(\mathbf{a}) \right) \right] \quad \text{(MAP-SqD)} \quad (1.27a)$$

$$-\ln p(\mathbf{z}|\mathbf{x}(\mathbf{a})) - \ln p(\mathbf{a}) = -\sum_{n=0}^{N-1} \ln p(z_n|x_n(\mathbf{a})) - \sum_{m=0}^{M-1} \ln p(a_m) \quad (1.27b)$$

which follows directly using the fact that the $-\ln p(\cdot)$ and $\max(\cdot)$ operations commute.

Expressing the MAP symbol decision rule in the metric domain is complicated by the fact that the $-\ln p(\cdot)$ and summation operations do not commute. However, defining the $\min^*(\cdot)$ operator [RoViHo95] as (see Problem 1.2)

$$\min{}^*(x,y) \triangleq -\ln\left(e^{-x} + e^{-y}\right) \tag{1.28a}$$

$$= \min(x,y) - \ln(1 + e^{-|x-y|}) \tag{1.28b}$$

$$\min{}^*(x,y,z) \triangleq -\ln\left(e^{-x} + e^{-y} + e^{-z}\right) \tag{1.28c}$$

$$= \min{}^*(\min{}^*(x,y),z) \tag{1.28d}$$

we can also express the MAP-SyD rule in the metric domain. Notice that $\min^*(x,y)$ is neither x nor y in general. Also, when $|x-y|$ is large $\min^*(x,y) \cong \min(x,y)$. Using this definition, the MAP-SyD rule in (1.19b) is obtained in the metric domain as

$$\hat{a}_m = \arg\min_{a_m} \left[\min_{\mathbf{a}:a_m}{}^* \left(-\ln p(\mathbf{z}|\mathbf{x}(\mathbf{a})) - \ln p(\mathbf{a}) \right) \right] \quad \text{(MAP-SyD)} \quad (1.29)$$

where $-\ln p(\mathbf{z}|\mathbf{x}(\mathbf{a})) - \ln p(\mathbf{a})$ may be computed as in (1.27b).

Example 1.4. ———————————————————————
The process described in Example 1.3 can be be carried out in the metric domain using the development above. For example, the second expression in (1.22) is already in the metric domain. Furthermore, (1.24) and (1.25) can be expressed, respectively, in the metric domain as

$$\min{}^* \left(\frac{[z-3A]^2}{N_0}, \frac{[z-A]^2}{N_0} \right) \mathop{\gtrless}_{\hat{a}_1 = 0}^{\hat{a}_1 = 1} \min{}^* \left(\frac{[z+3A]^2}{N_0}, \frac{[z+A]^2}{N_0} \right) \tag{1.30}$$

$$\frac{8A^2}{N_0} + \min{}^* \left(\frac{-6A}{N_0}|z|, \frac{6A}{N_0}|z| \right) \mathop{\gtrless}_{\hat{a}_0 = 1}^{\hat{a}_0 = 0} \min{}^* \left(\frac{-2A}{N_0}|z|, \frac{2A}{N_0}|z| \right) \tag{1.31}$$

To obtain the ML-SyD rule for a_1 directly from (1.30), see Problem 1.11.
——————————————————————————————— *End Example*

1.2.1   The General Combining and Marginalization Problem and Semi-Ring Algorithms

Expressing the MAP symbol and MAP sequence decision rules as thresholded[1] versions of some soft information on the conditional symbol value a_m has been accomplished in both the probability and metric domains. Inspecting (1.19a), (1.19b), (1.27a) and (1.29), four *marginalization operators* are identified: max, $\sum$, min, and min*, respectively. Each of these marginalization operators converts joint soft information on input-output pair $(\mathbf{a}, \mathbf{x}(\mathbf{a}))$ to marginal soft information on the conditional values of the symbols $\{a_m(\zeta)\}$. Furthermore, the *combining operators* used in the probability and metric domains are $\prod$ and $\sum$, respectively. Note that these combining operators combine marginal soft information on a_m and $x_n(\mathbf{a})$ to obtain soft information on the input-output pair $(\mathbf{a}, \mathbf{x}(\mathbf{a}))$ by either summing metrics or multiplying probabilities and/or likelihoods.

With this interpretation, it is useful to formally name the soft information that is thresholded to obtain the MAP-SqD and MAP-SyD solutions in the previous development. Specifically, let u be any quantity derived from the input-output pair $(\mathbf{a}, \mathbf{x}(\mathbf{a}))$, then[2]

$$\mathrm{APP}[u] \triangleq \sum_{\mathbf{a}:u} p(\mathbf{z}|\mathbf{a})p(\mathbf{a}) \qquad \text{(A-Posteriori Prob.)} \qquad (1.32a)$$

$$\mathrm{M^*SM}[u] \triangleq \min_{\mathbf{a}:u}{}^*[-\ln p(\mathbf{z}|\mathbf{a})p(\mathbf{a})] \quad \text{(Min}^* \text{ Sum Metric)} \qquad (1.32b)$$

$$\mathrm{GAP}[u] \triangleq \max_{\mathbf{a}:u} p(\mathbf{z}|\mathbf{a})p(\mathbf{a}) \qquad \text{(Generalized APP)} \qquad (1.32c)$$

$$\mathrm{MSM}[u] \triangleq \min_{\mathbf{a}:u} -\ln[p(\mathbf{z}|\mathbf{a})p(\mathbf{a})] \quad \text{(Min. Sequence Metric)} \qquad (1.32d)$$

where $\triangleq$ denotes definition to an equivalent quantity. For the case of $u = a_m$ the quantities in (1.32) give soft information on a_m which is thresholded in (1.14a) and (1.27a) for MAP-SqD and in (1.15a) and (1.29) for MAP-SyD, respectively. Furthermore, the joint quantities that are marginalized in (1.32) are obtained via the combining operations in (1.19c) and (1.27b) for the probability and metric domains, respectively. This is summarized in Table 1.1. The terminology arises from the fact

---

[1]Thresholding is a maximization in the probability domain and a minimization in the metric domain. This will be implicit in the following.

[2]These and related quantities are referred to in the literature using different terms. For example, what we refer to as an APP algorithm is often called a "MAP" algorithm. Similarly, an algorithm producing the negative of the MSM is referred to as a "log-max-MAP" algorithm by some authors.

that $\text{GAP}[a_m] = p(a_m)g(\mathbf{z}|a_m)$ where $g(\mathbf{z}|a_m)$ is the generalized likelihood of (1.8) with $\Theta(\zeta) = \{a_i(\zeta)\}_{i \neq m}$. Similarly, the APP in (1.32a) for $u = a_m$ is the product of the a-priori probability and the average likelihood over this nuisance set. These associated likelihood quantities are also listed in Table 1.1.

Marg.	Combining	Reliability Measure	Associated Likelihood
sum	product	APP	average likelihood (AL)
min*	sum	neg-log-APP (M*SM)	neg-log AL
max	product	generalized APP	generalized likelihood (GL)
min	sum	MSM	neg-log GL

Table 1.1. Summary of marginal soft information resulting from various combining and marginalization operators.

While the notation is a bit cumbersome, we carry it to emphasize the general nature of the soft information quantities defined and to encourage the development of intuition. The MSM soft information measure is especially intuitive. Specifically, $\text{MSM}[u]$ is the metric associated with the best sequence $\mathbf{a}$ that is consistent with the conditional quantity u. Thus, $\text{MSM}[u]$ is found by solving a constrained shortest-path problem. This concept is illustrated in Fig-1.7 for $u = a_k$. We have

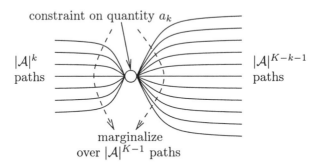

Figure 1.7. Soft information on a quantity corresponds to a marginalization of the joint soft information over all possible paths $\mathbf{a}$.

defined these quantities in terms of a general quantity u to allow future use in the most general sense. For example, one may consider the $\text{MSM}[x_n]$ – *i.e.*, the metric of the best sequence that is consistent with the output $x_n(\zeta) = x_n$. Similarly, this allows one to directly solve MAP decision problems on quantities related to $(\mathbf{a}, \mathbf{x}(\mathbf{a}))$. For example, the MAP decision on $u(\zeta) = (a_m(\zeta), a_{m+2}(\zeta))$ is obtained by maximizing $\text{APP}[a_m, a_{m+2}]$ with respect to (a_m, a_{m+2}). Similarly, the decision on

(a_m, a_{m+2}) consistent with MAP-SqD is obtained by thresholding the corresponding MSM.

Another important reason to carefully identify the soft-out that is produced is that if one can confirm that two algorithms are producing the same soft-output, then those two algorithms are equivalent. We will see that for many systems, one can avoid the exhaustive marginalization and combining implied in the above equations by exploiting local structure of the system. Furthermore, it will become clear that different methods for exploiting this structure can be followed which eventually produce the same soft information.

It is worth noting that the definitions in (1.32) are valid even in cases more general than the example system being considered. Specifically, only two components are required: the a-priori probability of the hypothesis or composite hypotheses, and the complete statistical description of the observation conditioned on the hypotheses. However, for the case of interest where $\mathbf{a}(\zeta)$ affects $\mathbf{z}(\zeta)$ only indirectly through $\mathbf{x}(\zeta) = \mathbf{x}(\mathbf{a}(\zeta))$, it is convenient to view the likelihood $p_{\mathbf{z}(\zeta)|\mathbf{x}(\zeta)}(\mathbf{z}|\mathbf{x}(\mathbf{a}))$ as soft information on $\mathbf{x}(\zeta) = \mathbf{x}(\mathbf{a})$. Specifically, denoting soft information on $u(\zeta)$ in the probability domain as P$[u]$ the combining in (1.19c) translates to

$$P[\mathbf{a}, \mathbf{x}(\mathbf{a})] \equiv p_{\mathbf{z}(\zeta)|\mathbf{x}(\zeta)}(\mathbf{z}|\mathbf{x}(\mathbf{a}))p_{\mathbf{a}(\zeta)}(\mathbf{a}) = P[\mathbf{x}(\mathbf{a})]P[\mathbf{a}] \qquad (1.33a)$$

$$\equiv \prod_{n=0}^{N-1} PI[x_n(\mathbf{a})] \prod_{m=0}^{M-1} PI[a_m] \qquad (1.33b)$$

where the notation PI[·] is used to denote marginal *soft-in* information in the probability domain: PI$[a_m] \equiv p(a_m)$ and PI$[x_n] \equiv p(z_n|x_n)$. Similarly, letting M[·] denote soft information in the metric domain, and MI[·] denote marginal soft-in metrics, the combining operation in (1.27b) may be written as

$$M[\mathbf{a}, \mathbf{x}(\mathbf{a})] \equiv -\ln p(\mathbf{z}|\mathbf{x}(\mathbf{a})) - \ln p(\mathbf{a}) = M[\mathbf{x}(\mathbf{a})] + M[\mathbf{a}] \qquad (1.34a)$$

$$\equiv \sum_{n=0}^{N-1} MI[x_n(\mathbf{a})] + \sum_{m=0}^{M-1} MI[a_m] \qquad (1.34b)$$

$$MI[a_m] \equiv -\ln p(a_m) \qquad (1.34c)$$

$$MI[x_n] \equiv -\ln p(z_n|x_n) \qquad (1.34d)$$

Note that the explicit dependence on the observation is lost in this notation. When important, we will stress the observation set using, for example, P$_i^j[a_m]$ to denote soft information based on $\mathbf{z}_i^j$.

The above relations make the correspondence between different decision criteria, executed in either probability or metric domain, clear.

Each involves a combination of marginal soft information on the inputs and the outputs of the system to obtain joint soft information, and a marginalization of this information over the structure of the system. Next we describe this similarity more formally.

1.2.1.1   Relations Between Soft Information

Let $S[u]$ and $S'[u]$ denote general soft measures on a quantity $u(\zeta)$ derived from the input-output pair $(\mathbf{a}(\zeta), \mathbf{x}(\mathbf{a}(\zeta)))$ – *i.e.*, $S[\cdot]$ may be in the metric, probability domain or even another domain. We define the following characteristics of a soft measure. First, we continue with the same use of the term *equivalent*. Specifically, $P[u] \equiv P'[u]$ in the probability domain if $P[u] = cP'[u]$ $\forall u$ for some $c > 0$, and in the metric domain if $M[u] = M'[u] + h$ $\forall u$ for some finite h. Essentially, two equivalent soft measures contain the same information and represent it in a compatible manner which implies that they are combined and marginalized using the same operations.

Second, two soft measures are *isomorphic* if there exists an invertible mapping from one measure to the other. More precisely, if one can always determine the equivalent of $S[u]$ from $S'[u]$ for all values of u and vice-versa, the measures are isomorphic. Isomorphic soft measures contain the same information, but may express this information in different forms. For example, the APP and $-$APP are isomorphic, with each using sum-product combining and marginalization, but with thresholding defined by $\max(\cdot)$ and $\min(\cdot)$, respectively. Similarly, the GAP and MSM are isomorphic via the $-\ln p(\cdot)$ mapping which changes not only the thresholding operation, but the marginalization and combining operations as well. Clearly, equivalent soft measures are isomorphic, but the converse does not hold in general.

Finally, $S[u]$ and $S'[u]$ are *threshold-consistent* if they both provide the same hard decisions when thresholded. Two isomorphic soft-outputs are threshold-consistent, but the converse does not hold. In fact, threshold-consistency is a very weak condition and says little about the relative quality of the soft information measures. For example, if $\hat{a}_m$ is the MAP-SyD, then the soft measure

$$P[a_m] = \begin{cases} 1 - \epsilon(|\mathcal{A}| - 1) & a_m = \hat{a}_m \\ \epsilon & a_m \neq \hat{a}_m \end{cases} \tag{1.35}$$

is threshold-consistent with $\text{APP}[a_m]$ for any $\epsilon \in [0, 1/|\mathcal{A}|)$. However, for small ϵ, the soft information in (1.35) provides little value beyond knowledge of $\hat{a}_m$.

Example 1.5.

From Example 1.3 we see that, for the specific 4-PAM case, MSM[a_1] and APP[a_1] are threshold-consistent, but MSM[a_0] and APP[a_0] are not. Note that MSM[a_1] and APP[a_1] are not isomorphic – *i.e.*, they contain distinct soft information.

End Example

1.2.1.2 Semi-Ring Algorithms

Since MAP-SqD and MAP-SyD generally provide different decisions on $a_m(\zeta)$, the APP and M*SM soft measures are not isomorphic to the MSM and GAP (*i.e.*, they are not even threshold-consistent). However, there is typically a correspondence between *algorithms* that compute the APP and, for example, the MSM. Specifically, as long as the marginalization and combining rules satisfy some properties, one need usually only derive one version of an algorithm and that algorithm can be converted to any other version by simply exchanging the marginalization and combining operations and modifying some initialization conditions. We refer to this as the *duality principle* between algorithms. Fortunately, the meaningful marginalization and combining operators in Table 1.1 satisfy these conditions. This duality principle is a powerful tool for two reasons: (i) only one version of an algorithm need typically be derived and specified; all other algorithms are implicitly specified, and (ii) one has flexibility to work with the most convenient set of marginalization and combining operations for the specific task.

The above advantages will become more clear as we develop algorithms and architectures in the following chapters. However, it is generally found that the APP ("sum-product") version is most useful when attempting to derive an algorithm based on formulas. Alternatively, the MSM ("min-sum") version can be used very effectively in deriving algorithms and architectures based on intuition. This intuition can then be formally verified using, for example, the APP version. Finally, for implementation, the metric domain is preferred for numerical stability. Thus, in simulation and implementation the min-sum and min*-sum versions are preferred and one is easily altered to provide the other.[3]

The condition underlying this duality principle is that the marginalization and combining operators considered (Ⓒ , Ⓜ), together with the ranges for the associated soft information ($\mathcal{F}$), form a *commutative*

---

[3]In software simulation this is as simple as selecting between two macros that define the min and min* operations.

semi-ring [AjMc00]. Specifically, $(\mathcal{F}, \copyright, \textcircled{m}, I_c, I_m)$ forms a *commutative semi-ring* if

(SR1) $\textcircled{m}$ and $\copyright$ are associative and commutative on $\mathcal{F}$

(SR2) Identity elements: $\exists\ I_c, I_m \in \mathcal{F}$ such that $f \copyright I_c = f$ and $f \textcircled{m} I_m = f$ for all $f \in \mathcal{F}$

(SR3) Distributive Law: $f \copyright (g \textcircled{m} h) = (f \copyright g) \textcircled{m} (f \copyright h)$

Note that, in general, there is no inverse for the marginalization or combining operator. However, for most cases of practical interest, the combining operation is invertible. Thus, throughout this book, we assume another property, namely

- Combining Inverse: $\forall\ f \in \mathcal{F}$ and $f \neq I_m$, there exists $\bar{f} \in \mathcal{F}$ such that $f \copyright \bar{f} = I_c$. We denote $g \copyright \bar{f}$ by $g \copyright^{-1} f$.

We use the inverse combining operator only to simplify the presentation of some operations. Furthermore, this operator is only applied in the form $(f \copyright g \copyright h) \copyright^{-1} f$ – *i.e.*, where it can be interpreted as operator that specifies a term be excluded from a stated combination.

The correspondence between the specific cases discussed and this general setting is summarized in Table 1.2. Since the combining operators

$S(\cdot)$	$\mathcal{F}$	$\textcircled{m}$	$\copyright$	I_m	I_c	$\bar{f}$	$\copyright^{-1}$	Threshold operation
APP	$[0, \infty)$	$+$	$\times$	0	1	$1/f$	$\div$	arg max
M*SM	$(-\infty, \infty]$	min*	$+$	∞	0	$-f$	$-$	arg min
GAP	$[0, \infty)$	max	$\times$	0	1	$1/f$	$\div$	arg max
MSM	$(-\infty, \infty]$	min	$+$	∞	0	$-f$	$-$	arg min

Table 1.2. Parameters of the semi-ring for each of the marginalization combining schemes discussed. The threshold operation is the method used to convert the given soft measure into a hard decision.

are associative and commutative, we use the notation

$$\overset{K-1}{\underset{k=0}{\copyright}} S[u_k] = S[u_0] \copyright S[u_1] \cdots \copyright S[u_{K-1}] \qquad (1.36a)$$

$$\overset{K-1}{\underset{k=0}{\textcircled{m}}} S[u_k] = S[u_0] \textcircled{m} S[u_1] \cdots \textcircled{m} S[u_{K-1}] \qquad (1.36b)$$

We can summarize all variations discussed thus far as a *general combining and marginalization (CM) problem*:

$$S[\mathbf{x}(\mathbf{a}), \mathbf{a}] = S[\mathbf{x}(\mathbf{a})] \, \copyright \, S[\mathbf{a}] \qquad (1.37a)$$

$$= \left(\overset{N-1}{\underset{n=0}{\copyright}} SI[x_n(\mathbf{a})] \right) \copyright \left(\overset{M-1}{\underset{m=0}{\copyright}} SI[a_m] \right) \qquad (1.37b)$$

$$S[a_m] = \underset{\mathbf{a}:a_m}{\textcircled{m}} S[\mathbf{x}(\mathbf{a}), \mathbf{a}] \qquad (1.37c)$$

where $SI[\cdot]$ has been used to denote input soft information ("soft-in") to distinguish it from the marginalized soft information (*e.g.*, $S[a_m]$). Specifically, it is straightforward to verify that computation of the four soft measures in (1.32) can be accomplished using (1.37) and the information in Table 1.2. This generality has been noted elsewhere. For example, based on the notion of min-sum processing, the problem of computing $S[a_m]$ in (1.37c) with the structure of (1.37c) is referred to as the *generalized shortest-path problem* [CoLeRi90, Sec. 26.4] or a *digital least metric* problem [ChChOrCh98]. Wiberg also noted the duality principle between certain min-sum and sum-product algorithms. Similarly, based on the sum-product intuition, the general CM problem has been called the *marginalized product function (MPF)* problem in [Aj99, AjMc00] with the commutative semi-ring properties of the marginalization and combining rules called the *generalized distributive law (GDL)*. A similar development may be found in [KsFrLo00]. The semi-ring property of the "max-sum" (*i.e.*, − min-sum) was identified and exploited to develop parallel architectures for the Viterbi Algorithm in [FeMe89].

We use the term *semi-ring* algorithm to describe an algorithm that solves (1.37) using *only* the semi-ring properties of the marginalization and combining operators. As such, a given semi-ring algorithm can be executed with different marginalization and combining operators to produce different marginal soft information. This is the *duality principle* referred to above. For example, if one has specified a min-sum semi ring algorithm to compute $MSM[a_m]$, then there is a corresponding APP algorithm that is defined by replacing all sums by products, all min operations by sums, and all occurrences of $I_c = 1$ by $I_c = 0$, and all occurrences of $I_m = 0$ by $I_m = \infty$. Thus, we may state a given semi-ring algorithm in any specific marginalization and combining semi-ring, or in the general notation used above.

Warning!   Care must be exercised in applying the duality principle. First, there are some meaningful combining and marginalization operations that do not satisfy the semi-ring property (*e.g.*, see [ChDeOr99] and Problem 1.6). Second, some algorithms based on semi-ring marginalization and combining operators may exploit characteristic of these operators that are beyond those guaranteed by the semi-ring property. For example, in the sum-product semi-ring there is an inverse under marginalization (subtraction) – *i.e.*, if $z = x + y$, then $y = z - x$. However, generally, there is no such inverse in a semi-ring and the min-sum semi-ring is an example of a semi-ring without a marginalization inverse – *i.e.*, if $z = \min(x, y)$, then y cannot be obtained from z and x. See Problem 1.10 for an non-semi-ring APP algorithm that has no corresponding MSM dual. Conversely, the Viterbi Algorithm, discussed in Section 1.3.2.1, is an MSM-based algorithm without an APP dual.[4] A commonly used non-semi-ring operation is the discarding of multiplicative constants in some min-sum algorithms. This point is illustrated in the next example and expounded upon in Section 2.4.1.

Example 1.6. ———————————————————————
From Examples 1.3 and 1.4 we may express the ML-SqD decision for $a_0(\zeta)$ as

$$\hat{a}_0 = \arg \min_{a_0} \mathrm{MSM}[a_0] \tag{1.38a}$$

$$\mathrm{MSM}[a_0] = \min_{a_1} \left[\frac{|z - x(a_1, a_0)|^2}{N_0} - \ln p(a_1) - \ln p(a_0) \right] \tag{1.38b}$$

$$= \frac{\min_{a_1} \left[|z - x(a_1, a_0)|^2 - N_0 \ln p(a_1) - N_0 \ln p(a_0) \right]}{N_0} \tag{1.38c}$$

The operations in (1.38b) and (1.38c) may be viewed as comprising a min-sum algorithm. It is tempting to simply replace the $\min(\cdot)$ operator in (1.38c) by the $\min^*(\cdot)$ operation. However, $\min^*(cx, cy) \neq c\min^*(x, y)$ in general. Thus, as expressed in (1.38c), this is not a semi-ring algorithm operating on the marginal soft-in information $\frac{|z-x(a_1,a_0)|^2}{N_0}$, $-\ln p(a_0)$, and $-\ln p(a_1)$. In fact, the corresponding $\min^*$-sum algorithm is

$$\hat{a}_0 = \arg \min_{a_0} \mathrm{M^*SM}[a_0] \tag{1.39a}$$

———————————————————
[4]The term semi-ring algorithm and GDL algorithm can basically be used interchangeably. However, we seek to emphasize the potential for a non-semi-ring algorithm based on marginalization and combining operators that form a semi-ring and thus, use this terminology to avoid inadvertently extending the term "GDL algorithm" in a manner not described in [AjMc00].

$$\text{M*SM}[a_0] = \min_{a_1}^* \left[\frac{|z - x(a_1, a_0)|^2}{N_0} - \ln p(a_1) - \ln p(a_0) \right] \tag{1.39b}$$

$$\neq \frac{1}{N_0} \min_{a_1}^* \left[|z - x(a_1, a_0)|^2 - N_0 \ln p(a_1) - N_0 \ln p(a_0) \right] \tag{1.39c}$$

Note that the factor $1/N_0$ in (1.38c) can be dropped for the purposes of making hard decisions. Moreover, if the a-priori probabilities are uniform, the ML-SqD rule does not require knowledge of N_0. In contrast, the ML-SyD rule does require knowledge of the noise variance.

—————————————————————————————— *End Example*

1.2.2   Detection with Imperfect CSI

When perfect channel state information (CSI) is not available at the receiver, *i.e.*, when the conditional observation depends on some additional parameter Θ, composite hypothesis tests can be formulated for the problem of sequence detection. The detection rule that minimizes the probability of sequence error for the case of a random parameter $\Theta(\zeta)$ is given by

$$\hat{\mathbf{a}} = \arg\max_{\mathbf{a}} p(\mathbf{z}, \mathbf{a}) = \arg\max_{\mathbf{a}} \mathbb{E}_{\Theta(\zeta)} \{ p(\mathbf{z}, \mathbf{a} | \Theta) \} \tag{1.40}$$

On the other hand, when a statistical description of Θ is not available, *generalized* tests can be used

$$\hat{\mathbf{a}} = \arg\max_{\mathbf{a}} \max_{\Theta} p(\mathbf{z}, \mathbf{a}; \Theta) \tag{1.41}$$

1.2.2.1   Deterministic Parameter Model

It was mentioned in Section 1.1.2 that the problem in (1.41) can be solved in two steps: first, conditioned on a hypothesized sequence $\mathbf{a}$, an appropriate estimate $\tilde{\Theta}(\mathbf{a})$ is formed, and then, $p(\mathbf{z}, \mathbf{a}; \tilde{\Theta}(\mathbf{a}))$ is maximized over all hypotheses, *i.e.*, all sequences $\mathbf{a}$. We are interested in the following special case for the observation

$$\mathbf{z}(\zeta) \triangleq \begin{bmatrix} z_{K-1}(\zeta) \\ \vdots \\ z_0(\zeta) \end{bmatrix} = \begin{bmatrix} x_{K-1}(\mathbf{a}(\zeta), \Theta) \\ \vdots \\ x_0(\mathbf{a}(\zeta), \Theta) \end{bmatrix} + \begin{bmatrix} w_{K-1}(\zeta) \\ \vdots \\ w_0(\zeta) \end{bmatrix}$$

$$= \mathbf{x}(\mathbf{a}(\zeta), \Theta) + \mathbf{w}(\zeta) \tag{1.42}$$

where $w_k(\zeta)$ is a complex circular AWGN with variance N_0. For this model, MAP-SqD simplifies to

$$\hat{\mathbf{a}} = \arg\min_{\mathbf{a}} \min_{\Theta} \left[-\ln p(\mathbf{a}) + \frac{1}{N_0} ||\mathbf{z} - \mathbf{x}(\mathbf{a}, \Theta)||^2 \right] \tag{1.43}$$

The inner minimization over the parameter Θ results in an estimate $\tilde{\Theta}(\mathbf{a})$, which in turn implies an estimate $\tilde{\mathbf{x}}(\mathbf{a}) = \mathbf{x}(\mathbf{a}, \tilde{\Theta}(\mathbf{a}))$. In the following example we further specialize the model in (1.42) and demonstrate that under this special case, the MAP sequence detector has a special structure, known as the Estimator-Correlator (EC) structure.

Example 1.7. ───
In this example, the unknown parameter Θ is assumed to be an unknown constant vector $(\Theta = \mathbf{f})$ of length $L + 1$. Furthermore, the mapping $\mathbf{x}(\mathbf{a}, \mathbf{f})$ is assumed to be decomposed as $\mathbf{x}(\mathbf{a}, \mathbf{f}) = \mathbf{Q}(\mathbf{a})\mathbf{f}$, where the $K \times (L + 1)$ matrix $\mathbf{Q}(\mathbf{a})$ represents an arbitrary mapping of the input sequence. The observation model in (1.42) can be written as

$$\mathbf{z}(\zeta) = \mathbf{Q}(\mathbf{a}(\zeta))\mathbf{f} + \mathbf{w}(\zeta) = \begin{bmatrix} \mathbf{q}_{K-1}(\mathbf{a}(\zeta))^T \\ \vdots \\ \mathbf{q}_0(\mathbf{a}(\zeta))^T \end{bmatrix} \mathbf{f} + \mathbf{w}(\zeta) \qquad (1.44)$$

and the sequence detection problem in (1.43) simplifies to

$$\hat{\mathbf{a}} = \arg\min_{\mathbf{a}} \; \min_{\mathbf{f}} \left[-\ln p(\mathbf{a}) + \frac{1}{N_0} \|\mathbf{z} - \mathbf{Q}(\mathbf{a})\mathbf{f}\|^2 \right] \qquad (1.45)$$

The inner minimization over $\mathbf{f}$ is a standard Least Squares (LS) problem, and the LS estimate of $\mathbf{f}$ is given by $\tilde{\mathbf{f}} = \mathbf{Q}^I \mathbf{z}$, where $\mathbf{Q}^I$ is the left pseudo-inverse of $\mathbf{Q}$, which for a full rank matrix $\mathbf{Q}$ is given by $\mathbf{Q}^I = (\mathbf{Q}^H \mathbf{Q})^{-1}\mathbf{Q}^H$. This estimate implies an LS estimate of $\mathbf{x}$ of the form $\tilde{\mathbf{x}} = \mathbf{Q}\tilde{\mathbf{f}} = \mathbf{Q}(\mathbf{Q}^H\mathbf{Q})^{-1}\mathbf{Q}^H\mathbf{z} = \mathbf{P}\mathbf{z}$, where $\mathbf{P} = \mathbf{Q}(\mathbf{Q}^H\mathbf{Q})^{-1}\mathbf{Q}^H$ is the matrix that projects to the column space of $\mathbf{Q}$. We note that the matrix $\mathbf{Q}$, and subsequently the quantities $\tilde{\mathbf{x}}$ and $\mathbf{P}$, depend on the hypothesized sequence $\mathbf{a}$. Substituting the last result back in (1.45), yields a quantity that explicitly involves only the data sequence

$$\hat{\mathbf{a}} = \arg\min_{\mathbf{a}} \left[-\ln p(\mathbf{a}) + \frac{1}{N_0} \|\mathbf{z} - \mathbf{P}(\mathbf{a})\mathbf{z}\|^2 \right] \qquad (1.46)$$

where the dependence on the data sequence is shown explicitly. Minimizing the projection error length is equivalent to maximizing the projection length, thus the above results in

$$\hat{\mathbf{a}} = \arg\max_{\mathbf{a}} \left[\ln p(\mathbf{a}) + \frac{\|\mathbf{P}(\mathbf{a})\mathbf{z}\|^2}{N_0} \right] = \arg\max_{\mathbf{a}} \left[\ln p(\mathbf{a}) + \frac{\mathbf{z}^H \tilde{\mathbf{x}}(\mathbf{a})}{N_0} \right]$$
$$(1.47)$$

where we have used the fact that an orthogonal projection matrix $\mathbf{P}$ satisfies $\mathbf{P} = \mathbf{P}^H$ and $\mathbf{P}\mathbf{P} = \mathbf{P}$. The last term in (1.47) is exactly the EC structure, with *correlation* being the inner product operation $\mathbf{z}^H \tilde{\mathbf{x}}(\mathbf{a})$.

End Example

It is now demonstrated that for the special case mentioned above, the EC can be implemented in an efficient forward recursive form. The only additional assumption required here is a causality condition for $\mathbf{q}_k(\mathbf{a})$, *i.e.*, $\mathbf{q}_k(\mathbf{a}) = \mathbf{q}_k(\mathbf{a}_0^k)$. The metric in (1.45) is slightly generalized by introducing an exponentially decaying window with forgetting factor ρ. This weighting provides increased numerical stability as well as the ability to track slow parameter variations. Minimization of this generalized metric over the unknown parameter $\mathbf{f}$, is equivalent to evaluating the following metric

$$
\mathrm{M}_0^{K-1}[\mathbf{a}, \mathbf{q}(\mathbf{a})] = \min_{\mathbf{f}} \left\{ \sum_{m=0}^{K-1} \left[\frac{|z_m - \mathbf{q}_m^T \mathbf{f}|^2}{N_0} - \ln p(a_m) \right] \rho^{K-1-m} \right\}
$$

(1.48)

By observing that the above quantity is the residual squared error of a weighted LS problem, a forward recursion can be derived as in [Ch95, Ha96]

$$
\mathrm{M}_0^k[\mathbf{a}_0^k, \mathbf{q}_0^k(\mathbf{a}_0^k)] = \rho \mathrm{M}_0^{k-1}[\mathbf{a}_0^{k-1}, \mathbf{q}_0^{k-1}(\mathbf{a}_0^{k-1})] - \ln p(a_k) +
$$
$$
\frac{\rho}{\rho + \mathbf{q}_k^T \tilde{\mathbf{P}}_{k-1} \mathbf{q}_k^*} \frac{|z_k - \mathbf{q}_k^T \tilde{\mathbf{f}}_{k-1}|^2}{N_0}
$$

(1.49a)

$$
\mathbf{r}_k = \frac{\tilde{\mathbf{P}}_{k-1} \mathbf{q}_k^*}{\rho + \mathbf{q}_k^T \tilde{\mathbf{P}}_{k-1} \mathbf{q}_k^*}
$$

(1.49b)

$$
\tilde{\mathbf{f}}_k = \tilde{\mathbf{f}}_{k-1} + \mathbf{r}_k(z_k - \mathbf{q}_k^T \tilde{\mathbf{f}}_{k-1})
$$

(1.49c)

$$
\tilde{\mathbf{P}}_k = \frac{1}{\rho}(\mathbf{I} - \mathbf{r}_k \mathbf{q}_k^T) \tilde{\mathbf{P}}_{k-1}
$$

(1.49d)

The method suggested by this set of equations, which have the form of a forward recursive Estimator-Correlator, with per-path recursive least squares (RLS) parameter estimators, *i.e.*, (1.49b)-(1.49d), is illustrated in Fig-1.8 and can be described as follows. Starting at time 0 a forward $|\mathcal{A}|$-ary tree is built, each node of which represents a sequence path. The quantity $\mathrm{M}_0^{k-1}[\mathbf{a}_0^{k-1}, \mathbf{q}_0^{k-1}(\mathbf{a}_0^{k-1})]$, together with $\tilde{\mathbf{f}}_{k-1}$ and $\tilde{\mathbf{P}}_{k-1}$ of that path are stored in each node. At each time k, the tree is expanded forward and the metrics corresponding to the newly generated branches are calculated using (1.49). In the context of the general soft quantities

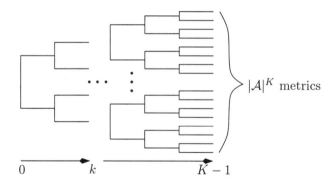

$\mathcal{|A|}^K$ metrics

0   k $K-1$

Figure 1.8.   Forward recursive Estimator-Correlator structure

of (1.37), the MAP-SqD problem can be expressed as

$$\hat{a}_k = \arg \max_{a_k} \mathrm{M}[a_k] \tag{1.50a}$$

$$\mathrm{M}[a_k] = \sum_{\mathbf{a}:a_k} \mathrm{M}[\mathbf{a}, \mathbf{q}(\mathbf{a})] \tag{1.50b}$$

where the quantity $\mathrm{M}[\mathbf{a}, \mathbf{q}(\mathbf{a})]$ is evaluated recursively by

$$\mathrm{M}[\mathbf{a}_0^k, \mathbf{q}_0^k(\mathbf{a}_0^k)] = \rho \mathrm{M}[\mathbf{a}_0^{k-1}, \mathbf{q}_0^{k-1}(\mathbf{a}_0^{k-1})] + \mathrm{MI}[a_k] + \mathrm{MI}[\mathbf{q}_k(\mathbf{a}_0^k)] \quad (1.50c)$$

Since we have a deterministic model for the nuisance parameter $\mathbf{f}$, and we operate in the logarithmic domain, the appropriate interpretation for the soft quantity $\mathrm{M}[a_k]$ is that of $\mathrm{MSM}[a_k]$.

We note that the innovations term in (1.49a) depends on the entire path history through the RLS recursions. This is the exact reason why sequence detection requires a maximization procedure over the entire path tree. This issue will be further discussed in 1.3.3.1.

1.2.2.2   Stochastic Parameter Model

It was mentioned that the MAP-SqD problem in (1.41) can be solved using a two step process. First, an estimate $\tilde{\Theta}(\mathbf{a})$ is obtained for each hypothesized sequence $\mathbf{a}$ and then $p(\mathbf{z}, \mathbf{a}; \tilde{\Theta}(\mathbf{a}))$ is maximized over all hypotheses. However, when a stochastic model for Θ is assumed, *i.e.*, $\Theta = \Theta(\zeta)$, the above two-step process is not applicable. This is due to the fact, that it is not always true that for each hypothesized sequence $\mathbf{a}$ there exists an estimate $\tilde{\Theta}(\mathbf{a})$, such that $\mathbb{E}_{\Theta(\zeta)}\{p(\mathbf{z}, \mathbf{a}|\Theta)\} = p(\mathbf{z}, \mathbf{a}|\tilde{\Theta}(\mathbf{a}))$. It can be shown though, that for an observation model similar to the one assumed in Example 1.7 (*i.e.*, $\mathbf{z}(\zeta) = \mathbf{Q}(\mathbf{a}(\zeta))\mathbf{f}(\zeta) + \mathbf{w}(\zeta)$) and under the assumption of joint Gaussian random vectors $\mathbf{f}(\zeta)$ and $\mathbf{w}(\zeta)$, an Estimator-Correlator structure exists (see Problem 4.1).

We now specialize to a linear observation model, similar to the one assumed in Example 1.7

$$\mathbf{z}(\zeta) = \begin{bmatrix} \mathbf{q}_{K-1}(\mathbf{a}(\zeta))^T \mathbf{f}_{K-1}(\zeta) \\ \vdots \\ \mathbf{q}_0(\mathbf{a}(\zeta))^T \mathbf{f}_0(\zeta) \end{bmatrix} + \mathbf{w}(\zeta) \qquad (1.51)$$

where the only difference between the above and (1.44) is a time varying random vector $\mathbf{f}_k(\zeta)$ instead of the constant $\mathbf{f}$. In addition, a first-order Gauss Markov (GM) random process $\{\mathbf{f}_k(\zeta)\}$ is assumed. This model is quite general, since it can describe higher order GM, as well as ARMA processes [AnMo79]. Assuming a time-invariant GM model for notational and expositional simplicity, the vector GM process $\{\mathbf{f}_k(\zeta)\}$ evolves in time according to the equations

$$\mathbf{f}_k(\zeta) = \mathbf{F}\mathbf{f}_{k-1}(\zeta) + \mathbf{u}_k(\zeta) \qquad \text{(forward)} \qquad (1.52a)$$

$$\mathbf{f}_k(\zeta) = \mathbf{F}^b \mathbf{f}_{k+1}(\zeta) + \mathbf{v}_k(\zeta) \qquad \text{(backward)} \qquad (1.52b)$$

where $\mathbf{u}_k(\zeta)$, $\mathbf{v}_k(\zeta)$ are zero-mean Gaussian vectors with covariance $\mathbf{K}_\mathbf{u}(m) = \mathbf{K}_\mathbf{u}\delta_m$ and $\mathbf{K}_\mathbf{v}(m) = \mathbf{K}_\mathbf{v}\delta_m$, respectively, where δ_k denotes the Kronecker delta function. The quantities $\mathbf{F}, \mathbf{F}^b, \mathbf{K}_\mathbf{u}$, and $\mathbf{K}_\mathbf{v}$ are selected such that the process is wide-sense stationary, with $\mathbb{E}\{\mathbf{f}_k(\zeta)\mathbf{f}_k(\zeta)^H\} = \mathbf{K}_\mathbf{f}$ (see Problem 1.29).

It is now shown that the MAP-SqD problem for the GM model above can be efficiently solved using a forward recursive Estimator-Correlator structure. Efficient evaluation of $p(\mathbf{z}_0^{K-1}, \mathbf{a}_0^{K-1})$ for each of the $|\mathcal{A}|^K$ input sequences can be based on the fact that the above joint density can be computed recursively as in [Il92]

$$p(\mathbf{z}_0^k, \mathbf{a}_0^k) = p(z_k|\mathbf{z}_0^{k-1}, \mathbf{a}_0^k)p(a_k)p(\mathbf{z}_0^{k-1}, \mathbf{a}_0^{k-1}) \qquad (1.53a)$$

$$p(z_k|\mathbf{z}_0^{k-1}, \mathbf{a}_0^k) = \mathcal{N}^{cc}(z_k; \mathbf{q}_k^T \tilde{\mathbf{f}}_{k|k-1}; N_0 + \mathbf{q}_k^T \tilde{\mathbf{F}}_{k|k-1}\mathbf{q}_k^*) \qquad (1.53b)$$

$$\mathbf{r}_k = \frac{\tilde{\mathbf{F}}_{k|k-1}\mathbf{q}_k^*}{N_0 + \mathbf{q}_k^T \tilde{\mathbf{F}}_{k|k-1}\mathbf{q}_k^*} \qquad (1.53c)$$

$$\tilde{\mathbf{f}}_{k|k} = \tilde{\mathbf{f}}_{k|k-1} + \mathbf{r}_k(z_k - \mathbf{q}_k^T \tilde{\mathbf{f}}_{k|k-1}) \qquad (1.53d)$$

$$\tilde{\mathbf{F}}_{k|k} = (\mathbf{I} - \mathbf{r}_k\mathbf{q}_k^T)\tilde{\mathbf{F}}_{k|k-1} \qquad (1.53e)$$

$$\tilde{\mathbf{f}}_{k+1|k} = \mathbf{F}\tilde{\mathbf{f}}_{k|k} \qquad (1.53f)$$

$$\tilde{\mathbf{F}}_{k+1|k} = \mathbf{F}\tilde{\mathbf{F}}_{k|k}\mathbf{F}^H + \mathbf{K}_\mathbf{u} \qquad (1.53g)$$

where $\mathcal{N}^{cc}(z; m; \sigma^2)$ denotes the probability density function of a complex circular Gaussian random variable with mean m, and variance $\sigma^2/2$

for the real and imaginary part, while $\tilde{\mathbf{f}}_{k|k-1}$ and $\tilde{\mathbf{F}}_{k|k-1}$ are the channel one-step prediction and corresponding covariance matrix updated by a sequence-conditioned Kalman filter (KF), *i.e.*, (1.53c)-(1.53g). This set of equations has an Estimator-Correlator structure similar to (1.49).

Again, the above processing can be put in the more generic notation of (1.37). Since we have a stochastic model for the nuisance parameter, we are able to find an exact expression for the quantity $P[\mathbf{a}, \mathbf{q}(\mathbf{a})] \equiv p(\mathbf{z}, \mathbf{a}) = \mathbb{E}_{\mathbf{f}(\zeta)} \{p(\mathbf{z}, \mathbf{a}|\mathbf{f})\}$, by averaging over $\{\mathbf{f}_k(\zeta)\}$. Therefore, marginalization of $P[\mathbf{a}, \mathbf{q}(\mathbf{a})]$ with respect to the nuisance sequence parameters $\{a_i\}_{i \neq k}$, results in the exact $GAP[a_k]$.

1.3   Data Detection for an FSM in Noise

In this section we consider the application of the preceding development to the important problem of data detection for systems that can be modeled as finite state machines (FSMs). This includes specialization of the MAP detection algorithms for perfectly known channels and extension to the case of imperfect CSI.

1.3.1   Generic FSM Model

We consider a system with input a_k and output x_k defined on the same time scale (*i.e.*, a *synchronous system*). For simplicity of the exposition, we assume that this is a causal system so that $x_k(\mathbf{a}) = x_k(\mathbf{a}_0^k)$. At any time k, the next output of the system x_k is determined by the current input a_k and the current system *state* s_k. For an arbitrary synchronous system, the state could be defined as all the previous inputs (*i.e.*, $s_k = \mathbf{a}_0^{k-1}$) but the number of states will grow exponentially with k. Thus, on a finite time interval, an arbitrary synchronous system can be represented by a degenerate case of a finite state machine, where s_k takes on $|\mathcal{A}|^k$ values for $k = 0, 1, \dots (K-1)$. However, the FSM term is typically reserved for the case when the number of states is bounded by a finite constant, on the index set $k = 0, 1, 2, \dots$ – *i.e.*, the system can only be in a finite number of states even when driven by an input sequence of indefinite length. Thus, while an arbitrary synchronous system can be represented by a *tree*, for an FSM, this tree *folds* into a *trellis* with nodes at depth k representing the conditional values of s_k. This is illustrated in Fig-1.9.

There are several equivalent representations of a given FSM, each of which characterizes the relations

$$x_k = x_k(a_k, s_k) = \text{out}_k(a_k, s_k) \qquad \text{(Output)} \qquad (1.54a)$$
$$s_{k+1} = s_{k+1}(a_k, s_k) = \text{ns}_k(a_k, s_k) \qquad \text{(Next State)} \qquad (1.54b)$$

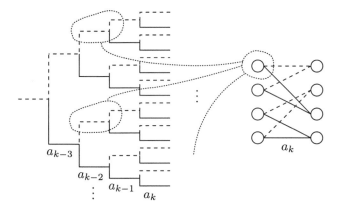

Figure 1.9. For an FSM, the tree of sequence hypotheses folds onto a trellis (shown for a memory 2 simple FSM with binary inputs).

Unless specified otherwise, we assume that the next-state and output functions are time invariant and that the number of possible states does not change with k. Thus, the FSM structure may be characterized by next-state and output tables, or an appropriately labeled state transition diagram. The trellis diagram is the latter with the time axis shown explicitly. We denote a state transition by $t_k = (s_k, a_k, s_{k+1})$.[5] Note that, in addition to $\mathbf{a}_0^{K-1}$, the FSM has the initial state s_0 as another implicit input. In the following, we may use $\mathbf{a}_0^k$ to implicitly represent $\{s_0, a_0, \ldots a_k\}$ for compactness.

This general FSM model, admits *parallel transitions* and complicated relationships between the state and the input sequence. Parallel transitions are transitions t_k that correspond to the same states s_k and s_{k+1}, but different inputs a_k. By "complicated" relations between the input and the state, we mean that at first glance the most efficient state representation for an FSM may not be clear. We refer to FSMs with no parallel transitions and $s_k = \mathbf{a}_{k-L}^{k-1}$ as *simple FSMs*. For a simple FSM, the transition is uniquely specified by (s_k, s_{k+1}) and the previous L symbols can be extracted from the state.

Example 1.8. ──

An equivalent discrete time model for a finite impulse response intersymbol interference (ISI) channel is a simple FSM. Specifically, the output is $x_k = a_k \circledast f_k = \sum_{m=0}^{L} f_m a_{k-m}$, where L is the memory (length) of the

---

[5]The transition is fully specified by (s_k, a_k), but we adopt the definition of $t_k = (s_k, a_k, s_{k+1})$ to make the notation $t_k : s_{k+1}$ more natural.

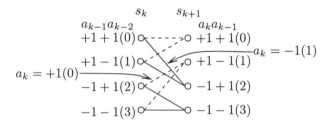

	0 input		1 input	
current state	next state	output	next state	output
0	0	0	2	4
1	0	1	2	5
2	1	2	3	6
3	1	3	3	7

Figure 1.10. Representing a 4-state FSM with a trellis diagram and next-state and output index tables.

channel. The trellis diagram for a 3-tap $(L = 2)$ ISI channel with BPSK input symbols (*i.e.*, $a_k \in \{-1, +1\}$) is illustrated in Fig-1.10. Note that, for this simple FSM, the state may be viewed as the contents of a shift register with inputs added on the left-hand side. Also, the value of the input a_k associated with a given transition is represented by the line style. We use this convention throughout.

It is also natural to represent the FSM by indexing the finite number of inputs, states, and outputs. For example, it is natural to consider $a_k = (-1)^{b_k}$ to be the result of BPSK modulation of the information bits $b_k \in \{0, 1\}$. Similarly, the states can be indexed naturally, as denoted in Fig-1.10. In some cases it is simpler to describe the FSM in terms of these indices while in others, it is easier to use the real (or complex) valued quantities. For example, in software implementation, indexing every variable starting from zero is most natural. Thus, the table shown in Fig-1.10 may be useful for describing an FSM in software with an additional table translating between the index and the numerical value of the variable (*i.e.*, see Problem 1.22). Note that, as shown, the output of the system is taken as the index of the transition. More generally, the number of outputs $|\mathcal{X}|$ may be much smaller than the number of transitions $|\mathcal{T}| = |\mathcal{A}| \times |\mathcal{S}|$. For example, if $f_0 = f_2 = 1$ and $f_1 = 2$ in the current example, there are only 5 distinct outputs.

Thus, one can describe an FSM as a system with integer-valued inputs, states, transitions, and outputs, followed by a mapper, or alternatively by the physical real-valued (or complex-valued) quantities that they represent. We do not introduce a notation to describe indexing of variables, which would require explicit mappers to be shown in the block diagrams, rather we use the convention that is most natural in the given context. It will be clear from the context when the indexing convention is being used.

———————————————————————————— *End Example*

An FSM model for a system is not unique. For example, many trellis coded modulation (TCM) encoders are most efficiently modeled as an FSM with parallel transitions, but a model with more states and no parallel transitions is possible (see Problem 1.14). Methods for obtaining FSM models or minimal state realizations are described in the literature (*e.g.*, [Gi62, BaCoJeRa74, BiDiMcSi91]) and are not addressed here.

When the structure of the FSM is described, in part, by some parameter set Θ, we may either view the parameter as part of the FSM, or add a mapper to map the FSM transitions to the real values. For example if the channel coefficients in Example 1.8 were unknown to the receiver, we may either model this internal to the FSM, or consider the FSM output to be a_{k-2}^k which is the input to a mapper producing $x_k = f^T a_{k-2}^k$. Again, the parameters may be modeled statistically, with a deterministic model, or with a mixture of these.

1.3.2 Perfect Channel State Information

Consider the special case of an FSM system in the development of Section 1.2. Specifically, assume that the output of an FSM drives a memoryless channel characterized by $p_{z_k(\zeta)|x_k(\zeta)}(z_k|x_k)$. It follows that $p(z_0^{K-1}|t_0^{K-1})$ factors into $\prod_k p(z_k|t_k)$ where t_0^{K-1} is a (valid) transition sequence. Thus, for the special case of the FSM system, the combining rule in the probability domain may be written as

$$p(z_0^{K-1}|a_0^{K-1}, s_0)p(a_0^{K-1}, s_0) = p(z_0^{K-1}|t_0^{K-1})p(t_0^{K-1}) \qquad (1.55a)$$

$$= \prod_{k=0}^{K-1} p(z_k|t_k) \prod_{k=0}^{K-1} p(t_k|t_{k-1}) \qquad (1.55b)$$

$$= \prod_{k=0}^{K-1} [p(z_k|x_k(t_k))p(a_k)]p(s_0) \qquad (1.55c)$$

where we use the one-to-one relation between the input sequence and the associated sequence of state transitions and $p(t_0) = p(a_0)p(s_0)$. In the following, this convention is used implicitly. Thus, as before, the soft-in information on x_k and a_k is

$$P_k[t_k] \triangleq \mathrm{PI}[x_k(t_k)]\mathrm{PI}[a_k] \tag{1.56a}$$

$$\mathrm{PI}[x_k(t_k)] \triangleq p(z_k|x_k(t_k)) \quad \text{and} \quad \mathrm{PI}[a_k] \triangleq p(a_k) \tag{1.56b}$$

In the metric domain (1.56) becomes

$$M_k[t_k] \triangleq \mathrm{MI}[x_k(t_k)] + \mathrm{MI}[a_k] \tag{1.57a}$$

$$\mathrm{MI}[x_k(t_k)] \triangleq -\ln p(z_k|x_k(t_k)) \quad \text{and} \quad \mathrm{MI}[a_k] \triangleq -\ln p(a_k) \tag{1.57b}$$

Notice that $P_k[t_k]$ and $M_k[t_k]$ include a term for the soft-in information of a_k that is implied by t_k. In cases where we wish to emphasize this relation, we will use $a_k(t_k)$. The sequence (path) metric is

$$M_{k_0}^{k_1}[a_{k_0}^{k_1}, x_{k_0}^{k_1}(\mathbf{a})] = M_{k_0}^{k_1}[\mathbf{t}_{k_0}^{k_1}] = \sum_{k=k_0}^{k_1} M_k[t_k] \tag{1.58}$$

where the term $\mathrm{MI}[s_0] \equiv -\ln p(s_0)$ is implicitly included in $M_0[t_0]$. Next we state two celebrated algorithms that exploit the fact that $\mathrm{SI}[x_k]$ depends only on t_k and not the entire hypothesized transition sequence.

1.3.2.1   The Viterbi Algorithm

The recursive form of the path metric in (1.58) immediately leads to a recursive algorithm for computing the MSM of s_{k+1} based on $\mathbf{z}_0^k$ (*i.e.*, the *forward state MSM*). Specifically,

$$\mathrm{MSM}_0^k[s_{k+1}] = \min_{\mathbf{t}_0^k:s_{k+1}} \sum_{i=0}^{k} M_i[t_i] \tag{1.59a}$$

$$= \min_{t_k:s_{k+1}} \left[\mathrm{MSM}_0^{k-1}[s_k] + M_k[t_k] \right] \tag{1.59b}$$

which is often referred to as a forward add-compare-select (ACS) computation. Furthermore, because the $\min(\cdot)$ marginalization operator has the property that $z = \min(x, y) \implies z = x$ or $z = y$, the minimizing values of the transitions t_k or a_k can be stored as this forward recursion proceeds. In other words, one can store the *survivor sequence* for state s_{k+1} which is the sequence $\breve{\mathbf{a}}_0^k(s_{k+1})$ with smallest metric entering that state

$$\breve{\mathbf{a}}_0^k(s_{k+1}) = \arg \min_{\mathbf{t}_0^k:s_{k+1}} M_0^k[\mathbf{a}_0^k, \mathbf{x}_0^k] \tag{1.60}$$

$$M_0^k[\breve{\mathbf{a}}_0^k(s_{k+1}), \mathbf{x}_0^k(\breve{\mathbf{a}}_0^k(s_{k+1}))] = \mathrm{MSM}_0^k[s_{k+1}] \tag{1.61}$$

Running the ACS recursions up to time k_1 and storing the survivors allows one to *trace back* to determine the a_k that minimizes $M_0^{k_1}[\cdot]$, which is for $k_1 = K - 1$, by definition, the estimate of $a_k(\zeta)$ implied by the MAP-SqD criterion. More specifically, one should trace back on the best survivor sequence – i.e., the survivor that minimizes $MSM_0^{K-1}(s_K)$ over all final states. It is also possible to modify this algorithm to perform a traceback after each forward ACS recursion. Specifically, suppose that the survivor sequences are stored only for D symbols in the past. Thus, after the ACS step that uses z_{k+D}, one has access to $MSM_0^{k+D}(s_k)$ so that the best path can be traced-back on yielding the MAP-SqD decision based on $\mathbf{z}_0^{k+D}$. We refer to algorithms that utilize $\mathbf{z}_0^{K-1}$ and $\mathbf{z}_0^{k+D}$ to obtain information on u_k (e.g., u_k is a_k or possibly some other quantity defined by t_k) as *fixed-interval* and *fixed-lag* algorithms, respectively. The above procedure defines the Viterbi algorithm (VA) and is illustrated in the following example.

Example 1.9. ————————————————————————
Consider the linear ISI-AWGN channel

$$z_k(\zeta) = \sum_{m=0}^{L} f_m a_{k-m}(\zeta) + w_k(\zeta) \qquad (1.62)$$

where $a_k(\zeta)$ is the independent input sequence, $\{f_m\}_{m=0}^L$ are the known equivalent discrete time channel coefficients, and $w_k(\zeta)$ is AWGN. In this example, we consider BPSK modulation (i.e., $\mathcal{A} = \{+1, -1\}$) and real channel coefficients so that the mean zero noise has $\mathbb{E}\{[w_k(\zeta)]^2\} = N_0/2$. Furthermore we consider an interval of length $K = 12$ and $p_{a_k(\zeta)}(-1) = p = 0.7$ – i.e., unequal a-priori probabilities for a stationary $a_k(\zeta)$. It follows that the transition metric defined in (1.57) is

$$M_k[t_k] = [z_k - x_k(t_k)]^2/N_0 + \ln\sqrt{\pi N_0} - \ln p(a_k) \qquad (1.63a)$$
$$= [z_k - \mathbf{f}^T \mathbf{a}_{k-2}^k]^2/N_0 + \ln\sqrt{\pi N_0} - \ln p(a_k) \qquad (1.63b)$$

with the $MI[a_k = -1] = -\ln(0.7) = 0.357$ and $MI[a_k = +1] = -\ln(0.3) = 1.20$. The FSM is started and terminated in the zero state. This has two implications. First, $MI[s_0 = (0)] = 0$ and $MI[s_0 \neq (0)] = \infty$ initialize the ACS recursion. Second, since it is known to the detector that the FSM was terminated into state $s_{12} = (0)$, a_{10} and a_{11} are both known to be $+1$ (i.e., index 0). This implies, for example, that $MI[a_{10} = -1] = \infty$.

A realization of $z_k(\zeta)$ was generated using $\mathbf{f}^T = [0.5\ 0.707\ 0.5]$ and $N_0 = 2$. Since $\|\mathbf{f}\| = 1$ and $a_k = \pm 1$, this is equivalent to a value of

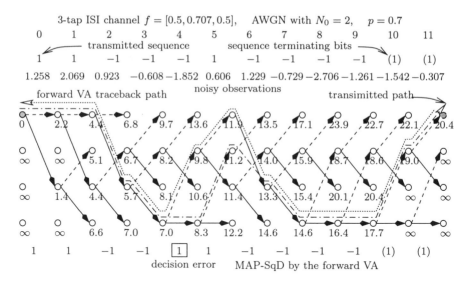

Figure 1.11. The Viterbi algorithm run in Example 1.9. The forward state metrics are shown under each state.

$E_b/N_0 = 1/2$ or -3 dB. This is an extremely low SNR, but the a-priori bias is also fairly strong. The realization of $\mathbf{z}$ and $\mathbf{a}$ along with the execution of the Viterbi algorithm for this example is shown in Fig-1.11 using the standard trellis convention in Fig-1.10. To give a concrete example of the ACS step, consider obtaining $\text{MSM}_0^5[s_6 = (0)] = 11.9$, in Fig-1.11.

There are two transitions t_5 that are consistent with $s_6 = (0)$; namely $t_5 = (s_5, a_5, s_6)$ of $((0), +1, (0))$ and $((1), +1, (0))$. These transitions have associated x_k values of $+f_0 + f_1 + f_2 = 1.707$ and $+f_0 + f_1 - f_2 = 0.707$, respectively. It follows that $\text{MI}[x_k] = [z_k - x_k(t_k)]^2/2$ takes the value 0.606 for the transition emanating from $s_5 = (0)$ and 0.0051 for the transition starting from $s_5 = (1)$. Since both of these correspond to $a_5 = +1$, the transition metric is obtained by adding $1.2 + \ln(\sqrt{\pi N_0}) = 2.12$ to each. The ACS for $s_6 = (0)$ therefore compares $(13.6 + 0.606 + 2.12) = 16.3$ (coming from $s_5 = (0)$) against $(9.8 + 0.0051 + 2.12) = 11.9$ (coming from $s_5 = (1)$) to select the survivor. Specifically, since the path coming from $s_5 = (1)$ has smaller metric, the survivor entering $s_5 = (1)$ is extended by $\breve{a}_5(s_6 = (0)) = +1$ to become the survivor entering $s_6 = (0)$. The state metric $\text{MSM}_0^5[s_6 = (0)]$ is set to the metric of the survivor, namely 11.9.

The procedure is repeated for each conditional state value at each time. The survivors at each step are shown in Fig-1.11. Since the last two bits are known with certainty, the incompatible transition metrics are infinite which yields infinite survivor metrics for all final states other than the zero state. The traceback is therefore conducted from this state yielding the decision shown below the trellis in Fig-1.11. Note that this results in an error of the form $\hat{a}_4 = +1$ when, in reality, $a_4 = -1$. The path in the trellis that indicates the transmitted sequence differs from that associated with $\hat{\mathbf{a}}$ over other locations, but the only error on the associated data value is at $k = 4$. More specifically, the transition sequences associated with the true sequence $\mathbf{a}$ and the MAP-SqD decision $\hat{\mathbf{a}}$ disagree for $k = 4, 5, 6$, but $\hat{a}_k \neq a_k$ only for $k = 4$.

There are several other notable properties shown in this example. Notice that *survivor path merging* occurs so that early decision can be made. Specifically, note that, after performing the ACS that incorporates z_4, all four survivors agree on $\check{\mathbf{a}}_0^2$. Thus, after, $k = 4$, further channel observations will not change the decision on a_0, a_1, and a_2. While this merging is a probabilistic event, a rule of thumb is that for a binary trellis associated with a simple FSM, a decoding lag of approximately $5L$ to $7L$ is sufficient to obtain performance near that of the fixed-interval Viterbi algorithm [HeJa71]. A similar result can be justified for arbitrary trellises as well (*i.e.*, see Problem 1.15). Also, notice the results remain unchanged if the term $\ln(\sqrt{\pi N_0})$ is subtracted from each metric.

$\text{——————————————————————} \textit{End Example}$

The Viterbi algorithm is not a semi-ring algorithm because the traceback operation uses a property of the $\min(\cdot)$ marginalizing operator that is not part of the semi-ring. The sum-product recursion corresponding to the forward ACS in (1.59b) is meaningful (*i.e.*, it produces the APP of s_{k+1} based on $\mathbf{z}_0^k$), however, there is no survivor sequence associated with the sum-product (min*-sum) version. A min-sum semi-ring algorithm that is very closely related to the Viterbi algorithm is constructed from a forward and backward ACS recursion. The next example builds intuition on the latter.

Example 1.10. ——————————————————————————

Notice that, while the FSM has a sense of temporal direction (*i.e.*, it is causal), the min-sum recursion, and the Viterbi algorithm in general, can be executed in the backwards direction. Specifically, the backward ACS recursion is defined as

$$\text{MSM}_k^{K-1}[s_k] = \min_{\mathbf{t}_k^{K-1}:s_k} \sum_{i=k}^{K-1} \text{M}_i[t_i] \tag{1.64a}$$

$$= \min_{t_k:s_k} \left[\text{MSM}_{k+1}^{K-1}[s_{k+1} + \text{M}_k[t_k] \right] \tag{1.64b}$$

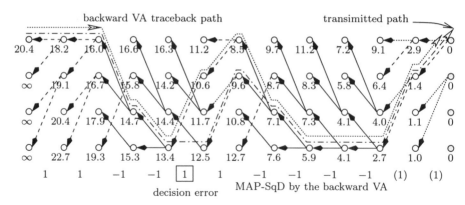

Figure 1.12. The backward Viterbi algorithm run used in Example 1.9. The forward state metrics are shown under each state.

Thus, the same dynamic programming principle that underlies the forward ACS recursion underlies the backward ACS and the implicit recursive solution to the shortest path problem. Notice that the transition metrics are identical in the forward and backward recursions. The backward Viterbi algorithm execution is shown in Fig-1.12 for the same scenario considered in Example 1.9.

For a concrete example, consider the backward ACS that incorporates the 5-th observation with a focus on $s_5 = (1)$. There are two transitions consistent with this value of s_5; namely $t_5 = ((1), +1, (0))$ and $t_5 = ((1), -1, (2))$. From Example 1.9, we know that the metric of the latter transition is 2.1. It follows that the path extending backward from $s_6 = (0)$ to $s_5 = (1)$ has metric $M_5^{11}[\cdot] = 8.5 + 2.1 = 10.6$. The transition metric of the $t_5 = ((1), -1, (2))$ is 1.68 (including the $-\ln(0.7)$ term), so that the comparison is between 10.6 and $10.8 + 1.68 = 12.5$. So, as shown, the backward survivor corresponds to the backward transition from $s_6 = (0)$ to $s_5 = (1)$.

Notice that, as in the forward version, merging can occur in the backward Viterbi algorithm. In this particular realization, after processing the observation z_5, merging occurs such that the decisions on a_k for $k \geq 7$ are finalized. The way that the backward Viterbi algorithm handles edge effects is also slightly different. Specifically, since (1.55a) is obtained by conditioning on the past, there is no a-priori term associated with s_K. In fact, a careful consideration of (1.55a) implies that the backward state metrics should be initialized to zero (*e.g.*, or equivalently, any finite constant). The edge information is enforced by $MI[a_{10}]$ and $MI[a_{11}]$ which are both infinite for the conditional -1 value. The input metric for s_0, however, must be included in the final backward step, which kills all survivors not consistent with $s_0 = (0)$. Once this left edge is reached the backward "traceback" (trace-left) operation can be performed. Note that, as it must, this yields the same final decision as the forward version.

—————————————————— *End Example*

1.3.2.2　The Forward Backward Algorithm

The total MSM of a transition t_k (*i.e.*, the metric based on all observations) can be obtained by marginalizing out over all other transitions t_i consistent with t_k. However, this consistency can be enforced by marginalizing over all $\mathbf{t}_0^{k-1}$ consistent with $s_k(t_k)$ and $\mathbf{t}_{k+1}^{K-1}$ consistent with $s_{k+1}(t_k)$. In other words, one can decouple the shortest path (MSM) problem into two such problems conditioned on the states: (i) from the initial time to each conditional value of s_k, yielding $\mathrm{MSM}_0^{k-1}[s_k]$, and (ii) from the final time (backwards) to each conditional value of s_{k+1}, yielding $\mathrm{MSM}_{k+1}^{K-1}[s_{k+1}]$. Together with the transition metric $\mathrm{M}_k[t_k]$, this yields the MSM of t_k based on the entire observation record

$$\mathrm{MSM}_0^{K-1}[t_k] = \mathrm{MSM}_0^{k-1}[s_k(t_k)] + \mathrm{M}_k[t_k] + \mathrm{MSM}_{k+1}^{K-1}[s_{k+1}(t_k)] \quad (1.65)$$

This concept is illustrated in Fig-1.13. More formally we have

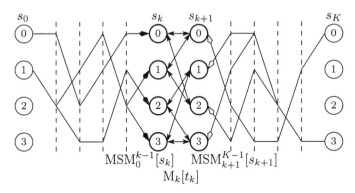

Figure 1.13.　The MSM for a given transition may be computed by summing the transition metric and the forward and backward state metrics.

$$\mathrm{MSM}_0^{K-1}[t_k] = \min_{\mathbf{t}_0^{K-1}:t_k} \sum_{i=0}^{K-1} \mathrm{M}_i[t_i] \quad (1.66a)$$

$$= \min_{\mathbf{t}_0^{K-1}:t_k} \left[\sum_{i=0}^{k-1} \mathrm{M}_i[t_i] + \mathrm{M}_k[t_k] + \sum_{i=k+1}^{K-1} \mathrm{M}_i[t_i] \right] \quad (1.66b)$$

$$= \min_{\mathbf{t}_0^k : t_k} \sum_{i=0}^{k-1} \mathrm{M}_i[t_i] + \mathrm{M}_k[t_k] + \min_{\mathbf{t}_{k+1}^{K-1} : t_k} \sum_{i=k+1}^{K-1} \mathrm{M}_i[t_i] \qquad (1.66c)$$

$$= \mathrm{MSM}_0^{k-1}[s_k(t_k)] + \mathrm{M}_k[t_k] + \mathrm{MSM}_{k+1}^{K-1}[s_{k+1}(t_k)] \qquad (1.66d)$$

where we have used $s_k(t_k)$ and $s_{k+1}(t_k)$ to emphasize that these states are determined by the conditional transition. Note that the term $\mathrm{M}_k[t_k]$ can be brought outside the minimization because it is a constant under the fixed t_k constraint. It is also worth noting that $\mathbf{t}_0^{K-1} : t_k$ implies minimization over all valid transitions sequences or, equivalently, that transitions not admitted by the FSM have infinite metric.

As a specific example, using the results from Examples 1.9 and 1.10, the MSM of $t_5 = ((1), +1, (0))$ is $9.8 + 2.1 + 8.5 = 20.4$ (*i.e.*, forward-MSM plus transition metric plus backward MSM). By definition this is the metric of the best path passing through this conditional transition. Since this transition is on the MAP-SqD decision path, this must be the smallest metric of any path – *i.e.*, $\mathrm{MSM}_0^{K-1} = \min_{t_k} \mathrm{MSM}_0^{K-1}[t_k]$. Similarly, any quantity $\hat{u}_k$ associated with this MAP-SqD path will also have a total MSM of 20.4. The key realization, however, is that the MSM of any quantity u_k derived from t_k can be computed by marginalizing over $\mathrm{MSM}_0^{K-1}[t_k]$.

$$\mathrm{MSM}_0^{K-1}[u_k] = \min_{t_k : u_k} \left[\mathrm{MSM}_0^{k-1}[s_k] + \mathrm{M}_k[t_k] + \mathrm{MSM}_{k+1}^{K-1}[s_{k+1}] \right] \quad (1.67)$$

We call the operation in (1.67), which may be interpreted as a bi-directional ACS, the *completion* operation. A specific example of interest is the $u_k = a_k$ since thresholding $\mathrm{MSM}_0^{K-1}[a_k]$ yields the MAP-SqD decision for a_k. This is described for continuation of the Viterbi algorithm example.

Example 1.11. ——————————————————————
The total MSM of the states s_k can be found by marginalizing the total MSM of the transition t_k over the right edge (s_{k+1}). Specifically,

$$\mathrm{MSM}_0^{K-1}[s_k] = \min_{t_k : s_k} \left[\mathrm{MSM}_0^{K-1}[s_k] + \mathrm{M}_k[t_k] + \mathrm{MSM}_{k+1}^{K-1}[s_{k+1}] \right] \quad (1.68a)$$

$$= \mathrm{MSM}_0^{k-1}[s_k] + \min_{t_k : s_k} \left[\mathrm{M}_k[t_k] + \mathrm{MSM}_{k+1}^{K-1}[s_{k+1}] \right] \quad (1.68b)$$

$$= \mathrm{MSM}_0^{k-1}[s_k] + \mathrm{MSM}_k^{K-1}[s_k] \qquad (1.68c)$$

which may be interpreted as simply adding the forward and backward state metrics for s_k. So, for example for $s_5 = (1)$ in Examples 1.9-1.10, we know that $\mathrm{MSM}_0^{11}[s_5 = (1)] = 9.8 + 10.6 = 20.4$.

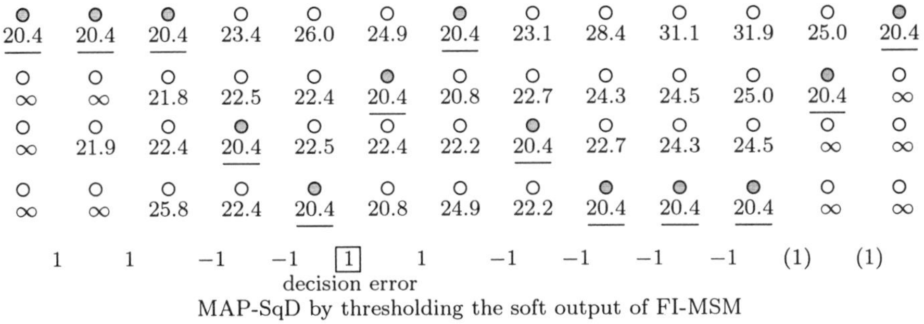

● 20.4	● 20.4	● 20.4	○ 23.4	○ 26.0	○ 24.9	● 20.4	○ 23.1	○ 28.4	○ 31.1	○ 31.9	○ 25.0	● 20.4
○ ∞	○ ∞	○ 21.8	○ 22.5	○ 22.4	● 20.4	○ 20.8	○ 22.7	○ 24.3	○ 24.5	○ 25.0	● 20.4	○ ∞
○ ∞	○ 21.9	○ 22.4	● 20.4	○ 22.5	○ 22.4	○ 22.2	● 20.4	○ 22.7	○ 24.3	○ 24.5	○ ∞	○ ∞
○ ∞	○ ∞	○ 25.8	○ 22.4	● 20.4	○ 20.8	○ 24.9	○ 22.2	● 20.4	● 20.4	● 20.4	○ ∞	○ ∞
1	1	−1	−1	[1]	1	−1	−1	−1	−1	(1)	(1)	

decision error

MAP-SqD by thresholding the soft output of FI-MSM

Figure 1.14. The total MSMs for the states for the realization in Examples 1.9-1.10.

The state MSMs have been computed for all states in the example realization and are shown in Fig-1.14. Notice that the best path can be identified by the constant MSM of 20.4. Because this example FSM is simple, the MSM of a_k can be obtained from s_{k+m} for $m = 1, 2$ – i.e., the state MSM can be marginalized in place of the transition MSM as in (1.67). For example, $\text{MSM}_0^{11}[a_3]$ may be obtained from the corresponding MSM of s_4 or s_5. Specifically, consider the conditional value $a_3 = +1$, which is consistent with $s_4 = (0), (1)$ so that $\text{MSM}_0^{11}[a_3 = +1] = \min(26.0, 22.4) = 22.4$. The condition $a_3 = +1$ is also consistent with $s_5 = (0), (2)$, so that $\text{MSM}_0^{11}[a_3 = +1] = \min(24.9, 22.4) = 22.4$. Notice that the $\text{MSM}_0^{11}[a_3 = -1] = 20.4$, so that thresholding, yields $\hat{a}_3 = -1$.

As another example, consider computing the metric of the shortest path consistent with $x_k = 0.707$. There are two values of t_k consistent with this output value: from $s_k = (1)$ to $s_{k+1} = (0)$ and from $s_k = (0)$ to $s_{k+1} = (2)$. Thus, minimizing over these two transition MSMs yields th desired quantity. For a specific example, consider $x_7 = 0.707$. This cannot be computed directly from the total state MSMs, but applying (1.67), yields the result $\text{MSM}_0^{11}[x_7 = 0.707] = 23.1$.

———————————————————————— *End Example*

While the Viterbi algorithm is not a semi-ring algorithm, the above *forward-backward algorithm* is a semi-ring algorithm. As a consequence, for example, there is a corresponding APP algorithm that performs sum-product operations. In fact, this APP forward-backward algorithm is generally more heavily referenced in the literature. The APP-version was discovered by a number of researchers in the late 1960s and early 1970s. Specifically, Chang and Hancock [ChHa66], Bahl, Cocke, Jelinek, and Raviv (BCJR) [BaCoJeRa74], McAdam [McWeWe72, Mc74] all de-

veloped a version of this APP algorithm. This is also a component of
the Baum-Welch algorithm (*e.g.*, [Ra89]).

Based on the duality principle, the completion operation in the sum-
product algorithm must compute $\text{APP}_0^{K-1}[u_k] \equiv p(\mathbf{z}_0^{K-1}, u_k)$ by summing $\text{APP}_0^{K-1}[t_k] \equiv p(\mathbf{z}_0^{K-1}, t_k)$ over all transitions consistent with u_k.
While a good deal of effort has been spent to convince the reader that
derivation of the sum-product version is not required with the min-sum
semi-ring version in place, we provide a direct derivation for this case
because this is a very important example. Using basic conditional probability relations, we obtain

$$p(\mathbf{z}_0^{K-1}, t_k) = p(\mathbf{z}_{k+1}^{K-1}|t_k)p(\mathbf{z}_0^k, t_k) \tag{1.69a}$$

$$= p(\mathbf{z}_{k+1}^{K-1}|t_k)p(z_k, a_k|s_k)p(\mathbf{z}_0^{k-1}, s_k) \tag{1.69b}$$

$$= [p(\mathbf{z}_0^{k-1}, s_k)][p(z_k|x_k(t_k))p(a_k)][p(\mathbf{z}_{k+1}^{K-1}|s_{k+1})] \tag{1.69c}$$

The three terms in (1.69c) correspond to $\text{APP}_0^{k-1}[s_k]$, $\text{P}_k[t_k]$, and
$\text{APP}_{k+1}^{K-1}[s_{k+1}]$, respectively. For any quantity u_k derived from t_k, the
APP can be obtained by summing over all t_k consistent with u_k (*i.e.*, the
completion operation). According to the duality principle, the forward
and backward state APPs should satisfy the sum-product dual of the
ACS operations in (1.59b) and (1.64b), respectively. This fact is simple
to verify directly yielding

$$p(\mathbf{z}_0^k, s_{k+1}) = \sum_{t_k:s_{k+1}} \left[p(\mathbf{z}_0^{k-1}, s_k)p(z_k|x_k(t_k))p(a_k) \right] \tag{1.70a}$$

$$p(\mathbf{z}_k^{K-1}|s_k) = \sum_{t_k:s_k} \left[p(\mathbf{z}_{k+1}^{K-1}|s_{k+1})p(z_k|x_k(t_k))p(a_k) \right] \tag{1.70b}$$

Many authors use the notation $\alpha(s_k)$, $\beta(s_{k+1})$, and $\gamma(t_k)$ to denote the
forward state APP, the backward state APP, and the transition proba-
bility. It may seem somewhat strange to refer to $p(\mathbf{z}_{k+1}^{K-1}|s_k)$ as an APP
– more precisely, it is a likelihood. This is actually the case in the MSM
version as well (*i.e.*, $\text{MSM}_{k+1}^{K-1}[s_{k+1}]$ is the generalized likelihood of s_{k+1}
based on $\mathbf{z}_{k+1}^{K-1}$). Furthermore, s_k is a "hidden variable," meaning that
it is not an input or output of the FSM. It will become clear in Chap-
ter 2 that one can always assume uniform a-priori information on such
hidden variables. Furthermore, the backward recursion is initialized by
the relation

$$p(\mathbf{z}_{K-1}^{K-1}|s_{K-1}) = p(z_{K-1}|s_{K-1}) \tag{1.71a}$$

$$= \sum_{t_{K-1}:s_{K-1}} p(z_{K-1}|t_{K-1})p(t_{K-1}|s_{K-1}) \tag{1.71b}$$

$$= \sum_{t_{K-1}:s_{K-1}} p(z_{K-1}|t_{K-1})p(a_{K-1}) \qquad (1.71c)$$

which is the standard backward recursion with the interpretation that $p(\mathbf{z}_K^{K-1}|s_K) = 1$ for all s_K.

Example 1.12. ──────────
The min*-sum version of the forward-backward algorithm was run for the same realization used in Examples 1.9-1.11 with the result shown in Fig-1.15. For this realization, the APP version yields the same final de-

1.00	1.08e-1	1.21e-2	1.76e-3	7.58e-5	1.98e-6	1.03e-5	3.81e-6	8.78e-8	1.53e-10	4.28e-10	9.15e-10	4.95e-9
1.24e-9	9.47e-9	6.72e-8	2.97e-8	2.72e-8	5.97e-6	5.31e-5	1.63e-5	3.32e-6	1.78e-4	2.66e-5	1.31e-2	0.25
1.24e-9	1.02e-9	8.15e-10	5.17e-11	2.06e-12	1.18e-11	5.48e-10	6.20e-11	2.92e-13	2.71e-14	1.14e-14	1.20e-11	1.24e-9
O	O	O	O	O	O	O	O	O	O	O	O	O
20.5	20.7	20.9	23.7	26.9	25.2	21.3	23.5	28.9	31.2	32.1	25.1	20.5

0.00	0.00	5.93e-3	1.22e-3	4.34e-4	8.57e-5	2.59e-5	1.51e-6	4.31e-7	2.48e-8	2.87e-8	2.05e-8	0.00
1.05e-9	3.48e-9	3.50e-8	6.72e-8	2.30e-7	8.00e-6	2.02e-5	4.16e-5	6.13e-5	7.72e-4	4.17e-4	5.96e-2	0.25
0.00	0.00	2.07e-10	8.18e-11	9.98e-11	6.86e-10	5.23e-10	6.28e-11	2.64e-11	1.92e-11	1.19e-11	1.22e-9	0.00
O	O	O	O	O	O	O	O	O	O	O	O	O
∞	∞	22.3	23.2	23.0	21.1	21.4	23.5	24.4	24.7	25.1	20.5	∞

0.00	2.40e-1	1.19e-2	4.10e-3	5.31e-4	3.67e-5	1.65e-5	4.79e-6	7.62e-7	6.61e-9	4.34e-9	0.00	0.00
2.65e-10	8.86e-10	1.75e-8	2.37e-7	2.48e-7	2.45e-6	9.01e-6	2.11e-4	1.63e-4	4.04e-3	4.41e-3	8.33e-2	0.00
0.00	2.12e-10	2.09e-10	9.71e-10	1.31e-10	9.00e-11	1.49e-10	1.01e-9	1.25e-10	2.67e-11	1.92e-11	0.00	0.00
O	O	O	O	O	O	O	O	O	O	O	O	O
∞	22.3	22.3	20.8	22.8	23.1	22.6	20.7	22.8	24.3	24.7	∞	∞

0.00	0.00	1.42e-3	8.96e-4	1.28e-3	4.30e-4	1.26e-5	7.56e-7	1.47e-6	2.78e-7	7.18e-8	0.00	0.00
6.07e-11	8.96e-11	3.42e-9	1.47e-7	7.86e-7	1.05e-6	1.27e-6	1.35e-4	7.39e-4	4.29e-3	1.67e-2	9.2e-2	0.25
0.00	0.00	4.86e-12	1.32e-10	1.00e-9	4.49e-10	1.62e-11	1.02e-10	1.09e-9	1.19e-9	1.21e-9	0.00	0.00
O	O	O	O	O	O	O	O	O	O	O	O	O
∞	∞	26.0	22.7	20.7	21.5	24.9	23.0	20.6	20.5	20.5	∞	∞
(1)	(1)	1	1	−1	−1	[1]	1	−1	−1	−1	−1	(1)

Figure 1.15. The APP forward-backward algorithm run on the realization from Examples 1.9-1.11. The value of $\mathrm{M^*SM}_0^{11}[s_k]$ is shown below each state, with the values of the forward, backward and total state APPs shown above the state (from top down).

cision as the MSM-version (*i.e.*, the MAP-SyD and MAP-SqD criterion for a_k yield the same results for this realization). The soft-information provided is, however different. This is most apparent from the fact that there is no common value of $\min_{s_k} \mathrm{M^*SM}_0^{11}[s_k]$ – *e.g.*, this is 20.8, 20.7, 21.1, ... for $k = 3, 4, 5 \dots$. This is expected since there is no survivor path (or best path) associated with the APP or M*SM soft information as these represent the average over all consistent paths. However, notice that the values of $\mathrm{M^*SM}_0^{K-1}[s_k]$ and $\mathrm{MSM}_0^{K-1}[s_k]$ for each k are close to each other. This is because $\min^*(x,y) \cong \min(x,y)$ when the SNR is moderate to high. Even in this fairly low SNR environment, the difference is not very large.

────────── *End Example*

Other Efficient Architectures   There are other structures for performing MAP symbol and sequence detection for an FSM system. These include fixed-lag algorithms and parallel tree-structured architectures. All of these structures can be derived in a similar intuitive manner. However, we postpone discussion of these structures until Chapter 2 since these algorithms play a key role in iterative detection.

One aspect that has not been considered in detail is the front-end processing of conversion from continuous time observation to the discrete time models assumed. In many cases, this is fairly straightforward (*e.g.*, the 4-PAM example). In the next example, we present two methods of obtaining metrics suitable for running the forward-backward or Viterbi algorithms for a continuous time ISI-AWGN channel.

Example 1.13. ——————————————————————

In this example we show two different methods of front-end processing for linear ISI-AWGN channels. The *Ungerboeck structure* [Un74] leads to colored noise discrete time model, but provides a suitable metric which can be used for the Viterbi or forward-backward algorithm. The *Forney structure* [Fo72b] yields the white noise model in (1.62).

Consider the baseband equivalent, continuous-time observation model for a known ISI channel with linear modulation given by

$$r(u,t) = y(t; \mathbf{a}(\zeta)) + n(u,t) \quad t \in \mathcal{T} \tag{1.72}$$

$$y(t; \mathbf{a}(\zeta)) = \sum_{i=0}^{K-1} a_i(u)h(t - iT) \tag{1.73}$$

where the support of $h(t)$ is assumed to be contained in $[0, (L+1)T)$. The received signal is assumed to be observed for some interval of the real line, $\mathcal{T}$, which contains the support of $y(t)$.[6]

By appealing to an orthogonal series expansion for the noise, the negative log-likelihood functional (*e.g.*, [Va68]) for the a conditional sequence **a** based on the observation $\{r(t) : t \in \mathcal{T}\}$ can be obtained. Including the a-priori probabilities,[7] the associated metric is

$$\mathrm{M}[\mathbf{a}] = -\ln p(\mathbf{a}) + \frac{1}{N_0} \left[\int_{\mathcal{T}} |y(t; \mathbf{a})|^2 dt - 2\Re \left\{ \int_{\mathcal{T}} r(t) y^*(t; \mathbf{a}) dt \right\} \right] \tag{1.74}$$

---

[6]In other words $\mathcal{T} \supseteq [0, (L+K)T)$. In fact, we may assume that $\mathcal{T} = [0, (L+K)T)$ since the observation outside this interval is irrelevant. This is not the case in colored noise.

[7]The initial state $s_0 = \{a_i\}_{i=-L}^{-1}$ is implicitly part of the hypothesized sequence and the a-priori probability $p(s_0)$ is included. We do not explicitly show this dependence in the following.

Substituting the assumed structure of the noise-free channel output yields

$$M[\mathbf{a}] = \sum_{i=0}^{K-1} -\ln p(a_i) + \frac{1}{N_0} \left[\sum_{i=0}^{K-1} \sum_{j=0}^{K-1} a_i a_j^* R(j-i) - 2\Re \left\{ \sum_{i=0}^{K-1} a_i^* r_i \right\} \right]$$

(1.75)

where $R(m)$ is the T-spaced sampled correlation of $h(t)$ and $\{r_i\}$ are the matched filter outputs

$$R(m) = [h(t) \circledast h^*(-t)]|_{t=mT} \tag{1.76a}$$

$$r_i = \int_T r(t) h^*(t - iT) dt = r(t) \circledast h^*(-t)|_{t=iT}. \tag{1.76b}$$

Note that, since the channel is FIR, the matched filter can be made causal by introducing a finite delay.

Ungerboeck Structure   The Ungerboeck metric [Un74], as defined in (1.74), can be computed recursively using the fact that Hermitian symmetric sequence $R(m)$ is nonzero only for $|m| \leq L$, which follows from the FIR assumption on $h(t)$. Specifically, the recursion is

$$M_0^k[\mathbf{a}_0^k] \triangleq \sum_{i=0}^{k} -\ln p(a_i) + \frac{1}{N_0} \left[\sum_{i=0}^{k} \sum_{j=0}^{k} a_i a_j^* R(j-i) - 2\Re \left\{ \sum_{i=0}^{k} a_i^* r_i \right\} \right]$$

$$= M_0^{k-1}[\mathbf{a}_0^{k-1}] + (-\ln p(a_k))$$

$$+ \frac{2}{N_0} \Re \left\{ a_k^* \left(\sum_{m=1}^{L} a_{k-m} R(m) + \frac{1}{2} a_k R(0) - r_k \right) \right\}$$

$$= M_0^{k-1}[\mathbf{a}_0^{k-1}] + \frac{2}{N_0} \Re \left\{ a_k^* (a_k \circledast g_k - r_k) \right\} - \ln p(a_k) \tag{1.77}$$

where

$$g_i = \begin{cases} 0 & i < 0, \ i > L \\ \frac{1}{2} R(0) & i = 0 \\ R(i) & i \in \{1, 2 \dots L\} \end{cases} \tag{1.78}$$

Thus, the total path metric for $\mathbf{a}_0^k$ based on the matched filter outputs $\mathbf{r}_0^k$ has been expressed as

$$M_0^k[\mathbf{a}_0^k] = \sum_{i=0}^{k} M_i[t_i] \tag{1.79a}$$

$$M_i[t_i] = MI[a_i] + \frac{2}{N_0} \Re \left\{ a_i^* (a_i \circledast g_i - r_i) \right\} \tag{1.79b}$$

where t_i is the state transition associated in the trellis with states $s_i = \{a_{i-m}\}_{m=1}^L$.

The recursive form of (1.79) allows one to perform MAP-SyD or MAP-SqD using the (min*-sum) forward-backward algorithm or the Viterbi algorithm, respectively. Also, the matched filter outputs clearly form a set of sufficient statistics for detection of the data sequence based on the observation waveform. However, the decision problem was never reformulated in terms of r_i, so there is no discrete-time system output directly associated with transitions t_k. More precisely, the stochastic model may be reformulated in terms of $r_i(\zeta)$ which yields

$$r_k(\zeta) = y_k(\mathbf{a}(\zeta)) + n_k(\zeta) \tag{1.80}$$

where $y_k(\mathbf{a}(\zeta))$ and $n_k(\zeta)$ are the matched filter outputs associated with the signal and noise, respectively. It can be shown that

$$y_k(\mathbf{a}(\zeta)) = y_k(\mathbf{a}_{k-L}^{k+L}) = a_k \circledast R(k) \tag{1.81}$$
$$\mathbb{E}\{n_{k+m}(\zeta)n_k^*(\zeta)\} = N_0 R(m) \tag{1.82}$$

Notice that, because y_k depends on a_k outside the range corresponding to t_k, it is difficult to interpret the $M_k[t_k] - MI[a_k]$ in (1.79) as input metric information on y_k.

Forney Structure   Since $R(m)$ is a non-negative definite sequence, it may be factored into $R(m) = f_m \circledast f_{-m}^*$, where f_m is nonzero only for $0 \le m \le L$. Assuming that f_{-m}^* is causally invertible (*i.e.*, no nulls in the Fourier spectrum), and denoting this inverse filter by f_m^{-H}, we obtain

$$z_k(\zeta) = r_k(\zeta) \circledast f_m^{-H} \tag{1.83a}$$
$$= a_k(\zeta) \circledast R(k) \circledast f_m^{-H} + n_k(\zeta) \circledast f_m^{-H} \tag{1.83b}$$
$$= a_k(\zeta) \circledast f_k + w_k(\zeta) \tag{1.83c}$$

where $w_k(\zeta)$ is a circular AWGN sequence with $\mathbb{E}\{|w_k(\zeta)|^2\} = N_0$. In other words, the output of this front-end processing is the model in (1.62).

Notice that both the Ungerboeck and Forney structure sample the output of a filter matched to $h(t)$. The Ungerboeck structure computes the likelihood functions directly, while the Forney structure uses a discrete-time whitening filter. The cascade of the T-spaced sampled, matched filter and the noise whitening filter, is referred to as a *whitened matched filter*. This is illustrated in Fig-1.16. Note that the value of the transition metrics $M_k[t_k]$ are *different* for the two structures. However,

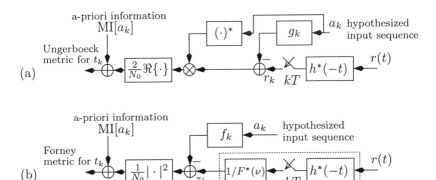

Figure 1.16. Computation of transition (a) Ungerboeck metrics and (b) Forney metrics for a linear ISI-AWGN channel.

the sum of these metrics along any sequence is the same using either convention.

The multiuser system example in Chapter 2 may be viewed as a vector input ISI channel. Inclusion of a time-varying whitening filter after the correlators/matched-filters may be problematic in this application, so the resulting trellis search algorithms are typically performed using the Ungerboeck metrics (*i.e.*, see Section 2.4.4).

——————————————————————— *End Example*

1.3.3 Detection with Imperfect CSI

1.3.3.1 The Inapplicability of the Viterbi Algorithm

The application of the Viterbi Algorithm for the problem of MAP-SqD is possible due to the special form of (1.55a). In particular, rewriting (1.55a) as

$$p(\mathbf{z}_0^{K-1}|\mathbf{a}_0^{K-1}, s_0)p(\mathbf{a}_0^{K-1}, s_0) = p(\mathbf{z}_0^{K-1}|\mathbf{t}_0^{K-1})p(\mathbf{t}_0^{K-1})$$

$$= p(z_0|t_0)p(a_0)p(s_0) \prod_{k=1}^{K-1} p(z_k|\mathbf{z}_0^{k-1}, \mathbf{t}_0^k)p(t_k|t_{k-1}) \quad (1.84)$$

it is observed that the application of the VA is enabled by the fact that

$$p(z_k|\mathbf{z}_0^{k-1}, \mathbf{t}_0^k) = p(z_k|\mathbf{z}_0^{k-1}, t_k) \quad (1.85)$$

The latter is usually referred to in the literature as the *folding* condition [Ch98], and in the case of memoryless channels is further simplified to $p(z_k|\mathbf{z}_0^{k-1}, t_k) = p(z_k|x_k(t_k))$.

Unfortunately, the innovation terms in (1.53), as well as in (1.49), do not possess the folding property, thus the Viterbi algorithm is not applicable in these cases. In general there are two approaches in designing suboptimal algorithms for MAP-SqD when folding does not hold.

In the first approach, an appropriate model for the unknown parameter is assumed, such that the innovation term depends only on a finite number of past input symbols. The linear predictive receiver in [LoMo90], and the innovations-based receiver in [YuPa95], follow exactly this approach and can be regarded as the optimal receivers under the assumed parameter model. In practice, this parameter model is viewed as an approximation, and folding is accomplished by substituting the infinite memory parameter estimators (*e.g.*, KF or RLS), by finite memory estimators. We point out that the complexity of such a suboptimal algorithm is determined by the memory of the estimator (and the size of the original trellis), which also determines the quality of the parameter estimate.

The second approach in constructing suboptimal MAP-SqD algorithms is based on the observation that, when folding does not hold, a search over the entire sequence tree is required, as mentioned in Section 1.2.2. The suboptimal receiver, referred to as the generalized Per-Survivor Processing (PSP) [RaPoTz95] receiver, searches only part of the tree, depending on the available resources. Many strategies are known to prune the sequence tree [AnMo84]. One such algorithm is the Viterbi algorithm, which maintains and updates – through the familiar ACS operations – a fixed number of paths in such a way that they are forced to have different recent paths. A KF (or RLS for deterministic modeling) parameter estimate is kept and updated for every trellis state. We emphasize that the trellis on which this algorithm operates is not tightly related to the FSM trellis. Its size is a design parameter that determines the amount of pruning, and eventually, the complexity of the algorithm.

We note that the two approaches mentioned above are not mutually exclusive, but can be combined to further reduce complexity. For instance, when the memory of the truncated estimator in the first approach results in a large trellis, PSP can be utilized to reduce the search effort.

Example 1.14. ───────────────────────────────
Consider the special case of a first order GM scalar process $\{f_k(\zeta)\}$ observed in no noise, *i.e.*, $z_k = a_k f_k$, with $\{a_k\}$ being a memoryless sequence. Equation (1.52) simplifies to $f_k(\zeta) = \alpha f_{k-1}(\zeta) + u_k(\zeta)$, with $\mathbb{E}\{|u_k(\zeta)|^2\} = \sigma_u^2$. For this special case, the forward filtered parameter estimate, and one step predictor are $f_{k|k} \triangleq \mathbb{E}\{f_k(\zeta)|\mathbf{z}_0^k, \mathbf{a}_0^k\} =$

z_k/a_k, $f_{k|k-1} \triangleq \mathbb{E}\{f_k(\zeta)|\mathbf{z}_0^{k-1}, \mathbf{a}_0^{k-1}\} = \alpha f_{k-1|k-1}$, respectively. The corresponding error variances are given by $\sigma_{k|k}^2 = 0$, and $\sigma_{k|k-1}^2 = \sigma_u^2$. The innovations term of the forward recursive EC in (1.53b) becomes

$$p(z_k|\mathbf{z}_0^{k-1}, \mathbf{a}_0^k)p(a_k) = \mathcal{N}^{cc}(z_k; a_k\alpha z_{k-1}/a_{k-1}; |a_k|^2\sigma_u^2)p(a_k) \qquad (1.86)$$

The above equation shows that the innovation term depends only on the symbols a_k and a_{k-1}, thus the problem of MAP-SqD can be solved using the VA, operating on an $|\mathcal{A}|$-state trellis.
—————————————— *End Example*

1.3.3.2 The Forward-backward Estimator-Correlator

An alternative exact expression for the calculation of $p(\mathbf{z}_0^{K-1}, \mathbf{a}_0^{K-1})$ for an FSM is now described for the GM parameter model (see Problem 4.4 for the deterministic case). As will be shown, these expressions have the form of two ECs operating in the forward and backward direction, respectively, hence the name *forward-backward* EC. Based on these expressions, several optimal and suboptimal practical algorithms will be developed in Chapter 4 for obtaining soft decisions in the presence of unknown parameters.

In evaluating $p(\mathbf{z}_0^{K-1}, \mathbf{a}_0^{K-1})$, we observe that due to the implicit presence of the parameter process $\{\mathbf{f}_k(\zeta)\}$, future observations depend on past observations conditioned on the state of the FSM. On the other hand, by conditioning on the parameter $\mathbf{f}_k$ as well, separation of the future and past observations occurs, yielding [An99]

$$p(\mathbf{z}_0^{K-1}, \mathbf{a}_0^{K-1}) = \underbrace{p(\mathbf{z}_0^k, \mathbf{a}_0^k)}_{\text{past/present}} \underbrace{p(\mathbf{z}_{k+1}^{K-1}, \mathbf{a}_{k+1}^{K-1}|s_{k+1})}_{\text{future}}$$

$$\underbrace{\int \frac{p(\mathbf{f}_k|\mathbf{a}_0^k, \mathbf{z}_0^k)p(\mathbf{f}_k|s_{k+1}, \mathbf{a}_{k+1}^{K-1}, \mathbf{z}_{k+1}^{K-1})}{p(\mathbf{f}_k)}d\mathbf{f}_k}_{\text{binding } b_p(\cdot)} \qquad (1.87)$$

The relation in (1.87) indicates that $p(\mathbf{z}_0^{K-1}, \mathbf{a}_0^{K-1})$ can be split into three factors, of which the first two depend on the past/present and future, respectively, while the third can be viewed as a weighting factor that binds them together. Indeed, the third factor quantifies the dependence of the future, present and past that is introduced due to the parameter process $\{\mathbf{f}_k(\zeta)\}$ and in the absence of parametric uncertainty would be eliminated. A closed form expression can be found for the binding factor since it involves an integral of Gaussian densities (see Problem 4.2), and although the expression is fairly complicated (it involves inverse matrices and matrix determinants), we emphasize that it does not require any re-processing of the observation record. The closed form expression for the binding factor is given by

$$b_p(\tilde{\mathbf{f}}_{k|k}, \tilde{\mathbf{F}}_{k|k}, \tilde{\mathbf{f}}^b_{k|k+1}, \tilde{\mathbf{F}}^b_{k|k+1}) = \frac{|\mathbf{K_f}||\mathbf{P}|}{|\tilde{\mathbf{F}}_{k|k}||\tilde{\mathbf{F}}^b_{k|k+1}|} \exp(\beta^H \mathbf{P}\beta - \gamma) \quad (1.88a)$$

with

$$\mathbf{P}^{-1} = \tilde{\mathbf{F}}^{-1}_{k|k} + (\tilde{\mathbf{F}}^b_{k|k+1})^{-1} - \mathbf{K_f}^{-1} \quad (1.88b)$$

$$\beta = \tilde{\mathbf{F}}^{-1}_{k|k}\tilde{\mathbf{f}}_{k|k} + (\tilde{\mathbf{F}}^b_{k|k+1})^{-1}\tilde{\mathbf{f}}^b_{k|k+1} \quad (1.88c)$$

$$\gamma = \tilde{\mathbf{f}}^H_{k|k}\tilde{\mathbf{F}}^{-1}_{k|k}\tilde{\mathbf{f}}_{k|k} + (\tilde{\mathbf{f}}^b_{k|k+1})^H(\tilde{\mathbf{F}}^b_{k|k+1})^{-1}\tilde{\mathbf{f}}^b_{k|k+1} \quad (1.88d)$$

where $\tilde{\mathbf{f}}_{k|k}$, $\tilde{\mathbf{f}}^b_{k|k+1}$ are the sequence-conditioned forward channel estimate, the one-step sequence-conditioned backward channel predictor and $\tilde{\mathbf{F}}_{k|k}$, $\tilde{\mathbf{F}}^b_{k|k+1}$ are the corresponding covariances. The first factor in (1.87) is recursively evaluated using (1.53), while the second is calculated through a similar backward recursion.

$$p(\mathbf{z}^{K-1}_{k+1}, \mathbf{a}^{K-1}_{k+1}|s_{k+1}) = p(z_{k+1}|\mathbf{z}^{K-1}_{k+2}, s_{k+1}, \mathbf{a}^{K-1}_{k+1})$$
$$p(a_{k+1})p(\mathbf{z}^{K-1}_{k+2}, \mathbf{a}^{K-1}_{k+2}|s_{k+2})$$
$$= \mathcal{N}^{cc}(z_{k+1}; \mathbf{q}^T_{k+1}\tilde{\mathbf{f}}^b_{k+1|k+2}; N_0 + \mathbf{q}^T_{k+1}\tilde{\mathbf{F}}^b_{k+1|k+2}\mathbf{q}^*_{k+1})$$
$$p(a_{k+1})p(\mathbf{z}^{K-1}_{k+2}, \mathbf{a}^{K-1}_{k+2}|s_{k+2}) \quad (1.89)$$

The scheme suggested by this forward-backward EC is illustrated in Fig-1.17 and can be described as follows: Starting at time 0 a forward $|\mathcal{A}|$-ary tree is built, exactly as in the case of forward EC. Similarly, starting at time $K - 1$ a backward tree is expanding according to the recursion (1.89), with the relevant channel estimates provided by a per-path backward running KFs. After $k+1$ forward and $K-k-1$ backward steps, the two trees meet each other. The metric $p(\mathbf{z}^{K-1}_0, \mathbf{a}^{K-1}_0)$ of each sequence $\mathbf{a}^{K-1}_0$ can now be evaluated as indicated by (1.87). The $|\mathcal{A}|^{k+1}$ metrics corresponding to the nodes of the forward tree are combined with the $|\mathcal{A}|^{K-k-1}$ metrics corresponding to the nodes of the backward tree (future) and weighted by the binding factor in (1.88). Note that the choice of k, the particular point in time when the past and future metrics are combined, is *completely arbitrary*. In fact, the two extreme values $k = K - 1$ and $k = 0$ correspond to a single forward or a single backward tree. Thus, while it may seem redundant to store and update

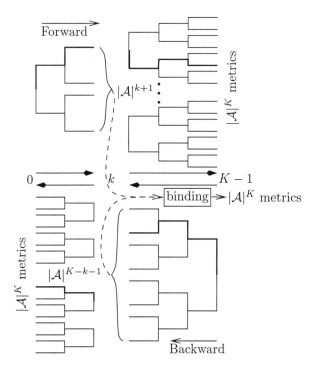

Figure 1.17.   Metric evaluation using a forward-backward estimator-correlator.

both a forward and a backward tree (*i.e.*, same result can be accomplished with a single forward tree), the fact that the two trees can be pruned independently, will result in practical algorithm, as discussed in Chapter 4.

1.4   Performance Bounds Based on Pairwise Error Probability

Consider an arbitrary deterministic decision rule for deciding among $|\mathcal{H}|$ hypotheses H_m based on some measurement $\mathbf{z} \in \mathcal{Z}$. Such a rule introduces a partition of the observation space as shown in Fig-1.2

$$\text{decide } H_m \quad \Longleftrightarrow \quad \mathbf{z} \in \mathcal{Z}_m \qquad (1.90)$$

For example, MAP detection for the $|\mathcal{H}|$-ary problem results in

$$\mathcal{Z}_m = \{\mathbf{z} : p(\mathbf{z}|H_m)p(H_m) > p(\mathbf{z}|H_j)p(H_j), \ \forall \ j \neq m\} \qquad (1.91)$$

The conditional probability of error is given by

$$P(\mathcal{E}|H_m) = \Pr\{\mathbf{z}(\zeta) \notin \mathcal{Z}_m|H(\zeta) = H_m\} = \Pr\{\mathbf{z}(\zeta) \in \mathcal{Z}_m^c|H(\zeta) = H_m\} \qquad (1.92)$$

Many useful bounds can be constructed by expressing $\mathcal{Z}_m^c$ in specific ways. For example, it is clear that

$$\mathcal{Z}_m^c = \mathcal{Z} - \mathcal{Z}_m = \bigcup_{j=0, j \neq m}^{|\mathcal{H}|-1} \mathcal{Z}_j \tag{1.93}$$

An expression that is typically more useful is obtained by constructing $\mathcal{Z}_m$ from pairwise decision regions $\mathcal{Z}_m^{PW}(j)$, defined as the region where H_m would be selected over H_j in a pairwise (binary) decision. For example, in the case of MAP $|\mathcal{H}|$-ary decisions,

$$\mathcal{Z}_m^{PW}(j) = \{\mathbf{z} : p(\mathbf{z}|H_m)p(H_m) > p(\mathbf{z}|H_j)p(H_j)\} \tag{1.94}$$

Comparing the definitions of the global decision region and the pairwise regions (*i.e.*, (1.91) and (1.94)) it is apparent that

$$\mathcal{Z}_m = \bigcap_{j=0, j \neq m}^{|\mathcal{H}|-1} \mathcal{Z}_m^{PW}(j) \tag{1.95}$$

The complement of this region is obtained by applying DeMorgan's Law yielding

$$\mathcal{Z}_m^c = \bigcup_{j=0, j \neq m}^{|\mathcal{H}|-1} \left[\mathcal{Z}_m^{PW}(j)\right]^c = \bigcup_{j=0, j \neq m}^{|\mathcal{H}|-1} \mathcal{Z}_j^{PW}(m) \tag{1.96}$$

where the fact that $\left[\mathcal{Z}_m^{PW}(j)\right]^c = \mathcal{Z}_j^{PW}(m)$ has been used.

Bounds can be constructed using simple Union Bounds and related techniques. Specifically, let $\{B_i\}$ be a set of events, then it follows that

$$\max_i P(B_i) \leq P\left(\bigcup_i B_i\right) \leq \sum_i P(B_i) \tag{1.97}$$

where the lower bound is constructed by obtaining the largest lower-bound from a family (*i.e.*, $P(B_i)$ is a lower bound for each value of i). Upper and lower bounds for the conditional error probability can then be constructed using either the union in (1.93) or (1.96). However, evaluation of $\Pr\{\mathbf{z}(\zeta) \in \mathcal{Z}_m^c | H_m\}$ is typically difficult (*i.e.*, if it can be obtained, then an exact expression for the error probability can be obtained), so a bound constructed from pairwise errors is more generally applicable. In this case, applying (1.97) using the expression in (1.96) yields

$$\max_j P_{PW}(j|H_m) \leq P(\mathcal{E}|H_m) \leq \sum_{j=0, j \neq m}^{|\mathcal{H}|-1} P_{PW}(j|H_m) \tag{1.98}$$

where the *pairwise error probability* is

$$P_{PW}(j|H_m) = \Pr\left\{ \mathbf{z}(\varsigma) \in \mathcal{Z}_j^{PW}(m)|H_m \right\} \qquad (1.99)$$

1.4.1 An Upper Bound using Sufficient Neighborhood Sets

In practice the expression for $\mathcal{Z}_m^c$ in (1.96) is overly conservative owing to the fact that a subset of terms in the union may actually fully define the complement of the decision region. Suppose that a set $\mathcal{F}_m \subset \{0, 1, \ldots |\mathcal{H}| - 1\}$ defines the region $\mathcal{Z}_m^c$ in the sense that

$$\mathcal{Z}_m^c = \bigcup_{j=0, j \neq m}^{|\mathcal{H}|-1} \mathcal{Z}_j^{PW}(m) = \bigcup_{j \in \mathcal{F}_m} \mathcal{Z}_j^{PW}(m) \qquad (1.100)$$

We refer to any such set as a *sufficient set* and the corresponding pairwise error events as *sufficient pairwise error events*. There are many such sufficient sets with smaller sets yielding tighter upper bounds. A key property of such sets is that if a global error occurs, then some sufficient pairwise error event must have occurred. One may view the sufficient set as a set of "nearest neighbors" that determine the decision region; this concept is illustrated in Fig-1.18 for the case of minimum distance decisions. As a result, the upper bound in (1.98) can be tightened by replacing the sum over $j \neq m$ with the sum over $j \in \mathcal{F}_m$.

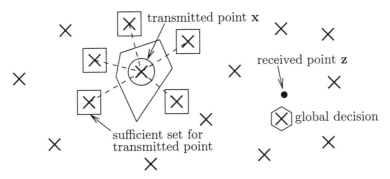

Figure 1.18. An example of a sufficient set for a given signal space. Note that a global error cannot occur without a sufficient pairwise error event occurring.

Bounds on the unconditional error probability can then be obtained by averaging these conditional bounds over $p(H_m)$ yielding

$$P(\mathcal{E}) \geq \sum_{m=0}^{|\mathcal{H}|-1} p(H_m) \left[\max_j P_{PW}(j|H_m) \right] \qquad (1.101a)$$

$$P(\mathcal{E}) \le \sum_{m=0}^{|\mathcal{H}|-1} p(H_m) \sum_{j \in \mathcal{F}_m}^{|\mathcal{H}|-1} P_{PW}(j|H_m) \qquad (1.101b)$$

Example 1.15. ――――――――――――――――――――――――――――――――

There are four hypotheses for the PAM sequence detection examples 1.2-1.6. Denote these by the shorthand $(a_1, a_0) \iff H_m$ with $m = 2a_1 + a_0$ (*i.e.*, the decimal value of the corresponding bit labels). For equal a-priori probabilities and ML-SqD, it follows from Fig-1.4 that sufficient sets are $\mathcal{F}_0 = \{1\}$, $\mathcal{F}_1 = \{0, 3\}$, $\mathcal{F}_2 = \{3\}$, and $\mathcal{F}_3 = \{1, 2\}$. In this decision problem, the global decision regions have the unique property that, for each hypothesis H_m, the pairwise decision regions comprising $\mathcal{F}_m$ are disjoint.

Alternately, one may view detection of the two symbols a_1 and a_0 separately (under either the MAP-SyD or MAP-SqD criterion) as a binary test. One can consider the probability of sequence error $P(\mathcal{Q}) = \Pr\{\hat{\mathbf{a}}(\zeta) \ne \mathbf{a}(\zeta)\}$ and/or the probability of symbol error at location n $P(\mathcal{Y}_n) = \Pr\{\hat{a}_n(\zeta) \ne a_n(\zeta)\}$ for any rule. Clearly, an error in any symbol location implies a sequence error – *i.e.*, $\mathcal{Q} = \bigcup_n \mathcal{Y}_n$.

――――――――――――――――――――――――――――――――――――― *End Example*

1.4.1.1   Special Cases for AWGN Channels

A common special case for the application of the bounds developed above is that of a-priori equally-likely signaling over an AWGN channel where $\mathbf{z}(\zeta) = \mathbf{s}_m + \mathbf{w}(\zeta)$ under H_m with $\mathbf{w}(\zeta)$ being AWGN. In this case, the MAP detection rule is the *minimum distance rule* and the pairwise error is

$$P_{PW}(j|H_m) = Q\left(\sqrt{\frac{d^2(j, m)}{2N_0}}\right) \qquad (1.102a)$$

$$d(j, m) = \|\mathbf{s}_j - \mathbf{s}_m\| \qquad (1.102b)$$

$$Q(x) \triangleq \int_x^\infty \mathcal{N}(z; 0; 1)dz = \frac{1}{\sqrt{2\pi}} \int_x^\infty \exp(-\frac{z^2}{2})dz \qquad (1.102c)$$

In this case, the bound simplifies to

$$\frac{1}{|\mathcal{H}|} \sum_{m=0}^{|\mathcal{H}|-1} Q\left(\sqrt{\frac{d_{\min}^2(m)}{2N_0}}\right) \le P(\mathcal{E}) \le \frac{1}{|\mathcal{H}|} \sum_{m=0}^{|\mathcal{H}|-1} \sum_{j \in \mathcal{F}_m} Q\left(\sqrt{\frac{d^2(j, m)}{2N_0}}\right)$$

$$(1.103)$$

where

$$d_{\min}(m) = \min_{j \neq m} d(j, m) \tag{1.104}$$

A simple set of bounds can be obtained in terms of the *global minimum distance*

$$d_{\min} = \min_{m} d_{\min}(m) \tag{1.105}$$

Specifically, using all other hypotheses as the sufficient set for the upper bound and constructing the lower bound by taking only a single term from the sum yields

$$\frac{1}{|\mathcal{H}|} Q \left(\sqrt{\frac{d_{\min}^2}{2N_0}} \right) \leq P(\mathcal{E}) \leq (|\mathcal{H}| - 1) Q \left(\sqrt{\frac{d_{\min}^2}{2N_0}} \right) \tag{1.106}$$

which implies that at high SNR, the error probability must decay proportionally to $Q(\sqrt{d_{\min}^2/(2N_0)})$ – *i.e.*, the error probability of a binary test with only the nearest neighbor.

With some book-keeping, these bounds can be improved. Specifically, consider the *distance spectrum* of the signal set – *i.e.*, the values of $d(m, j)$ that can occur for the specific set of signals. Order these distances so that $d_{\min} = d_1 < d_2 < d_3 \ldots$. Let $N_i(\{\mathcal{F}_m\})$ be the number of times that d_i occurs in a listing of all sufficient pairwise error events. Let K_i be the number of hypotheses for which $d_{\min}(m) = d_i$. The basic bound in (1.103) then simplifies to

$$\sum_i \frac{K_i}{|\mathcal{H}|} Q \left(\sqrt{\frac{d_i^2}{2N_0}} \right) \leq P(\mathcal{E}) \leq \sum_i \frac{N_i(\{\mathcal{F}_m\})}{|\mathcal{H}|} Q \left(\sqrt{\frac{d_i^2}{2N_0}} \right) \tag{1.107}$$

Note that one could use any single term from the lower bound in (1.107) as a lower bound. This may be much easier to compute and only slightly looser. For example, this yields a lower bound in $d_{\min}$ of the form

$$P(\mathcal{E}) \geq \frac{K_1}{|\mathcal{H}|} Q \left(\sqrt{\frac{d_{\min}^2}{2N_0}} \right) \tag{1.108}$$

Note that, depending on the SNR, this may not be the best single-term lower bound. For example, the analogous bound based on d_2 may be larger at low SNR if $K_2 > K_1$. This motivates the lower bound

$$P(\mathcal{E}) \geq \max_i \frac{K_i}{|\mathcal{H}|} Q \left(\sqrt{\frac{d_i^2}{2N_0}} \right) \tag{1.109}$$

At moderate to high SNR, the $d_{\min}$ term in the upper bound dominates and may be used as an approximation to the error probability.

Example 1.16. ────────────────────────────────────

Continuing with the 4-PAM example, consider analysis of the MAP-SqD rule with equal a-priori probabilities. To emphasize the decision criterion used, we use $P_{Sq}(\cdot)$ in denoting the symbol and sequence error probabilities – i.e., $P_{Sq}(\cdot)$ and $P_{Sy}(\cdot)$ will be used to denote error probabilities under the MAP-SqD and MAP-SyD criteria, respectively.

For probability of sequence error $P_{Sq}(\mathcal{Q})$, we consider the 4-ary ML-SqD rule. The expressions in (1.106) and (1.103)/(1.107) both yield bounds of the form

$$C_L Q\left(\gamma\right) \le P_{Sq}(\mathcal{Q}) \le C_U Q\left(\gamma\right) \tag{1.110}$$

where $\gamma = \sqrt{2A^2/N_0}$. Specifically, (1.106) yields $(C_L, C_U) = (1/4, 3)$ and (1.103)/(1.107) yield the tighter bounds $(C_L, C_U) = (1, 3/2)$. Note that, for this example $d_{\min}(m) = d_{\min} = 2A$ – i.e., each signal has its closest neighbor at distance $2A$.

Because of the special disjoint structure of the pairwise decision regions in $\mathcal{F}_m$, the upper bound from (1.107) coincides with the exact value of $P_{Sq}(\mathcal{Q})$. The symbol error probabilities for the ML-SqD can also be found exactly as

$$P_{Sq}(\mathcal{Y}_0) = Q\left(\gamma\right) + \frac{1}{2}Q\left(3\gamma\right) - \frac{1}{2}Q\left(5\gamma\right) \tag{1.111a}$$

$$P_{Sq}(\mathcal{Y}_1) = \frac{1}{2}Q\left(\gamma\right) + \frac{1}{2}Q\left(3\gamma\right) \tag{1.111b}$$

── *End Example*

1.4.2   Lower Bounds Based on Uniform Side Information

A simple lower bound for any problem can be obtained using side information. This is often described as a "genie" that aids a receiver. The reasoning is that the optimal genie-aided receiver must perform at least as well as the non-aided receiver. Thus, by choosing the genie's rules carefully, one can obtain a good lower bound that is easy to evaluate.

To formalize this notion, denote the side information by $\mathcal{V}(\zeta)$. Let us focus on the case where a lower bound is desired for $P(\mathcal{E})$ for an $|\mathcal{H}|$-ary decision when the MAP $|\mathcal{H}|$-ary decision rule is used. In that case, the decision rule of a genie-aided receiver is

$$\max_m p(\mathbf{z}, \mathcal{V}|H_m)p(H_m) \iff \max_m p(\mathbf{z}|H_m)p(\mathcal{V}|H_m)p(H_m) \tag{1.112}$$

where it has been assumed that $\mathbf{z}(\zeta)$ is independent of $\mathcal{V}(\zeta)$ given the hypothesis. Consider the special case where the genie provides the receiver with the index of the correct hypothesis and some other hypotheses according to a given probability. For example, if H_m is the true hypothesis, the genie will provide the receiver with $\mathcal{V}_{i,j,m} = \{i, j, m\}$ with probability $p(\mathcal{V}_{i,j,m}|H_m)$. The genie never gives incorrect information – *e.g.*, the genie will never reveal $\mathcal{V}_{3,4} = \{3, 4\}$ if H_2 is correct.

We define a special type of side information scheme as *uniform side information (USI)* which has the property that

$$p_{\mathcal{V}(\zeta)|H(\zeta)}(\mathcal{V}|H_m) = \begin{cases} b(\mathcal{V}) & m \in \mathcal{V} \\ 0 & m \notin \mathcal{V} \end{cases} \tag{1.113}$$

where $b(\mathcal{V}) > 0$ is a constant for a given side information realization. As a consequence, the genie-aided MAP detector executes

$$\max_{m \in \mathcal{V}} p(\mathbf{z}|H_m)p(H_m) \tag{1.114}$$

which is simply a MAP decision rule over the subset of the hypotheses revealed by the side information. Notice that the $p(\mathcal{V}|H_m)$ terms have been canceled due to the USI property.

A *pairwise uniform side information* scheme reveals either one or two elements. That is, when $H(\zeta) = H_m$, the genie reveals either $\mathcal{V}_m = \{m\}$ or $\mathcal{V}_{j,m} = \{j, m\}$. In the former case, the error probability is zero since $p(\mathcal{V}_m|H_j) = \delta_{j,m}$. In the latter case, the MAP rule in the presence of the side information is a binary MAP test between H_m and H_j. Thus, the conditional probability of error for a USI scheme is the error probability of a MAP detector that must select only between H_j and H_m – *i.e.*,

$$P(\mathcal{E}|\mathcal{V} = \{m, j\}, H_m) = P_{PW}(j|H_m) \tag{1.115}$$

An example of how this may be used is to reveal pairs that correspond to signals at a given distance, thus obtaining a bound similar to that in (1.109).

Example 1.17. ———————————————————————————

To illustrate the subtleties associated with development of lower bounds based on side information, consider obtaining a lower bound on the sequence error probability associated with the MAP-SqD $P_{Sq}(\mathcal{Q})$ when the a-priori probabilities are uniform. Table 1.3 shows three different side-information schemes, each revealing a single hypothesis or a pair of hypotheses. Schemes A and B are USI schemes, but scheme C is not. The pairwise USI schemes provide lower bounds immediately. For example,

(a_1, a_0)	$x(a_1, a_0)$	H_m	(A) $\mathcal{V}(\varsigma)$	(B) $\mathcal{V}(\varsigma)$	(C) $\mathcal{V}(\varsigma)$
$(0,0)$	$-3A$	H_0	$\mathcal{V}_{0,1}$ wp 1	$\mathcal{V}_0$ wp 1	$\mathcal{V}_{0,1}$ wp 1
$(0,1)$	$-A$	H_1	$\mathcal{V}_{0,1}$ wp 1	$\mathcal{V}_{0,1}$ wp 1	$\mathcal{V}_{0,1}$ wp 0.5 $\mathcal{V}_{1,3}$ wp 0.5
$(1,1)$	$+A$	H_3	$\mathcal{V}_{2,3}$ wp 1	$\mathcal{V}_{2,3}$ wp 1	$\mathcal{V}_{1,3}$ wp 0.5 $\mathcal{V}_{2,3}$ wp 0.5
$(1,0)$	$+3A$	H_2	$\mathcal{V}_{2,3}$ wp 1	$\mathcal{V}_2$ wp 1	$\mathcal{V}_{2,3}$ wp 1

Table 1.3. Three different side-information schemes for the 4-PAM sequence detection problem ("wp" is "with probability"). The notation $\mathcal{V}_{i,j}$ is used to denote the set with x under H_i and H_j. For example, $\mathcal{V}_{2,3} = \{+3A, +A\}$ and $\mathcal{V}_2 = \{+3A\}$.

consider the MAP-SqD rule given $\mathbf{z}$ and the $\mathcal{V}(\varsigma) = \{-3A, -A\} = \mathcal{V}_{0,1}$ under scheme A. In this case $p_{\mathcal{V}(\varsigma)|H(\varsigma)}(\{-3A, -A\}|H_m) = 0$ for $m \notin \{0, 1\}$. Thus, the MAP rule in (1.112) reduces to

$$p(\mathbf{z}|H_0)p(\mathcal{V}_{0,1}|H_0)p(H_0) \underset{H_1}{\overset{H_0}{\gtrless}} p(\mathbf{z}|H_1)p(\mathcal{V}_{0,1}|H_1)p(H_1) \qquad (1.116)$$

Since the a-priori probabilities are uniform and the scheme is USI, this reduces to a pairwise ML test between H_0 and H_1 so that

$$P_{Sq}^{(A)}(\mathcal{Q}|H_0, \mathcal{V}_{0,1}) = P_{Sq}^{(A)}(\mathcal{Q}|H_1, \mathcal{V}_{0,1}) = Q(\gamma) \qquad (1.117)$$

Considering the same exact situation under side-information scheme C, the rule in (1.116) reduces to

$$p(\mathbf{z}|H_0) \times (1) \underset{H_1}{\overset{H_0}{\gtrless}} p(\mathbf{z}|H_1) \times (1/2) \qquad (1.118)$$

since $p(\mathcal{V}_{0,1}|H_m)$ is 1 and 1/2 for $m = 0, 1$, respectively. Thus, MAP-SqD in the presence of this nonuniform side-information must take into account the statistics of the side information (*i.e.*, the "genie" is biased). The conditional error probabilities for the binary MAP test in (1.118) are (see Problem 1.7)

$$P_{Sq}^{(C)}(\mathcal{Q}|H_0, \mathcal{V}_{0,1}) = Q\left(\gamma + \gamma^{-1}\ln(2)\right) \qquad (1.119)$$

$$P_{Sq}^{(C)}(\mathcal{Q}|H_1, \mathcal{V}_{0,1}) = Q\left(\gamma - \gamma^{-1}\ln(2)\right) \qquad (1.120)$$

Regardless of the (accurate) side information scheme considered, the MAP-SqD rule in the presence of this additional information can perform no worse than the MAP-SqD rule without access to the side information. Thus, after considering all possible conditional values on the sequence and side information and averaging over $p(\mathcal{V}|H_m)p(H_m)$, a lower bound is obtained for each scheme considered. For the schemes listed in Table 1.3, we obtain the error probabilities of the aided receivers

$$P_{Sq}^{(A)}(\mathcal{Q}) = Q(\gamma) \tag{1.121a}$$

$$P_{Sq}^{(B)}(\mathcal{Q}) = \frac{1}{2}Q(\gamma) \tag{1.121b}$$

$$P_{Sq}^{(C)}(\mathcal{Q}) = \frac{3}{4}\left[\frac{2}{3}Q\left(\gamma + \frac{\ln 2}{2\gamma}\right) + \frac{1}{3}Q\left(\gamma - \frac{\ln 2}{2\gamma}\right)\right] + \frac{1}{4}Q(\gamma) \tag{1.121c}$$

The bound associated with scheme C has been written in a form such that it is apparent that it is looser than the lower bound obtained by scheme A (see Problem 1.7). In general, for equally likely hypotheses, a tighter lower bound will be obtained by revealing as little information as possible. For pairwise side-information, this means a good scheme is a USI scheme that reveals the true hypothesis with certainty as infrequently as possible (see Problem 1.8). For example, scheme A is preferred over scheme B because H_0 and H_2 are revealed with certainty in scheme B whenever they are true. However, the main advantage of a USI scheme is that the evaluation of the resulting bound is substantially simplified.

———————————————————————————— *End Example*

Obviously the PAM example is not representative of a practical application of the bounding techniques because the exact error analysis can be obtained directly. A common practical application is the use of the upper-bounding techniques applied to the MAP-SqD and the USI-based lower bounds to the MAP-SyD. Together, these provide lower and upper bounds on the symbol error probability (SEP) using either MAP sequence or symbol detection rules. Specifically, a lower bound on the SEP for MAP-SyD is a valid lower bound for *any* receiver and the upper bound on SEP for the MAP-SqD is also valid for the MAP symbol detector. The next example concludes the 4-PAM example and illustrates this typical use.

Example 1.18. ————————————————————————————
Since the ML-SqD and ML-SyD rules are identical for $a_1(\zeta)$, we focus on the analysis of the MAP-SyD rule for $a_0(\zeta)$ with equal a-priori probabilities. In this case, the exact error probability can be found based in

terms of the parameter ϵ defined in Example 1.3 as

$$P_{Sy}(\mathcal{Y}_0) = \frac{1}{2}\left[Q(\gamma - \epsilon\sqrt{N_0/2}) + Q(\gamma + \epsilon\sqrt{N_0/2})\right]$$
$$+ \frac{1}{2}Q(3\gamma + \epsilon\sqrt{N_0/2}) - \frac{1}{2}Q(5\gamma + \epsilon\sqrt{N_0/2}) \quad (1.122)$$

which still requires solving a transcendental equation for ϵ.

A pairwise USI scheme can be designed so that MAP-SyD of a_0 in the presence of the side information is an ML binary test. Specifically, when a hypothesis H_m in the 4-ary problem is true, the genie should reveal H_j where the value of $a_0(\zeta)$ associated with H_m and H_j are different. In fact, the scheme A from Table 1.3 has this property. The MAP-SyD receiver for $a_0(\zeta)$ based on $\mathbf{z}$ and the side information from scheme A always performs a minimum distance test between two signals at distance $d_{\min}$ so that

$$P_{Sy}(\mathcal{Y}_0) \geq P_{Sy}^{(A)}(\mathcal{Y}_0) = Q(\gamma) \quad (1.123)$$

A lower bound on $P_{Sy}(\mathcal{Y}_1)$ can be found using a pairwise USI scheme that coincides with the exact SEP (see Problem 1.9).

The exact expressions for the SEP associated with MAP-SyD and MAP-SqD for both bit labels is plotted in Fig-1.19. Also shown is the lower bound in (1.123). Notice that the only noticeable difference be-

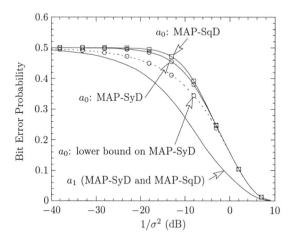

Figure 1.19. Performance comparison for various decision rules for the 4-PAM example.

tween the SEP associated with the two detectors for a_0 occurs at very low SNR as implied by Fig-1.6. Furthermore, the simple lower bound in (1.123) is relatively tight for all reasonable values of the SNR.

————————————————————————— *End Example*

1.4.3   An Upper Bound for MAP-SqD

An upper bound on the probability that the MAP-SqD decision is in error at location k can be obtained using the methods introduced in Section 1.4.1 by properly defining the global error event and a sufficient set of pairwise error events. This may be done by conditioning on the transmitted sequence being $\mathbf{a}(\zeta) = \mathbf{a}$. Let $\mathcal{G}_k(\mathbf{a})$ be the set of all valid input sequences that differ from $\mathbf{a}$ at position k. There are two different error events of interest for a given transmitted sequence $\mathbf{a}$ and another allowable sequence $\tilde{\mathbf{a}} \in \mathcal{G}_k(\mathbf{a})$. The *global error event* $\hat{\mathcal{E}}(\hat{\mathbf{a}}|\mathbf{a})$ occurs if and only if $\hat{\mathbf{a}} \in \mathcal{G}_k(\mathbf{a})$ and $\mathrm{M}[\hat{\mathbf{a}}] \leq \mathrm{M}[\tilde{\mathbf{a}}]$ for all $\tilde{\mathbf{a}}$, where we are using $\mathrm{M}[\mathbf{a}]$ as shorthand for the sequence metric $\mathrm{M}[\mathbf{a}, \mathbf{x}(\mathbf{a})]$ in (1.34). In words, $\hat{\mathcal{E}}(\hat{\mathbf{a}}|\mathbf{a})$ is the event that the MAP-SqD decision $\hat{\mathbf{a}}(\zeta) = \hat{\mathbf{a}}$ disagrees with the transmitted sequence $\mathbf{a}(\zeta) = \mathbf{a}$ at location k. The *pairwise error event* $\mathcal{E}_{PW}(\tilde{\mathbf{a}}|\mathbf{a})$ is the event that $\tilde{\mathbf{a}} \in \mathcal{G}_k(\mathbf{a})$ is more likely than the transmitted sequence $\mathbf{a}$ – *i.e.*, $\mathrm{M}[\tilde{\mathbf{a}}] < \mathrm{M}[\mathbf{a}]$. Note that a symbol error occurs at location k when $\mathbf{a}$ is transmitted if and only if $\hat{\mathcal{E}}(\hat{\mathbf{a}}|\mathbf{a})$ occurs, but the occurrence of a pairwise error $\mathcal{E}_{PW}(\tilde{\mathbf{a}}|\mathbf{a})$ does not necessarily mean that there will be a symbol error at location k.

Assuming a unique tie-breaking strategy is employed for the MAP-SqD rule, the error events $\{\hat{\mathcal{E}}(\tilde{\mathbf{a}}|\mathbf{a})\}$ are disjoint for different $\tilde{\mathbf{a}} \in \mathcal{G}_k(\mathbf{a})$. Thus, the symbol error probability, given $\mathbf{a}$ is transmitted, is the probability of $\hat{\mathcal{E}}(\mathbf{a})$, defined as

$$\hat{\mathcal{E}}(\mathbf{a}) \triangleq \bigcup_{\tilde{\mathbf{a}} \in \mathcal{G}_k(\mathbf{a})} \hat{\mathcal{E}}(\tilde{\mathbf{a}}|\mathbf{a}) \tag{1.124}$$

More precisely, the conditional SEP is $P_{Sq}(\mathcal{Y}_k|\mathbf{a}) = P(\hat{\mathcal{E}}(\mathbf{a}))$.

An upper bound may be obtained, as in Section 1.4.1, by identifying a set of sufficient pairwise error events. Specifically, a sufficient set of sequences $\mathcal{F}_k(\mathbf{a})$ is defined by the property

$$\hat{\mathcal{E}}(\mathbf{a}) \subseteq \bigcup_{\tilde{\mathbf{a}} \in \mathcal{F}_k(\mathbf{a})} \mathcal{E}_{PW}(\tilde{\mathbf{a}}|\mathbf{a}) \tag{1.125}$$

or, as before, a global error implies at least one pairwise error with a member of the sufficient set. An upper-bound on the symbol error probability at index k for the MAP-SqD is then

$$P_{Sq}(\mathcal{Y}_k|\mathbf{a}) \leq \sum_{\tilde{\mathbf{a}} \in \mathcal{F}_k(\mathbf{a})} P(\mathcal{E}_{PW}(\tilde{\mathbf{a}}|\mathbf{a})) = \sum_{\tilde{\mathbf{a}} \in \mathcal{F}_k(\mathbf{a})} P_{PW}(\tilde{\mathbf{a}}|\mathbf{a}) \tag{1.126}$$

for any sufficient set $\mathcal{F}_k(\mathbf{a})$. The pairwise error probability $P_{PW}(\tilde{\mathbf{a}}|\mathbf{a})$ is implicitly defined in (1.126) as $P(\mathcal{E}_{PW}(\tilde{\mathbf{a}}|\mathbf{a}))$.

The most commonly used sufficient set is the set of *simple* sequences $\mathcal{S}_k(\mathbf{a})$ [Ve87]. A simple sequence $\tilde{\mathbf{a}} \in \mathcal{S}_k(\mathbf{a})$ is one in $\mathcal{G}_k(\mathbf{a})$ with associated state sequence $\tilde{\mathbf{s}}$ that differs from that associated with $\mathbf{a}$ in a connected pattern.[8] In other words, the paths in the trellis defined by the transmitted sequence $\mathbf{a}$ and the associated simple sequence $\tilde{\mathbf{a}}$ may remerge to the left or right of the location k at most once. Note that since $\mathcal{S}_k(\mathbf{a}) \subseteq \mathcal{G}_k(\mathbf{a})$, the sequences differ at index k (*i.e.*, $a_k \neq \tilde{a}_k$) so that the associated transitions also differ at this location (*i.e.*, $t_k \neq \tilde{t}_k$). Moreover, for any sequence $\tilde{\mathbf{a}} \in \mathcal{G}_k(\mathbf{a})$ we define its simple component $\tilde{\mathbf{a}}_\mathcal{S}$. Specifically, let the $k_0 \leq k$ and $k_1 + 1 > k$ be the locations closest state agreements between the paths $\tilde{\mathbf{a}}$ and $\mathbf{a}$ to the left and right of k, respectively, and $\mathcal{X}_k = \{k_0, \ldots k \ldots k_1\}$, then

$$\tilde{a}_{\mathcal{S},n} = \begin{cases} \tilde{a}_n & n \in \mathcal{X}_k \\ a_n & n \notin \mathcal{X}_k \end{cases} \tag{1.127}$$

Note that k_0 (k_1) may extend to the left (right) edge of the observation interval. For example, referring to Fig-1.20, the simple component of the MAP-SqD decision $\hat{\mathbf{a}}$ is shown.

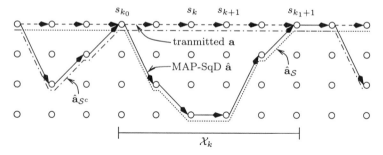

Figure 1.20. The MAP-SqD decision must be the best path between any two states through which it passes. Thus, $\mathbf{a}_\mathcal{S}$ is more likely than the transmitted sequence $\mathbf{a}$.

It follows that the set of all simple sequences $\mathcal{S}_k(\mathbf{a})$ is sufficient because, referring to Fig-1.20, it is not possible that $\hat{\mathbf{a}}$ is the globally shortest path, yet $\mathbf{a}$ is more likely than $\hat{\mathbf{a}}_\mathcal{S}$. Specifically, the MAP-SqD must be the shortest path through any two states that it passes. If this were not the case, another sequence $\hat{\mathbf{a}}_{\mathcal{S}^c}$ could be constructed which would be more likely than $\hat{\mathbf{a}}$ by replacing that segment.

More formally, we claim that $\hat{\mathcal{E}}(\hat{\mathbf{a}}|\mathbf{a}) \subseteq \mathcal{E}_{PW}(\hat{\mathbf{a}}_\mathcal{S}|\mathbf{a})$. This may be shown by contradiction. Specifically, assume that $\hat{\mathcal{E}}(\hat{\mathbf{a}}|\mathbf{a})$ has occurred

---

[8]When referring to the input sequence $\mathbf{a}$ to an FSM, we adopt the convention that the initial state is also included in this sequence.

and $\mathcal{E}_{PW}(\hat{\mathbf{a}}_S|\mathbf{a})$ has not. This means that $\mathrm{M}[\hat{\mathbf{a}}_S] > \mathrm{M}[\mathbf{a}]$. Since the transitions associated with $\hat{\mathbf{a}}_S$ and $\mathbf{a}$ agree on $\mathcal{X}_k^c$, it follows that $\mathrm{M}_{\mathcal{X}_k}[\hat{\mathbf{a}}_S] > \mathrm{M}_{\mathcal{X}_k}[\mathbf{a}]$ holds. Recall that $\mathrm{M}_{\mathcal{J}}[\cdot]$ corresponds to the sequence metric summed over indices in $\mathcal{J}$. However, $\mathrm{M}_{\mathcal{X}_k}[\hat{\mathbf{a}}_S] = \mathrm{M}_{\mathcal{X}_k}[\hat{\mathbf{a}}]$ since the transitions of $\hat{\mathbf{a}}_S$ and $\hat{\mathbf{a}}$ agree on $\mathcal{X}_k$. Thus, a sequence $\hat{\mathbf{a}}_{S^c}$ that agrees with $\mathbf{a}$ on $\mathcal{X}_k$ and $\hat{\mathbf{a}}$ on $\mathcal{X}_k^c$ will have smaller total metric than $\hat{\mathbf{a}}$ – *i.e.*, $\mathrm{M}[\hat{\mathbf{a}}_{S^c}] < \mathrm{M}[\hat{\mathbf{a}}]$ – contradicting the assumption that $\hat{\mathcal{E}}(\hat{\mathbf{a}}|\mathbf{a})$ occurred.

The unconditional upper bound may then be expressed as

$$P_{Sq}(\mathcal{Y}_k) \leq \sum_{\tilde{\mathbf{a}} \in \mathcal{S}_k(\mathbf{a})} P_{PW}(\tilde{\mathbf{a}}|\mathbf{a}) p_{\mathbf{a}(\varsigma)}(\mathbf{a}) \tag{1.128}$$

Note that the proof of the upper bound relies only on the fact that the metric decouples (*i.e.*, as in (1.85)) so the bound is valid whenever the Viterbi algorithm or forward-backward algorithm apply. Thus, this applies to non-uniform a-priori probabilities on the symbols, time-varying systems, systems with terminated initial and final states, etc.

A common special case of (1.128), however, is for a simple, linear, time-invariant FSM, driven by an iid sequence, and observed in AWGN. For this special case, (1.128) is typically simplified in a manner similar to the that in Section 1.4.1.1. Specifically, if $z_k = x(t_k) + w_k$ is the observed realization where $w_k(\varsigma)$ is an AWGN (complex circular) sequence with variance $N_0/2$ per dimension, and $x(t_k) = \mathbf{f}^T \mathbf{a}_{k-L}^k$, then the pairwise error probability simplifies to

$$P_{PW}(\tilde{\mathbf{a}}|\mathbf{a}) = Q\left(\sqrt{\frac{d^2(\mathbf{e})}{2N_0}}\right) \tag{1.129a}$$

$$d^2(\mathbf{e}) = \|\mathbf{x}(\tilde{\mathbf{a}}) - \mathbf{x}(\mathbf{a})\|^2 = \|\mathbf{x}(\tilde{\mathbf{a}} - \mathbf{a})\|^2 \tag{1.129b}$$

$$= \|\mathbf{x}(\tilde{\mathbf{e}})\|^2 = \sum_n |x_n(\tilde{\mathbf{e}})|^2 \tag{1.129c}$$

Thus, one can translate the sum over sequences $\tilde{\mathbf{a}} \in \mathcal{S}_k(\mathbf{a})$ in (1.128) into a sum over associated *error sequences*. Specifically, error sequences are digital sequences with elements in the difference set $\Delta \mathcal{A}$, which contains all possible values obtained by differences of the elements in $\mathcal{A}$. For example, if $\mathcal{A} = \{+1, -1\}$, then the error alphabet is $\Delta \mathcal{A} = \{-2, 0, +2\}$. The set of *simple error sequences* associated with $\{\mathcal{S}_k(\mathbf{a})\}_{\mathbf{a}}$ are those with $e_k \neq 0$ and all nonzero elements separated by no more than $L - 2$ zero entries.[9]

---

[9]Thus, a single simple error sequence corresponds to multiple pairs of $\mathbf{a}$ and $\tilde{\mathbf{a}} \in \mathcal{S}_k(\mathbf{a})$.

It is common to focus on a location k that is near the center of a long sequence and assume that all error sequences with significant pairwise error probability may occur for that location. Under that assumption, the SEP is not a function of k for those points sufficiently far from the edges of the observation interval. Collecting terms, it is then possible to express (1.128) as

$$P_{Sq}(\mathcal{Y}) \le \sum_{d \in \mathcal{D}} K_{UB}(d) Q\left(\sqrt{d^2/(2N_0)}\right) \qquad (1.130a)$$

$$K_{UB}(d) = \sum_{\tilde{e} \in \mathcal{S}: d(\tilde{e})=d} P_C(\tilde{e}) w(\tilde{e}) \qquad (1.130b)$$

where $\mathcal{D}$ is the *distance spectrum* of the FSM. The set $\mathcal{S}$ is the set of all simple error patterns with all shift equivalences removed – *i.e.*, if $\tilde{e}_k$ is in $\mathcal{S}$, then $\tilde{e}_{k-n}$ for $n \ne 0$ is not in $\mathcal{S}$. The number of nonzero elements in $\tilde{e}$ or *weight* is denoted by $w(\tilde{e})$. The inclusion of $w(\tilde{e})$ accounts for not counting different shift equivalent sequences. The term $P_C(\tilde{e})$ accounts for the sum of a-priori probabilities of all transmitted sequences consistent with the error sequence $\tilde{e}$. This factors into a product of terms $P_C(\tilde{e}_k)$ which is the sum of the probabilities of all the values $a_k \in \mathcal{A}$ that could yield $\tilde{e}_k$. Thus, for example, since all transmitted sequence values are consistent with $\tilde{e}_k = 0$, $P_C(0) = 1$. For the case of $\mathcal{A} = \{+1, -1\}$, the error value $+2$ can only be achieved if $a_k = -1$ (and $\tilde{a}_k = +1$), so $P_C(+2) = 1/2$. Similarly, $P_C(-2) = 1/2$. Thus, $P_C(\tilde{e}) = 2^{-w(\tilde{e})}$ for this special case. This is further illustrated in the following example.

Example 1.19. ────────────────────────────────
Consider the QPSK alphabet $\mathcal{A} = \{+1, +j, -1, -j\}$. The associated difference set can be obtained by forming $\tilde{a}_k - a_k$ for all possible values of $\tilde{a}_k, a_k \in \mathcal{A}$, as shown in Table 1.4. It follows from this table that

		a_k			
		$+1$	$+j$	-1	$-j$
	$+1$	0	$+1-j$	$+2$	$+1+j$
$\tilde{a}_k$	$+j$	$-1+j$	0	$+1+j$	$+2j$
	-1	-2	$-1-j$	0	$-1+j$
	$-j$	$-1-j$	$-2j$	$+1-j$	0

Table 1.4.   The difference alphabet for QPSK signaling

$\Delta \mathcal{A} = \{0, +2, -2, +(1+j), -(1+j), +(1-j), -(1-j)\}$. Summing over all values of a_k consistent with $\tilde{e}_k$ and noting that $p_{a_k(\varsigma)}(a_k) = 1/4$ for all conditional values, we can obtain the $P_C(\tilde{e}_k)$ terms. Specifically, this yields $P_C(0) = 1$, $P_C(+2) = P_C(-2) = P_C(+2j) = P_C(-2j) = 1/4$, and

$$P_C(+1+j) = P_C(-1-j) = P_C(+1-j) = P_C(-1+j) = 1/2.$$

End Example

Error analysis of the form presented in this section dates back to the introduction of the Viterbi algorithm [Vi67]. Forney stated the bound in the form of (1.130) in [Fo72]. Foschini noted some errors in the development of [Fo72] and proved that the upper bound in (1.130) converged [Fo75]. Foschini's proof of the upper bound, however, is also flawed (see [ChAn00] and Problem 1.24). The development above is similar to that of Mazo [Ma75] and Verdú [Ve87], and Sheen and Stüber [Sh91, St96], slightly generalized. For the linear FSM channel, Verdú [Ve87] also introduced a set of sufficient sequences that is strict subset of simple sequences, thus tightening the bound (see Problem 1.25).

Evaluation of the upper bound can be approximated by searching simple sequences up to some maximum length. For linear convolutional codes, the upper bound can be evaluated by determining the graph determinant for an error state diagram [ViOm79, LiCo83]. This extends conceptually to more general linear FSMs, but the evaluation becomes tedious [ViOm79]. A branch and bound algorithm for evaluation of the upper bound is given in [Ve87].

At moderate to high SNR, the union bound sum is dominated by the value associated with the maximum pairwise error probability. For the special case of (1.130), this corresponds to the minimum distance term. The *normalized minimum distance* is the minimum distance under the normalization convention $\mathbb{E}\{|a_k(\zeta)|^2\} = 1$ and $\|\mathbf{f}\| = 1$. For example, with BPSK signaling, the pairwise error term associated with the minimum distance is $Q(\sqrt{2E_b/N_0}d_{\min})$. It may be shown that the normalized minimum distance for linear FSM channels is no more than one (see Problem 1.26). Thus, $d_{\min}$ is a measure of the degradation in the SNR associated with ISI. Finding the value of $d_{\min}$ may be viewed as a special case of the CM problem. Specifically, $d_{\min}$ is found by minimizing

$$M[\mathbf{e}] = \sum_k M_k[e_{k-L}^k] = \sum_k |x(e_{k-L}^k)|^2 = \sum_k |f_k \circledast e_k|^2 \qquad (1.131)$$

with the condition that one location in **e** is nonzero. Thus, the minimum distance is the overall associated metric of the best nonzero error path through a trellis with $|\Delta\mathcal{A}|^L$ states. This may be found in most cases by running the Viterbi algorithm on this trellis using the metric defined in (1.131), initialized to the all-zero error state, with the first transition forced to correspond to a nonzero error symbol.

The minimum distance can be found more efficiently by checking only a small subset of simple error sequences. This is shown in [AnFo75] with some insight to these clever observations given in Problem 1.27.

1.4.4   A Lower Bound for MAP-SyD

Lower bounds on the SEP associated with MAP-SyD can be constructed using the method of uniform side information as described in Section 1.4.2. Since it is usually relatively simple to evaluate pairwise error probabilities, the most common approach is to reveal pairs of sequences in a USI manner. The performance of the MAP-SyD receiver aided by this USI is a lower bound on the SEP for any receiver. A simple scheme is to group sequences associated with a given pairwise error probability – *i.e.*, a *consistency set*. For transmitted sequences not in that set, the transmitted sequence is revealed. For sequences in that consistency set, the transmitted sequence is revealed with another sequence which yields the specific pairwise error probability. If this is done in a USI manner, a lower bound is obtained which is simply the given pairwise error probability times the probability of the consistency set. In the following, we demonstrate this for the linear ISI-AWGN channel, for which the distance is a proxy for the pairwise error probability.

Let $\mathcal{C}_k(d)$ be the set of all possible sequences $\mathbf{a}$, that are consistent with an error sequence $\tilde{\mathbf{e}} = \tilde{\mathbf{a}} - \mathbf{a}$ having $\tilde{e}_k \neq 0$ and distance d (*i.e.*, $\|\mathbf{x}(\tilde{\mathbf{a}} - \mathbf{a})\| = d$ for some valid input sequence $\tilde{\mathbf{a}}$). Let $\mathcal{B}_k(d) \subseteq \mathcal{C}_k(d)$ have a pairwise USI scheme defined. Specifically, the side information $\mathcal{V}(\zeta)$ is defined as follows. For transmitted sequences not in $\mathcal{B}_k(d)$, reveal the transmitted sequence

$$p(\mathcal{V} = \{\mathbf{a}\}|\mathbf{a}) = 1 \quad \text{for } \mathbf{a} \notin \mathcal{B}_k(d) \tag{1.132}$$

For transmitted sequences in $\mathcal{B}_k(d)$, reveal the transmitted sequence along with one other sequence in $\mathcal{B}_k(d)$ in a USI manner. With this, errors are only made when sequences in $\mathcal{B}_k(d)$ are transmitted and it follows that the lower bound is

$$P_{Sy}(\mathcal{Y}_k) \geq P_{Sy}^{(\text{aided})}(\mathcal{Y}_k) = P(\mathcal{B}_k(d))Q\left(\sqrt{d^2/(2N_0)}\right) \tag{1.133}$$

where $P(\mathcal{B}_k(d))$ is shorthand for $\Pr\{\mathbf{a}(\zeta) \in \mathcal{B}_k(d)\}$. Note that if a pairwise partitioning of the set $\mathcal{B}_k(d)$ exists, then there is a natural pairwise USI scheme defined. Specifically, for each sequence $\mathbf{a} \in \mathcal{B}_k(d)$, define the associated sequence $\mathbf{a}^{(\text{si})} \in \mathcal{B}_k(d)$ (superscript (si) denotes side information) in such a way that each sequence in $\mathcal{B}_k(d)$ is paired with exactly one other sequence in $\mathcal{B}_k(d)$. Then the pairwise USI scheme is simply to reveal $\mathcal{V} = \{\mathbf{a}, \mathbf{a}^{(\text{si})}\}$ if and only if either $\mathbf{a}$ or $\mathbf{a}^{(\text{si})}$ is transmitted – *i.e.*,

$$p(\mathcal{V} = \{\mathbf{a}, \mathbf{a}^{(\text{si})}\} | \mathbf{a}) = p(\mathcal{V} = \{\mathbf{a}, \mathbf{a}^{(\text{si})}\} | \mathbf{a}^{(\text{si})}) = 1 \qquad (1.134)$$

and $p(\mathcal{V} = \{\mathbf{a}, \mathbf{a}^{(\text{si})}\} | \tilde{\mathbf{a}}) = 0$ for all other values of $\tilde{\mathbf{a}}$. In most cases of interest, again, the index k can be dropped by the assumption of a sufficiently large observation record.

A lower bound based on a pairwise partition of a subset of $\mathcal{C}_k(d_{\min})$ was suggested by Mazo [Ma75b]. A more well-referenced lower bound is that suggested by Forney [Fo72b] which is $P(\mathcal{C}_k(d_{\min}))Q(\sqrt{d_{\min}^2/(2N_0)})$. The development of this claim is, however, not correct. Specifically, the "genie" used in the derivation of [Fo72b] operates as follows. If the transmitted sequence has no $d_{\min}$ neighbor, then it is revealed. If the transmitted sequence $\mathbf{a}$ is in $\mathcal{C}_k(d_{\min})$, then it is revealed along with another sequence that is randomly (uniformly) selected from all sequences $d_{\min}$ away from $\mathbf{a}$. This is not a USI scheme, and therefore does not yield the bound claimed. This is illustrated by reinterpreting Example 1.17 as in the next example.

Example 1.20. ———————————————————————————
Consider detection of a 4-PAM symbol sequence that has been corrupted by an ISI channel and AWGN. Specifically, consider the trivial ISI channel (*i.e.*, no ISI) so that $z_k = x_k + w_k$ where x_k is a sequence of 4-PAM symbols uniformly distributed over the 4-ary constellation of Fig-1.4, and w_k is AWGN with variance $N_0/2$. Note that, in this context, the symbols x_k are the symbols and $\mathbf{x}$ is the sequence. The minimum distance for this problem is $d_{\min} = 2A$ and every transmitted sequence is in $\mathcal{C}(d_{\min})$. A USI scheme is to reveal $\mathcal{V} = \{\mathbf{a}, \tilde{\mathbf{a}}\}$ with probability one when $\mathbf{a}$ is transmitted, where $\tilde{\mathbf{a}}$ agrees with $\mathbf{a}$ in all locations except k. At location k, the scheme is defined as genie A in Example 1.17. Specifically at location k, the side information is $\mathcal{V} = \{-3A, -A\}$ if $a_k \in \{-3A, -A\}$ and $\mathcal{V} = \{+A, +3A\}$ otherwise. This yields a lower bound on the 4-PAM symbol error probability of $Q(\gamma)$ (recall $\gamma = \sqrt{(2A^2)/(N_0)}$).

The side information scheme of [Fo72] applied to this trivial problem yields a lower bound on the 4-PAM SEP the expression in (1.121c). The lower bound stated in [Fo72], however, reduces to $Q(\gamma)$ in this case.
————————————————————————————— *End Example*

Example 1.20 shows that the development of Forney's lower bound is flawed. However, the bound claimed in [Fo72], is, in fact, valid for this example (*i.e.*, it was obtained using the pairwise USI scheme). Another such example is given in [Ve98, Problem 4.24]. In general, the bound stated in [Fo72] itself is invalid [ChAn00] (*i.e.*, see Problem 1.28), although it appears to hold for most cases of practical interest.

Note that the set $\mathcal{B}_k(d)$ may be any subset of $\mathcal{C}_k(d)$ for which a USI scheme is properly defined. Thus, in practice, one need not find the entire distance spectrum to obtain a useful lower bound. Useful lower bounds can be based on incomplete searches of error sequences. Also, the value of d was never specified, so lower bounds can be constructed based on any distance in the distance spectrum. If a search has been performed to obtain the sets $\mathcal{B}_k(d_i)$ for different d_i, each of which has a pairwise USI scheme defined on it, then a valid lower bound of the form in (1.133) can be obtained for each value of d_i. Which of these lower bounds is tightest depends on the SNR. Thus, for a given SNR, the lower bound can be taken as

$$P_{Sy}(\mathcal{Y}_k) \geq \max_i P(\mathcal{B}_k(d_i)) Q\left(\sqrt{d_i^2/(2N_0)}\right) \qquad (1.135)$$

This may be useful, for example, when the minimum distance error sequences have large weight. In this case, $P(\mathcal{B}_k(d_{\min}))$ will be small, making it useful at high SNR but very loose at low SNR. In a similar manner, if the sets $\{\mathcal{B}_k(d_i)\}$ are designed such that they are disjoint, then a multi-d bound can be obtained

$$P_{Sy}(\mathcal{Y}_k) \geq \sum_i P(\mathcal{B}_k(d_i)) Q\left(\sqrt{d_i^2/(2N_0)}\right) \qquad (1.136)$$

An algorithm for constructing such a multi-d bound was described in [MoAu99].

The USI scheme does not need to reveal only pairs. The next example illustrates how the USI scheme can be useful for establishing an intuitive lower bound for ISI channel.

Example 1.21. ───────────────────────────────

Consider the linear ISI-AWGN channel with $z_n(\zeta) = \mathbf{f}^T \mathbf{a}_{n-L}^n(\zeta) + w_n(\zeta)$ where $w_n(\zeta)$ is (complex circular) AWGN with variance $N_0/2$ in each dimension. A USI scheme is defined as follows. When the sequence $\mathbf{a}$ is transmitted, reveal a set of $|\mathcal{A}|$ sequences with probability one. These sequences are those that agree with $\mathbf{a}$ at all locations other than k, with each of the $|\mathcal{A}|$ sequences corresponding to a different conditional value of $a_k(\zeta)$. At the receiver, using the side information, construct a sequence $\bar{x}_n$ by convolving the channel with a sequence that is a_n at all locations except $n = k$, for which it is zero. Subtracting the sequence $\bar{x}_n$ from z_n, we obtain the sufficient statistic y_n which has the model

$$y_n(\zeta) = \begin{cases} w_n(\zeta) & n < k \\ a_k(\zeta)f_{n-k} + w_n(\zeta) & n = k, k+1, \ldots k+L \\ w_n(\zeta) & n > k+L \end{cases} \qquad (1.137)$$

Because the noise is white, a set of sufficient statistics is provided by $\{y_n\}_{n=k}^{k+L}$ or, in vector form $\mathbf{y}_k^{k+L} = a_k\mathbf{f} + \mathbf{w}_k^{k+L}$. This can further be reduced without loss of sufficiency by forming $r = (\mathbf{f}^T\mathbf{y}_k^{k+L})/\|\mathbf{f}\|$ which has the simple model $r(\zeta) = a_k(\zeta)\|\mathbf{f}\| + n(\zeta)$, where $n(\zeta)$ is a circular complex Gaussian random variable with variance $N_0/2$ in each dimension.

Thus, this example shows that a lower bound for the SEP of a linear ISI-AWGN channel is the SEP for an ISI-free channel with equivalent signal energy. In other words, linear ISI can never improve the minimum achievable SEP (*i.e.*, the performance of a MAP-SyD receiver).

———————————————————————— *End Example*

1.5 Chapter Summary

In this chapter we have summarized data detection for fairly general systems and memoryless channels. Effort has been taken to present the most commonly used optimality criteria in a common framework. For example, regardless of the exact system structure MAP-SyD and MAP-SqD may be viewed as special cases of the general Combine and Marginalize (CM) problem. Upper and lower bounds for MAP detection can be constructed from pairwise error probabilities using sufficient sets of neighbors and uniform side information schemes, respectively.

The important special system of a Finite State Machine was considered in detail. With perfect channel state information, the forward-backward algorithm provides MAP sequence or symbol decisions. For MAP sequence detection, the backward recursion can be replaced by a store and trace-back operation which yields the Viterbi algorithm. The Viterbi algorithm produces exactly the same decisions as the min-sum forward-backward algorithm, but it does not explicitly compute the input symbol MSM and then threshold it. In Chapter 2, we seek algorithms to compute soft information and, instead of thresholding that information, pass it on to other subsystem detectors for further refinement.

For channels with nuisance parameters (*i.e.*, imperfect CSI), the FSM structure does not yield an efficient MAP detector structure. In fact, even for very simple systems with imperfect CSI, the optimal detector is, strictly speaking, an exhaustive combining and marginalization processor. Fortunately, however, approximations to this prohibitively complex processing perform well in practice. Our presentation of these practical receivers for imperfect CSI emphasized the view of a forced folding approximation to the optimal processing. The bidirectional recursive estimator-correlator developed in this chapter serves as the basis for the practical adaptive iterative detection methods presented in Chapter 4.

1.6   Problems

1.1. Consider the system with $z_k(\zeta) = \sqrt{E_s}a_k(\zeta)e^{j\phi} + w_k(\zeta)$, where $a_k(\zeta)$ is a PSK modulation sequence so that $|a_k|^2 = 1$ for all $a_k \in \mathcal{A}$ and $w_k(\zeta)$ is AWGN.

(a) Assuming that ϕ is an unknown, deterministic constant in $[-\pi, +\pi)$, find the generalized likelihood of $\mathbf{a}_0^{K-1}$ based on $\mathbf{z}_0^{K-1}$.

(b) Assume a random model $\phi(\zeta)$ which is uniform on $[-\pi, +\pi)$. Find both the generalized likelihood and the average likelihood for this model.

(c) Compare your results and discuss the impact of the assumption that $|a_k|^2$ is constant.

1.2. Verify that the computation of the min*$(\cdot)$ operator defined in (1.28a) and (1.28c) can be computed as shown in (1.28b) and (1.28d). To show (1.28b), consider the cases $x > y$ and $x < y$ separately.

1.3. Consider natural labeling of a PSK constellation. Specifically, let $x(\mathbf{b}_0^{K-1}) = \sqrt{E_s}\exp[j\theta(\mathbf{b}_0^{K-1})]$, where

$$\theta(\mathbf{b}_0^{K-1}) = \frac{1}{2\pi}2^{-K}\sum_{k=0}^{K-1}b_k 2^k$$

and each b_k takes on values in $\{0, 1\}$. In this problem, we compare the ML-SqD and ML-SyD decision rules. Specifically, assume that $b_k(\zeta)$ is an iid sequence with each equally likely to take 0 or 1, and that the observation is $z_k = x(\mathbf{b}_0^{K-1}) + w$, where w is a realization of a circular complex Gaussian random variable with mean zero and $\mathbb{E}\{|w(\zeta)|^2\} = N_0$.

(a) For $K = 2$ (*i.e.*, QPSK), show that the ML-SqD and ML-SyD rules are the same.

(b) For $K = 3$ (*i.e.*, 8PSK), show that the ML-SqD and ML-SyD rules are the same.

(c) Can you find a labeling for which the two criteria yield different decisions?

1.4. Show that, in an AWGN channel, the ML-SyD decision regions for naturally labeled 2^K-PSK are identical (modulo $2\pi/2^K$) to those for a Gray indexed 2^K-PSK. **Hint**: the Gray labeling can be obtained from the natural labeling via the mapping $b_1 b_2 \cdots b_l \rightarrow (b_1 \oplus b_2) \cdots (b_{l-1} \oplus b_l) b_l$.

1.5. Define a soft information measure as $V[x] = -I_0^{-1}(P[x])$, where $I_0(\cdot)$ is the modified Bessel function of the first kind. Is there a combining and marginalizing semi-ring for this soft information measure? If so, identify the associated elements in Table 1.2, if not, explain why.

1.6. [ChDeOr99] Consider the soft information measure that computes the marginal metric $M[a_k]$ using sum combining with marginalizing defined as follows. Let $\hat{a}$ be the path with shortest path and $\bar{a}(a_k)$ be the same sequence with $\hat{a}_k$ replaced by a_k, then $M[a_k] = M[\bar{a}(a_k)] - M[\hat{a}]$.

(a) Show that this soft output is threshold consistent with the Viterbi algorithm with appropriately defined metrics.

(b) If a_k is the input to a simple FSM with output x_k, give a simple algorithm to augment the Viterbi algorithm to produce this soft information.

(c) Consider the case where y_k is the input to a quantizer and $a_k(\mathbf{y})$ is the output. When the inverse quantizer is driven by a_k, the reconstructed estimate of y_k is $x_k(\mathbf{a})$. In this case, if $M[\mathbf{a}] = \sum_k M[y_k, x_k(\mathbf{a})]$ represents the reconstruction distortion associated with quantizing $\mathbf{y}$ to $\mathbf{a}$, what does the marginal soft-information computed represent?

1.7. Consider a binary hypothesis test defined by $\mathbf{z} = \mathbf{s}_m + \mathbf{w}(\zeta)$ under H_m, with $m = 0, 1$. The noise $\mathbf{w}(\zeta)$ is AWGN. Show that the error probability for the MAP test is

$$P(\mathcal{E}) = p_1 Q\left(\frac{d}{2} + \frac{\ln(p_1/p_0)}{d}\right) + p_0 Q\left(\frac{d}{2} - \frac{\ln(p_1/p_0)}{d}\right) \quad (1.138)$$

where $p_m = p(H_m)$ and

$$d = \sqrt{\frac{\|\mathbf{s}_1 - \mathbf{s}_0\|^2}{N_0/2}} \quad (1.139)$$

Show that this is maximized when the priors are equal.

1.8. Reconsider the lower bound development for the 4-PAM example with the a-priori probabilities: $p_{a_0}(0) = 2/3$, $p_{a_1}(0) = 1/2$. Does scheme C yield a tighter lower bound?

1.9. Consider the 4-PAM example. Find the lower bound on the SEP for a_1 based on the side information scheme: $p(\mathcal{V}_{0,3}|H_0) = 1$, $p(\mathcal{V}_{0,3}|H_3) = 1$, $p(\mathcal{V}_{1,2}|H_1) = 1$, and $p(\mathcal{V}_{1,2}|H_1) = 1$. Show that this lower bound coincides with the exact value of the SEP in this example. Generally, under what conditions will a pairwise USI scheme yield the exact SEP?

1.10. [Ch99] Consider computing the soft-information $\text{APP}_{k+1}^{k+D}[s_{k+1}]$ that is the output of the backward recursion of the forward-backward algorithm started at observation z_{k+D}. Denote the $|\mathcal{S}| \times 1$ vector of these state APPs by $\mathbf{b}_{k+1}^{k+D}$

   (a) Show that sum-product backward recursion can be written as a matrix multiplication of the form $\mathbf{b}_k^{k+D} = \mathbf{P}_k \mathbf{b}_{k+1}^{k+D}$. Identify the elements of $\mathbf{P}_k$.

   (b) Show that $\mathbf{b}_{k+1}^{k+D} = \mathbf{P}_{k+1}\mathbf{P}_{k+2}\cdots\mathbf{P}_{k+D}\mathbf{1}$, where $\mathbf{1}$ is the all one's vector.

   (c) Describe the details of a fixed-lag APP algorithm computing $p(a_k|\mathbf{z}^{k+D})$ using the standard sum-product forward recursion and competition operation, but with the backward recursion replaced by $\mathbf{b}_k^{k+D-1} = \mathbf{P}_k\mathbf{G}[\mathbf{P}_{k+D+1}]^{-1}$. Identify $\mathbf{G}$.

   (d) Is this algorithm a semi-ring algorithm? If so, state the min-sum version. If not, explain why. Is there a min*-sum version?

   (e) Consider applying the above backward operation to the 4-state trellis in Fig-1.10. Can you exploit the sparse connections between states to implement the matrix inversion in a simple form?

1.11. Give a sufficient condition on x_0 and y_0 so that $\min^*([z-x_0]^2, [z-y_0]^2) \equiv \min([z-x_0]^2, [z-y_0]^2)$ for all z. Generalize this to real-valued vectors.

1.12. Consider a binary sequence $\mathbf{b}$ that labels a signal set via the mapping $\mathbf{x}(\mathbf{b})$. Let $\bar{\mathbf{b}}_m$ denote $\mathbf{b}$ with the m-th element complemented. Assume that the signal set satisfies $\mathbf{x}(\bar{\mathbf{b}}_m) = -\mathbf{x}(\mathbf{b})$ for all $\mathbf{b}$ and all m. Use the result of Problem 1.11 to show that MAP-SyD and MAP-SqD yield the same decisions for this case.

1.13. Consider a binary convolutional code with input b_k and outputs $c_k(0)$ and $c_k(1)$ defined by

$$c_k(0) = b_{k-1} \oplus b_{k-2} \qquad (1.140a)$$
$$c_k(1) = b_k \oplus b_{k-1} \oplus b_{k-2} \qquad (1.140b)$$

where $\oplus$ represents modulo two addition and all variable are in $\{0,1\}$.

(a) Sketch a block diagram of the encoder using two delay elements and two adders.

(b) Show that this system is an FSM that can be represented by the trellis diagram in Fig-1.10.

1.14. Consider the TCM scheme illustrated in Fig-1.21 where the convolutional code is the toy code from Problem 1.13. The 8PSK con-

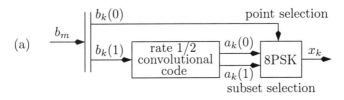

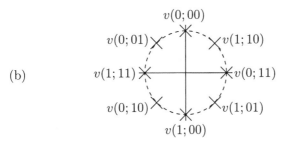

Figure 1.21. The (a) toy TCM encoder for Problem 1.14 with the (b) bit labeling convention

stellation is labeled according to the principle of Ungerboeck set-partitioning. Specifically, let $x_k = \sqrt{E_S} v(b_k(0); a_k^{(0)} a_k^{(1)})$ where $v(\cdot) \in \left\{ \exp\left[\frac{\pi}{4} m\right] \right\}_{m=0}^{7}$ is as illustrated in Fig-1.21(b). Draw the trellis diagram based on:

(a) a state definition of $s_k = (b_{k-1}(1), b_{k-2}(1))$.

(b) a state definition of $s_k = (b_{k-1}(0); b_{k-1}(1), b_{k-1}(1))$.

(c) the simple FSM resulting from the 4-ary input vector $\mathbf{b}_k = [b_k(0) \; b_k(1)]$.

1.15. Consider the standard trellis representing a simple $|\mathcal{A}|$-ary input FSM which has $|\mathcal{A}|^L$ states, and $|\mathcal{A}|$ transitions leaving and entering each state. Consider a Viterbi-like algorithm performs the ACS in a completely random manner – *i.e.*, the survivor is selected randomly from the $|\mathcal{A}|$ paths entering the state. Find an expression for the probability of all paths merging after D trace-back steps. You may want to consider the L-step trellis to simplify the computation. Do you expect the merging for this algorithm to be sooner or later than the Viterbi algorithm?

1.16. Confirm that $\text{MSM}_0^{11}[x_7 = 0.707] = 22.9$ in Example 1.11.

1.17. Based on the information in Fig-1.15, compute the M*SM version of all MSM quantities considered in Examples 1.9-1.12.

1.18. Let $z(\zeta) = f_k(\zeta)a_k(\zeta)$ be an observation sequence where $a_k(\zeta)$ is an independent data sequence and $f_k(\zeta)$ is a complex circular Gaussian process. Furthermore, assume that $f_k(\zeta)$ is a stationary AR process of order P that is independent of the data.

 (a) Show that this problem satisfies the folding condition (1.85).

 (b) Describe the transition metrics used in the associated Viterbi algorithm.

 (c) Describe the forward backward APP algorithm for this special case.

 (d) Is the folding condition still satisfied if there is additional AWGN which is independent of both the channel and data?

1.19. Consider an ISI-AWGN channel with an iid binary input sequence b_0^K. Denote $p[b_k = 1] = p$. Show that when the input binary symbols are not uniformly distributed (i.e., $p \neq 0.5$), the ML-SqD obtained without the knowledge of p (i.e., assume $p = 0.5$) asymptotically equals to the ML-SqD obtained with the perfect knowledge of p as SNR$\to \infty$.

1.20. Consider an algorithm for performing data detection for the FSM system and memoryless channel case. Specifically, the algorithm runs a forward sum-product recursion of the form in (1.70b). For each state s_{k+1}, the algorithm notes the value t_k that maximizes $p(z_k|x_k(t_k))p(a_k)p(z_0^{k-1}, s_k)$ – i.e., contributes most to the sum. It stores this sequence of transitions $\check{t}_i(s_{i+1})$ for $i = 0, 1, 2, \ldots$ $(K-1)$. At the end of the trellis, a traceback is done for the state s_{K-1} with maximum APP to infer $\hat{a}_k$. Is this a MAP algorithm? Is this a semi-ring algorithm? Do you expect this algorithm to perform well?

1.21. Consider an arbitrary time-invariant channel. Given the transmitted sequence $\check{a}$, we observe z through this channel.

 (a) Show that if there exists a sequence $\tilde{a}$ such that $P[\tilde{a}|z] > 0.5$, the MAP-SqD based on z is equivalent to the MAP-SyD.

 (b) Assume the channel is an ISI-AWGN (with variance σ^2) channel. Show that $P[z|\check{a}] \to 1$ in probability as $\sigma^2 \to 0$.

 (c) Assume the channel is an ISI-AWGN (with variance σ^2) channel. Show that for $a \neq \check{a}$, $P[z|a] \to 0$ in probability as $\sigma^2 \to 0$.

(d) Show that for an ISI-AWGN channel the MAP-SqD is equivalent to the MAP-SyD in probability as $\sigma^2 \to 0$.

1.22. This problem is aimed at software implementation of the forward-backward algorithm (*e.g.*, in C or C++).

(a) Create a data structure for a generic FSM – *i.e.*, type FSM. This should have the next-state and next-out tables in terms of symbols indices. Make the values of $|\mathcal{A}|$, $|\mathcal{S}|$, $|\mathcal{X}|$ part of the data structure so that they may be read at run time. In addition, you may want to create a previous-state table, $s_k = \mathrm{ps}(u_k, s_{k+1})$ with u_k having $|\mathcal{U}| = |\mathcal{A}|$.

(b) Write subroutines to read-in the defining size and table variables for the FSM data type. This routine should allocate the appropriate memory. Also, write a subroutine that can free the memory allocated.

(c) Write a subroutine that executes the forward-backward algorithm. Make the subroutine accept pointers to MI$[a_k]$, MI$[x_k]$, and M$[a_k]$ as well as a pointer to an FSM-type variable. Use a #define statement to create a mymin (x,y) macro that can be easily toggled between the standard $\min(x,y)$ and $\min^*(x,y)$.

(d) Use the numerical values given in Examples 1.9-1.12 to verify your code.

1.23. Consider an ISI channel with

$$h(t) = \begin{cases} e^{-t/T} & t \in [0, 3T) \\ 0 & \text{otherwise} \end{cases} \tag{1.141}$$

(a) Plot $h(t)$ and $R_h(\tau) = h(t) \circledast h(-t)$.

(b) Determine and plot $R(m)$, the whitening filter, and the equivalent channel f_k.

(c) Consider the design of an approximate whitening filter based on an FIR predictor. Specifically, if the colored noise $n_k(\zeta)$ is predicted (*i.e.*, MMSE) from $\{n_{k-m}(\zeta)\}_{m=1}^{N}$ to obtain $\hat{n}_k^{(N)}(\zeta) = p_k \circledast n_k(\zeta)$, then the show that the corresponding error $e_k^{(N)}(\zeta) = n_k(\zeta) - \hat{n}_k^{(N)}(\zeta) = (\delta_k - p_k) \circledast n_k(\zeta)$ is approximately white. More precisely, as $N \to \infty$, e_k becomes AWGN, and g_k becomes an auto-regressive whitening filter. Perform this design procedure for various values of N; determine the output noise correlation and the overall equivalent ISI channel coefficients. Based on these results (in comparison with those of part 1.23b), what is a reasonable choice for N?

1.24. [AnCh00] Consider the upper bound development in Section 1.4.3. Using a sketch similar to that in Fig-1.20, give examples to show the following:

(a) The MAP-SqD path based on all observations need not be the best path on a partial interval. Given the scenario in Fig-1.20, show another path that may have a smaller metric on X_k than the MAP-SqD path. **Hint:** this better path cannot go through the same states as the MAP-SqD at times k_0 and k_1.

(b) A global error causing a symbol error at time k does not imply a simple pairwise error with symbol error at time k. Specifically, that for simple sequences, the relation in (1.125) is not equality.

1.25. [Ve87] For the special case of a linear ISI-AWGN channel show that a condition for a sufficient set $\mathcal{F}_i(\mathbf{a}) \subset \mathcal{S}_i(\mathbf{a})$ is

$$\Delta M[\mathbf{a}; \mathbf{a} + \hat{\mathbf{e}}_\mathcal{F}] = \Delta M[\mathbf{a} + \hat{\mathbf{e}}_{\mathcal{F}^c}; \mathbf{a} + \mathbf{e}] + 2\mathbf{x}^T(\hat{\mathbf{e}}_{\mathcal{F}^c})\mathbf{x}(\hat{\mathbf{e}}_\mathcal{F})$$
$$\geq 2\mathbf{x}^T(\hat{\mathbf{e}}_{\mathcal{F}^c}), \mathbf{x}(\hat{\mathbf{e}}_\mathcal{F}) \geq 0 \qquad (1.142a)$$

where $\hat{\mathbf{e}} = \hat{\mathbf{e}}_\mathcal{F} + \hat{\mathbf{e}}_{\mathcal{F}^c}$ is the decomposition of $\hat{\mathbf{e}} = \hat{\mathbf{a}} - \mathbf{a}$ into $\mathbf{a} + \mathbf{e}_\mathcal{F} \in \mathcal{F}_i(\mathbf{a})$ and the remainder and $\Delta M[\mathbf{b}, \mathbf{c}] = M[\mathbf{b}] - M[\mathbf{c}]$. Does this extend to nonlinear mappings $\mathbf{x}(\mathbf{a})$?

1.26. Use the results of Example 1.21 to show that the normalized minimum distance of a linear ISI channel is no more than one.

1.27. [AnFo75] Consider the problem of finding $d_{\min}$ for the special case of a linear ISI-AWGN channel, $i.e.$, $z_k = \mathbf{f}^T \mathbf{a}_{k-L}^k + w_k$, with $||\mathbf{f}|| = 1$. For an error sequence of length K, $\mathbf{e} = (\ldots, 0, e_0, e_1, \ldots, e_{K-1}, 0, \ldots)$ define the following quantities

$$d^2(\mathbf{e}, \mathbf{f}) \triangleq \sum_{k=0}^{K+L-1} |\mathbf{f}^T \mathbf{e}_{k-L}^k|^2 = \sum_{k=0}^{K+L-1} \mathbf{f}^H \mathbf{E}_k \mathbf{f} = \mathbf{f}^H \mathbf{E} \mathbf{f}$$

$$d_{\min}^2(\mathbf{f}) \triangleq \min_{\mathbf{e}} d^2(\mathbf{e}, \mathbf{f})$$

where $\mathbf{E}_k = (\mathbf{e}_{k-L}^k)^*(\mathbf{e}_{k-L}^k)^T$, $\mathbf{E} = \sum_{k=0}^{K+L-1} \mathbf{E}_k$, and the last minimization is over all simple error sequences.

(a) For a given error sequence $\mathbf{e}$, show that if a subsequence of length L is repeated ($i.e.$, $\mathbf{e}_{j-L+1}^j = \pm \mathbf{e}_{i-L+1}^i$ for some $i < j$,) then there exists an error sequence $\mathbf{e}'$, such that $d^2(\mathbf{e}', \mathbf{f}) \leq d^2(\mathbf{e}, \mathbf{f})$. Thus, in finding $d_{\min}$, the sequence $\mathbf{e}$ need not be extended.

(b) Define $b(e_{k-L+1}^{k}) \triangleq [e_{k-L+1}, \dots, e_{k}]^{T}$. For a given error sequence $\mathbf{e}$, show that if $e_{j-L+1}^{j} = \pm b(e_{i-L+1}^{i})$ for some $i \leq j$, then, in finding $d_{\min}$ no sequence with prefix e_{0}^{j} need to be extended, except the one having $e_{j-L+2}^{j+1} = \pm b(e_{i-L}^{i-1}), \dots,$ $e_{i+j-L}^{i+j-1} = \pm b(e_{2-L}^{1})$.

(c) Given a simple error sequence $\mathbf{e}_0$ (with corresponding matrix $\mathbf{E}_0$), show that, in finding $d_{\min}$, there is no need to extend an error pattern $\mathbf{e}$ (with corresponding matrix $\mathbf{E}$) if the minimum eigenvalue of $\mathbf{E} - \mathbf{E}_0$ is positive.

1.28. [ChAn00] In this problem we demonstrate a case where the bound stated in [Fo72] is invalid. Consider a binary antipodal length-3 sequence $\mathbf{a} = a_0^2$, $a_i \in \{-1, 1\}$ as the input to a linear time-varying mapping producing the noise-free sequence $\mathbf{x} = x_0^2 = \mathbf{Fa} = \mathbf{SRa}$, where $\mathbf{R}$ and $\mathbf{S}$ represent a rotation and a scaling, respectively. Choose

$$\mathbf{R} = \frac{1}{\sqrt{6}} \begin{bmatrix} \sqrt{2} & \sqrt{2} & \sqrt{2} \\ -1 & 2 & -1 \\ -\sqrt{3} & 0 & \sqrt{3} \end{bmatrix} \qquad \mathbf{S} = \begin{bmatrix} \Delta & 0 & 0 \\ 0 & 1/\sqrt{8} & 0 \\ 0 & 0 & 1/\sqrt{8} \end{bmatrix}$$

with $\Delta \geq 2/\sqrt{8}$ being a parameter that controls the stretching. The resulting constellation is shown in Fig-1.22.

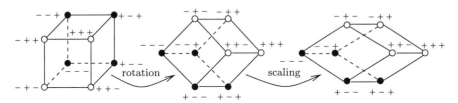

Figure 1.22. Counter-example of Forney's lower bound using a linear mapping of bit triplets used in Problem 1.28.

(a) Show that the probability of error of the optimal ML-SqD receiver for the middle bit a_1 can be evaluated as

$$P_{Sq}(\mathcal{Y}_1) = \frac{1}{2\pi} \int_0^{2\pi/3} \exp\left(-\frac{1}{8\sigma^2 \sin^2 \phi}\right) d\phi + \epsilon(\Delta)$$

where σ^2 is the observation noise variance per dimension and $\epsilon(\Delta) \to 0$ as $\Delta \to \infty$ for a fixed σ^2.

(b) Show that the bound suggested by [Fo72] is based on a non-USI scheme, resulting in $P_{Sq}^{(Fo)}(\mathcal{Y}_1) = \frac{6}{8}Q(\frac{1}{2\sigma})$. Consider the limiting case of $\Delta \to \infty$, such that $\Delta/\sigma^2 \to \infty$, and show that there exists a noise variance and a large enough Δ, such that $P_{Sq}^{(Fo)}(\mathcal{Y}_1) > P_{Sq}(\mathcal{Y}_1)$, thus Forney's lower bound is incorrect. Verify the above claim using simulation with $\sigma^2 = 2$dB and $\Delta = 10$.

(c) Show that the pairwise USI scheme suggested in [Ma75b] results in $P_{Sq}^{(Ma)}(\mathcal{Y}_1) = \frac{4}{8}Q(\frac{1}{2\sigma})$. For the special case of $\Delta = 10$, show that a tighter multi-d lower bound can be constructed as in (1.136), resulting in $P_{Sq}(\mathcal{Y}_1) \geq \frac{4}{8}Q(\frac{1}{2\sigma}) + \frac{4}{8}Q(\frac{d}{2\sigma})$, where $d = 23.1$ is the distance between $[+++]$ and $[+--]$ and the distance between $[---]$ and $[-++]$.

1.29. A zero-mean Gauss Markov vector process $\{\mathbf{f}_k(\zeta)\}$ is described by the equation

$$\mathbf{f}_k(\zeta) = \mathbf{F}\mathbf{f}_{k-1}(\zeta) + \mathbf{u}_k(\zeta)$$

where $\mathbf{u}_k(\zeta)$ is a zero-mean Gaussian vector with covariance $\mathbf{K_u}(m) = \mathbf{K_u}\delta_m$.

(a) Show that a necessary and sufficient condition for stationarity of $\{\mathbf{f}_k(\zeta)\}$ is that the covariance matrix of $\mathbf{f}_k(\zeta)$ satisfies the equation

$$\mathbf{K_f} \triangleq \mathbb{E}\{\mathbf{f}_k(\zeta)\mathbf{f}_k(\zeta)^H\} = \mathbf{F}\mathbf{K_f}\mathbf{F}^H + \mathbf{K_u}$$

(b) Show that the time-reversed process $\{\mathbf{f}_{-k}(\zeta)\}$ is also a GM process, described by the backward equation

$$\mathbf{f}_k(\zeta) = \mathbf{F}^b\mathbf{f}_{k+1}(\zeta) + \mathbf{v}_k(\zeta)$$

where $\mathbf{v}_k(\zeta)$ is a zero-mean Gaussian vector with covariance $\mathbf{K_v}(m) = \mathbf{K_v}\delta_m$. In particular, show that the matrices $\mathbf{F}^b$ and $\mathbf{K_v}$ satisfy the equations

$$\mathbf{F}^b = \mathbf{K_f}\mathbf{F}^H\mathbf{K_f}^{-1}$$
$$\mathbf{K_v} = \mathbf{K_f} - \mathbf{K_f}\mathbf{F}^H\mathbf{K_f}^{-1}\mathbf{F}\mathbf{K_f}$$

Chapter 2

PRINCIPLES OF ITERATIVE DETECTION

In this chapter we are concerned with detecting the input to a system which is expressed as a concatenated network of subsystems. In particular, the goal is to mimic the implicit decomposition of the system in the detector processing. First, we consider sufficient conditions for when this detector comprised of subsystem-detectors yields the optimal decision as defined in Chapter 1. Second, we summarize the suboptimal method of iterative (turbo) detection which is often a very effective, low-complexity approximation to the optimal detector. This description, which is based on the block diagram system descriptions developed in Chapter 1, consists of a simple subsystem "soft-inversion" process which, given the system model, implies the iterative detector for the most part. Several detailed application examples are given.

Viewing system block diagrams using explicit indexing as a method of graphical system modeling, the various graphical modeling and iterative detection methods in the literature are described. Finally, we conclude by noting that system models are generally not unique and, when used to imply an iterative detector, yield different performance and complexity. We describe a set of guidelines for selecting among system modeling options. All channel parameters are assumed to be known in this chapter unless otherwise specified.

2.1 Optimal Detection in Concatenated Systems

Consider a system comprising a concatenated network of subsystems as illustrated in Fig-2.1. In such a system, the output of each subsystem (except H) is the input to another subsystem. The only assumptions made about the subsystem structure are those from Chapter 1. Namely that all inputs/outputs are digital, and that the sequence of inputs to a

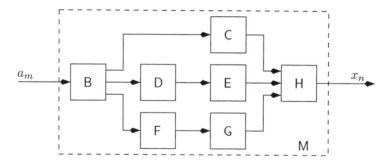

Figure 2.1.   The block diagram of a generic concatenated network.

particular subsystem uniquely determines its output sequence. The concatenated network is assumed to have no explicit feedback loops. More precisely, for a given subsystem (*e.g.*, E), the output cannot be fed back to another subsystem (*e.g.*, B) that affects its input. The overall or *global* system M is defined by the mapping from the sequence of inputs $\{a_m\}$ to the sequence of outputs $\{x_n\}$ as in Fig-1.3. Note that it is somewhat arbitrary how one decomposes a given system into a concatenated network of subsystems since many such representations that maintain the global system structure are possible. Thus, given a concatenated network with feedback, one can attempt to combine subsystems to eliminate the explicit feedback. For example, the subsystem E may actually represent a grouping of several other subsystems between which there exist feedback loops. In many applications, however, a natural decomposition exists because the global system is actually obtained by a particular concatenation.

Based on the development in Chapter 1, optimal detection of a_m directly from the global structure is conceptually clear. For most concatenated systems of interest, however, this direct approach is prohibitively complex. Thus, we consider the *concatenated detector* of Fig-2.2 in which a network of *soft-output algorithm (SOA)* modules is constructed according to the system network. At this point in the development, we do not specify the exact form of the SOA and simply denote it as a shaded block corresponding to the block diagram. We do assume, however, that the SOA produces soft information on the inputs to the system as implicitly shown in the system block diagram. For example, since the block diagram shown in Fig-2.1 is in implicit index form, the corresponding SOA in the detection network in Fig-2.2 passes soft information on the individual elements of the sequence, or *marginal soft information*. For a concrete example, if the output of subsystem D is d_k, then the soft information passed on the corresponding connection between SOA E and

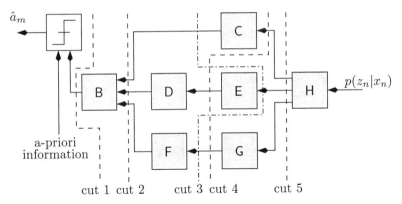

Figure 2.2. The concatenated detector structure considered for the system in Fig-2.1. This structure is optimal if the soft information passed over the cut boundaries comprises a set of sufficient statistics.

SOA D in the detector of Fig-2.2 is $S[d_k]$. More precisely, for each value of k, the SOA E produces $|\mathcal{D}|$ values to describe the reliability or belief for each conditional value of $d_k(\zeta)$ and the sequence of this marginal soft information $\{S[d_k]\}$ is passed to the SOA D. This is the same as the explicit index form. However, if the same system is expressed in the vector mapping convention of Fig-1.3(c), then the corresponding concatenated detector would implicitly pass soft information on the entire sequence **d**.

The issue to be addressed is: under what conditions is the concatenated detector equivalent to the optimal detector based on the global detector? The answer is fairly straightforward, and follows from two simple facts. Consider the example in Fig-2.2. First, the soft information delivered by SOA B must be a set of sufficient statistics for optimal detection of a_m based on the observation **z**. Second, given some soft information that does not provide a set of sufficient statistics, additional processing of this soft information which is independent of the observation **z** cannot yield a set of sufficient statistics. Intuitively, this second fact means that once sufficiency is lost, it cannot be recovered. This is straightforward to show formally (see Problem 2.1). Together, these two simple facts imply that the set of soft information that is input to SOA B must also form a set of sufficient statistics for detecting a_m from **z**. Since the boundaries between subsystems is only conceptual, one can *cut* the concatenated detector into two distinct parts and the soft information passing over this cut boundary must form a set of sufficient statistics for deciding on a_m from the observation.

In general, therefore, the concatenated detector shown in Fig-2.2, which implicitly passes marginal soft information, is suboptimal. The concatenated detector implied by the vector mapping system diagram would pass soft information on the entire sequence **d**. While this is considerably more complex, it is also generally suboptimal. This is because, for example, in order to obtain sufficiency at cut 4, it may be necessary that SOAs E and G perform joint processing. These subtleties are illustrated in the examples that follow.

Two simple types of concatenated networks are the *serial* and *parallel* concatenations shown in Fig-2.3. For the serial concatenation in

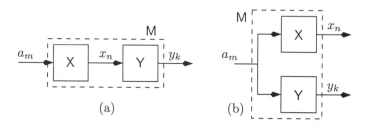

Figure 2.3. (a) Serial and (b) parallel concatenation of two systems.

Fig-2.3(a), the associated concatenated detector passes marginal soft information on x_n which is generally suboptimal. However, for this simple network, the concatenated detector based on vector mapping convention of Fig-1.3(c) passes soft information on **x** which can be sufficient. For example, $p(\mathbf{z}|\mathbf{x})$ for each of the $|\mathcal{X}|^N$ possible values of **x** is a set of sufficient statistics for detection of a_m. Moreover, this essentially decouples the SOA for Y in the associated concatenated detector since S[**x**] (*i.e.*, the soft information on the entire sequence) may be computed without considering the structure of the system X. In other words, the soft information for each of $|\mathcal{X}|^N$ possibilities can be computed even though only $|\mathcal{A}|^M$ values are possible. While optimal, this may be more complex than necessary depending on the structure of the systems X and Y. This is illustrated in the following example.

Example 2.1. ————————————————————————
Consider the special case of Fig-2.3(a) where the system X is a mapping of the independent binary sequence a_m onto a 4-PAM constellation as described in Fig-1.4 and the system Y is a finite state machine. This example is shown in Fig-2.4 in the three conventions introduced in Fig-1.3. The concatenated detector implied by each is shown in Fig-2.5. Consider performing MAP-SyD of a_m. Clearly, from the above general discussion, the detector in Fig-2.5(c) provides optimal processing when S[**x**] is equivalent to $p(\mathbf{z}|\mathbf{x})$. In this case, the soft information S[**a**] $\equiv$ $p(\mathbf{z}|\mathbf{a}) = p(\mathbf{z}|\mathbf{x}(\mathbf{a}))$ can be marginalized to $p(a_m|\mathbf{z})$ using $\{p(a_i)\}$.

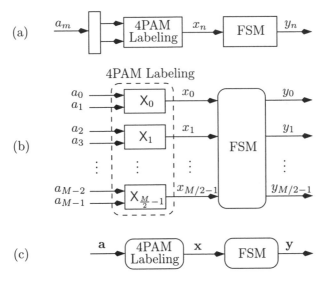

Figure 2.4. The specific example of the serially concatenated system in Example 2.1 shown in the conventions of (a) implicit index block diagram, (b) explicit index block diagram, and (c) vector mappings. The left-most block in (a) is a serial to parallel conversion that maps the bits a_m into non-overlapping blocks of length two.

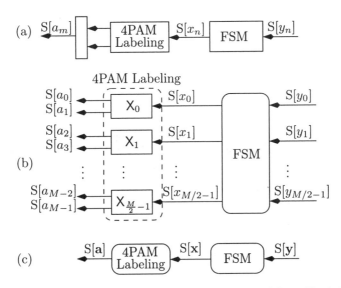

Figure 2.5. The concatenated detectors implied by the models in Fig-2.4. The soft information $S[\mathbf{x}]$ is extracted from the observation $\mathbf{z}$.

Since the implicit and explicit index block diagram notations are equivalent, the associated concatenated detectors in Fig-2.5(a) and (b) are also equivalent. The explicit index version makes the local memory structure of the PAM labeling clear. Specifically, it is clear from Fig-2.4(b) that the PAM mapping can be decomposed into $M/2$ decoupled systems (*i.e.*, denoted by X_n in Fig-2.4(b)) which make passing soft information on the entire sequence $\mathbf{x}$ unnecessary. The concatenated detectors in Fig-2.5(a)-(b) can be used to obtain optimal MAP-SyD decisions if $S[x_n]$ is equivalent to $p(\mathbf{z}|x_n)$. For example, consider detection of $a_1(\zeta)$ which can be based on $p(\mathbf{z}|x_0)$ and $p(a_m)$ for $m = 0, 1$. Specifically, the soft information on x_0 can first be marginalized via

$$p(a_1|\mathbf{z}) \equiv \sum_{\mathbf{a}:a_1} p(\mathbf{z}|x_0(\mathbf{a}))p(\mathbf{a}) \tag{2.1a}$$

$$= \sum_{\mathbf{a}_0^1:a_1} p(\mathbf{z}|x_0(\mathbf{a}_0^1))p(\mathbf{a}_0^1) \tag{2.1b}$$

$$= p(a_1)[p(\mathbf{z}|x_0(a_1, a_0 = 0))p(a_0 = 0) + p(\mathbf{z}|x_0(a_1, a_0 = 1))p(a_0 = 1)] \tag{2.1c}$$

Thus, for this simple example, the concatenated detector in Fig-2.5(a)-(b) can be used for optimal detection. Furthermore, this structure is substantially less complex than the detector shown in Fig-2.5(c). Note that one can treat the a-priori information $p(a_1)$ separately by marginalizing the likelihood $p(\mathbf{z}|x_0)$ to obtain the equivalent of $p(\mathbf{z}|a_0)$ using the a-priori information of the other symbols.

If the channel is memoryless so that $p(\mathbf{z}|y_n) = p(z_n|y_n)$, then the required soft information $p(\mathbf{z}|x_n)$ can be computed using the APP version of the forward-backward algorithm with the assumption of uniform a-priori information on x_n. The equivalent processing can be performed in the metric domain using a min*-sum forward-backward algorithm and replacing the sum marginalization operations in (2.1) by the min* operation of the corresponding metric quantities. Similarly, if the MAP-SqD criterion is desired, one should use an MSM-version of the forward-backward algorithm and use min marginalization of the soft information on x_n to obtain the MSM of a_m.

——————————————————————————————— *End Example*

Example 2.1 shows that for a system with non-overlapping block memory one may marginalize "early" in a concatenated detector. For example, while $p(\mathbf{x}|\mathbf{z})$ is always sufficient soft information for the serially concatenated system in Fig-2.3(a), the simple structure of the outer system in Example 2.1 (*i.e.*, the PAM labeling) allows this information to be marginalized down to the corresponding soft information on x_n without loss of sufficiency. The next example shows that this property does not extend to two finite state machines in serial concatenation.

Example 2.2. ───

Consider the serial concatenation shown in Fig-2.3(a) where both X and Y are FSMs and the output is passed through a memoryless channel. Thus, all sequences are on the same time scale (*e.g.*, a_k, x_k, y_k). Again, the concatenated detector implied by the vector-mapping block diagram is optimal when passing $p(\mathbf{z}|\mathbf{x})$ for all possible sequences $\mathbf{x}$. However, the concatenated detectors implied by the block diagram in Fig-2.3(a) (and its explicit index version) is suboptimal. Specifically, the concatenated detector will produce marginal soft information on x_k (*e.g.*, $p(x_n|\mathbf{z})$) which is not sufficient. There is a detector with processing much simpler than that associated with the vector-mapping concatenated detector. Specifically, denoting the state of X (Y) as $s_k^{(X)}$ ($s_k^{(Y)}$) the sufficient set of statistics can be expressed as

$$p(\mathbf{z}|\mathbf{a}) = \prod_k p(z_k|y_k(\mathbf{x})) \tag{2.2a}$$

$$= \prod_k p(z_k|y_k(x_k, s_k^{(Y)})) \tag{2.2b}$$

$$= \prod_k p(z_k|y_k(x_k(a_k, s_k^{(X)}), s_k^{(Y)})) \tag{2.2c}$$

$$= \prod_k p(z_k|a_k, s_k^{(X)}, s_k^{(Y)}) \tag{2.2d}$$

The relation in (2.2) implies that the global system is also an FSM with $s_k^{(M)} = (s_k^{(X)}, s_k^{(Y)})$. Thus, the marginalization of the likelihood in (2.2) can be executed efficiently using the forward-backward algorithm with at most $|\mathcal{S}^{(X)}| \times |\mathcal{S}^{(Y)}|$ states.

Although the concatenated detector passing soft information on x_k is suboptimal, it is, intuitively at least, a reasonable receiver. More precisely, suppose that the SOA for Y computes the soft information $p(\mathbf{z}|x_k)$ and passes it to a SOA that exploits the structure of X. One may use $p(\mathbf{z}|x_k)$ as the "soft-in" information on x_k and run a forward-backward algorithm based on the trellis describing X. This results in a detector with complexity on the order of $|\mathcal{S}^{(X)}| + |\mathcal{S}^{(Y)}|$ since a forward-backward algorithm is run separately on a trellis for X and a trellis for Y. Thus, while suboptimal, the concatenated detector may be viewed as a reduced-complexity approximation to the optimal receiver. Computer simulations are typically required to characterize the associated performance degradation. This is performed for this type of serial concatenation in an upcoming example.

――――――――――――――――――――――――― *End Example*

For the simple parallel concatenation in Fig-2.3(b) the detector processing also does not generally decouple. Specifically, assume that a memoryless channel maps the output sequence x_n to $z_n(1)$ and a similar memoryless channel maps y_k to $z_k(2)$. Furthermore, assume that these two parallel channels are independent. Then the sufficient statistic is $p(\mathbf{z}|\mathbf{a})$, where $\mathbf{z} = \{\mathbf{z}(1), \mathbf{z}(2)\}$. This can be obtained using the structure of the constituent systems separately

$$p(\mathbf{z}|\mathbf{a}) = p(\mathbf{z}(1), \mathbf{z}(2)|\mathbf{a}) = p(\mathbf{z}(1), \mathbf{z}(2)|\mathbf{x}(\mathbf{a}), \mathbf{y}(\mathbf{a})) \tag{2.3a}$$
$$= p(\mathbf{z}(1)|\mathbf{x}(\mathbf{a}))p(\mathbf{z}(2)|\mathbf{y}(\mathbf{a})) \tag{2.3b}$$
$$= \prod_n p(z_n(1)|x_n(\mathbf{a})) \times \prod_k p(z_k(2)|y_k(\mathbf{a})) \tag{2.3c}$$

Thus, the combining of marginal soft-information from the channel can be done separately. However, the marginalization of the joint soft information in (2.3) over $\mathbf{a}$ cannot be performed separately because addition (marginalizing) does not distribute over multiplication (combining). However, note that the concatenated detector implied by the vector mapping system model is optimal under the above assumptions. As usual, however, considering explicit combining and marginalization over all possible $\mathbf{a}$ is not typically a viable solution.

Example 2.3. ―――――――――――――――――――――――――
For the special case of two FSMs in parallel concatenation, the concatenated detector implied by Fig-2.3(b) is not optimal. However, like the case of serial concatenation of FSMs, there is a relatively simple optimal receiver. Specifically, in this case (2.3) can be expressed as

$$p(\mathbf{z}|\mathbf{a}) = \prod_k p(z_k(1)|x_k(\mathbf{a})) \times \prod_k p(z_k(2)|y_k(\mathbf{a})) \tag{2.4a}$$
$$= \prod_k p(z_k(1)|x_k(a_k, s_k^{(X')})) \times \prod_k p(z_k(2)|y_k(a_k, s_k^{(Y')})) \tag{2.4b}$$
$$= \prod_k p(z_k(1)|x_k(a_k, s_k^{(X')}))p(z_k(2)|y_k(a_k, s_k^{(Y')})) \tag{2.4c}$$

Again, the overall system can be viewed as a single FSM with no more than $|\mathcal{S}^{(X)}| \times |\mathcal{S}^{(Y)}|$ states.

――――――――――――――――――――――――― *End Example*

In the next section, we develop some notational conventions that simplify the exposition.

2.2 The Marginal Soft-Inverse of a System

For a given system M, which maps the sequences $\{a_m\}$ to $\{x_n\}$ we found that, for a memoryless channel, optimal detection of a_m can be achieved, conceptually at least, by two steps. First the marginal *soft-in information* on a_m and x_n, $\{SI[a_m]\}$ and $\{SI[x_n]\}$, is *combined* to obtain joint soft-information on the input-output pair $S[\mathbf{a}, \mathbf{x}(\mathbf{a})]$. Then, this joint soft information is marginalized to obtain soft information on a_m. This process, as summarized in (1.37), yields soft information that, when thresholded, produces the MAP-SyD or MAP-SqD decision for a_m with the proper choices for the marginalizing and combining operators and soft-in information measures. For example, the measures $SI[a_m] \equiv -\ln p(a_m)$ and $SI[x_n] \equiv -\ln p(z_n|x_n)$ and min-sum processing yields $S[a_m] = MSM[a_m]$ which may be thresholded to obtain the optimal MAP-SqD decision.

In order to describe concatenated detectors of the form in Fig-2.2 and the iterative variation described in the next section, it is useful to slightly extend this general processing in two ways. First, instead of producing the soft information shown in (1.37) we remove the direct effects of $SI[a_m]$ from $S[a_m]$ to produce *soft-output* information. More precisely, the soft-output produced is $SO[a_m] \equiv S[a_m] \,\mathbb{C}^{-1} SI[a_m]$. Referring to Table 1.1, it is clear that this process converts the soft information $S[a_m]$ into the corresponding *likelihood* quantity. For example, in the MSM version, the soft-out information is $MSM[a_m]\text{-}MI[a_m]$, which is the negative-log of the generalized likelihood $g(\mathbf{z}|a_m)$ with nuisance parameters $\Theta(\zeta) = \{a_i(\zeta)\}_{i \neq m}$. Second, we produce the analogous marginal soft-out information for the system output. The quantity $S[x_n]$ is well-defined (*e.g.*, $APP[x_n]$, $MSM[x_n]$, etc.), but since there is no direct a-priori information on the random variable $x_n(\zeta)$, the associated likelihood quantity for the system output is more nebulous to define. However, this is done in the same manner – *i.e.*, $SO[x_n] = S[x_n] \,\mathbb{C}^{-1} SI[x_n]$ where $S[x_n]$ is obtained via marginalization of $S[\mathbf{a}, \mathbf{x}(\mathbf{a})]$.

We refer to the processing unit that performs the above operation as the *marginal soft inverse*[1] of the system M and denote this by M^{-s}. In the general ($\mathbb{m}$, $\mathbb{C}$) semi-ring notation, this soft inverse is defined by

$$S[\mathbf{a}, \mathbf{x}(\mathbf{a})] = S[\mathbf{x}(\mathbf{a})] \,\mathbb{C}\, S[\mathbf{a}] \qquad (2.5a)$$

---

[1]We also use the term *soft inverse* for brevity.

$$= (\overset{N-1}{\underset{n=0}{\textcircled{c}}} \mathrm{SI}[x_n(\mathbf{a})]) \textcircled{c} (\overset{M-1}{\underset{m=0}{\textcircled{c}}} \mathrm{SI}[a_m]) \qquad (2.5b)$$

$$\mathrm{SO}[a_m] = (\underset{\mathbf{a}:a_m}{\textcircled{m}} \mathrm{S}[\mathbf{a}, \mathbf{x}(\mathbf{a})]) \textcircled{c}^{-1} \mathrm{SI}[a_m] \qquad (2.5c)$$

$$\mathrm{SO}[x_n] = (\underset{\mathbf{a}:x_n}{\textcircled{m}} \mathrm{S}[\mathbf{a}, \mathbf{x}(\mathbf{a})]) \textcircled{c}^{-1} \mathrm{SI}[x_n] \qquad (2.5d)$$

The marginal soft inverse takes in $\mathrm{SI}[a_m]$ and $\mathrm{SI}[x_n]$ and produces $\mathrm{SO}[a_m]$ and $\mathrm{SO}[x_n]$ for all index values m and n, and for each conditional value of a_m and x_n. We use the convention shown in Fig-2.6(a) and (b) to denote the soft-inverse for the implicit and explicit index block diagrams. To emphasize processing in the metric (probability) domain, we will use $\mathrm{MI}[\cdot]$, $\mathrm{MO}[\cdot]$, and $\mathrm{M}[\cdot]$ ($\mathrm{PI}[\cdot]$, $\mathrm{PO}[\cdot]$, and $\mathrm{P}[\cdot]$) in place of general notation $\mathrm{SI}[\cdot]$, $\mathrm{SO}[\cdot]$, and $\mathrm{S}[\cdot]$.

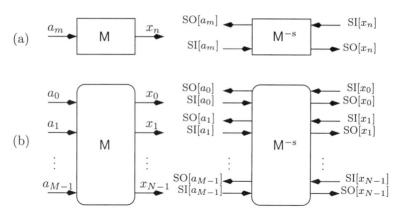

Figure 2.6.   The marginal soft inverse of a system in the (a) implicit and (b) explicit index block diagram conventions.

The marginal soft-inverse of a system is based on the implicit assumption of independent inputs and a memoryless channel.[2] The only structure exploited in the process is the structure of the (sub)-system (*i.e.*, the allowable input output pairs). The soft output information $\mathrm{SO}[\cdot]$ is the soft information typically passed to other system soft inverses in a concatenated detector. Thus, $\mathrm{SO}[\cdot]$ is often referred to as

---

[2]The soft-inverse notion also implicitly assumes that the combining and marginalization operators form a semi-ring. This is not necessary – *i.e.*, the soft inverse still simply combines marginal soft-in information and then marginalizes this joint information to produce marginal soft-out information.

extrinsic information while S[·] is referred to as *intrinsic information* since it is used only internally in the soft inverse. The conversion from intrinsic to extrinsic information is accomplished by the inverse combining operation in (2.5c)-(2.5d). The use of likelihoods for extrinsic information may be motivated by considering that, for example, the a-priori information on a_m is available at the left side of the concatenated detector in Fig-2.2 so that obtaining the likelihood information on a_m is all that is required to obtain a reliable decision on a_m (see also Problem 2.2). The most reliable soft information is generally the intrinsic information S$[a_m]$ so that, while the extrinsic information is passed to other soft inverse processors, final decisions are made based on the intrinsic information S$[a_m]$ = SI$[a_m]$ ⓒ SO$[a_m]$. For the isolated system in Fig-2.6, if the outputs x_n are observed through a memoryless channel, the MAP-SyD (MAP-SqD) for a_m can be obtained by thresholding S$[a_m]$ = SI$[a_m]$ ⓒ SO$[a_m]$ with properly defined, metric-domain, soft-in information and min*-sum (min-sum) processing.

The terminology we adopt for the remainder of the book is as follows. A soft-output algorithm (SOA) is any algorithm producing any type of soft information. The marginal soft inverse is an algorithm that computes quantities equivalent to those in (2.5). A *soft-in/soft-out* (SISO) module is any algorithm that computes any kind of marginal soft-out information for the system inputs and outputs in *extrinsic* form. Thus, we may speak of a "sub-optimal" or ad-hoc SISO that is not the marginal soft-inverse, or SISOs that are the soft inverses.[3]

Finally, the conversion to the extrinsic information can be done inside of the marginalization and combining operation. Specifically, note that, for example, a_m is fixed in the marginalization of (2.5c) so that the effect of SI$[a_m]$ can be removed in this process. Thus, the operations in (2.5) are often written equivalently as

$$SO[a_m] = \underset{\mathbf{a}:a_m}{\textcircled{m}} \left[\left(\overset{N-1}{\underset{n=0}{\textcircled{c}}} SI[x_n(\mathbf{a})] \right) \textcircled{c} \left(\overset{M-1}{\underset{j=0,j\neq m}{\textcircled{c}}} SI[a_j] \right) \right] \quad (2.6a)$$

$$SO[x_n] = \underset{\mathbf{a}:a_m}{\textcircled{m}} \left[\left(\overset{N-1}{\underset{i=0,i\neq n}{\textcircled{c}}} SI[x_i(\mathbf{a})] \right) \textcircled{c} \left(\overset{M-1}{\underset{m=0}{\textcircled{c}}} SI[a_m] \right) \right] \quad (2.6b)$$

We will generally use the form in (2.5) for clarity with the understanding that the conversion to extrinsic information can usually be done more efficiently in a manner analogous to that in (2.6).

---

[3]The term SISO is often used in the literature as the soft-inverse of an FSM.

The *marginal soft inverse is the key concept in understanding iterative detection techniques.* Specifically, the remainder of this chapter, and iterative detection by implication, simply repeatedly applies this notion to specific systems. For a given system or subsystem, it is desirable to avoid direct evaluation of operations in (2.5), instead using local dependencies between input and output variables whenever possible (*e.g.*, much in the manner used in the forward-backward algorithm). Finally, it is important to keep in mind that the marginal soft inverse of a system is found by determining the corresponding optimal soft-output algorithm (*e.g.*, MSM for MAP-SqD, APP for MAP-SyD) assuming independent inputs and a memoryless channel. Thus, once the system structure and the optimality criterion is specified, the system soft-inverse can be specified.

2.2.1   Some Common Subsystems

For the implicit block diagram convention, Benedetto et. al. [BeDiMoPo98] defined the marginal soft inverse for a variety of systems commonly encountered. These are shown in Fig-2.7 with minor modification.

The *interleaver* is a system that reorders the components of $\mathbf{a}_0^{K-1}$ according to some known permutation so that the output is $x_k = a_{I(k)}$ where $I(k)$ is the permutation. Considering (2.5), the soft-inverse of the interleaver is trivial. Specifically, it is the interleaver/deinterleaver pair shown in Fig-2.7 which sets $\mathrm{SO}[x_k] = \mathrm{SI}[a_{I(k)}]$ and $\mathrm{SO}[a_k] = \mathrm{SI}[x_{I^{-1}(k)}]$. The *serial to parallel* and *parallel to serial* converters also have trivial soft inverses. The parallel to serial converter inputs I streams of information $\{a_k(i)\}_{i=0}^{I-1}$ and outputs the sequence $x_{nI+i} = a_n(i)$ for $n = 0, 1, 2 \ldots$. The soft-inverse simply rearranges the soft information to account for the different time scales of the input and output.

The *memoryless mapper* maps the inputs $\{a_k(i)\}_{i=0}^{I-1}$ to a set of outputs $\{x_k(j)\}_{j=0}^{J-1}$ for a fixed index (time) k by some known function. More precisely $\mathbf{x}_k = f(\mathbf{a}_k)$ where f is a deterministic mapping from $\mathcal{A}^I$ to $\mathcal{X}^J$. The soft inverse of the mapper simply carries out the marginalization and combining operations of (2.5) exhaustively over the structure $f(\cdot)$. Note that this is different from exhaustive combining and marginalization over the *entire* set of input sequences $\{\mathbf{a}_0^{K-1}(i)\}_{i=0}^{I-1}$. This is illustrated by the next example.

Example 2.4. ——————————————————————————————

Consider the generalization of the simple 4-PAM labeling described in Example 1.2 where a sequence of bits a_m are converted to non-overlapping bit-pairs and then mapped into 4-PAM signals and sent across an AWGN channel. This is illustrated in Fig-2.8(a). The optimal (MAP) receiver may be viewed as a concatenated detector as illustrated in Fig-

System	Marginal Soft Inverse

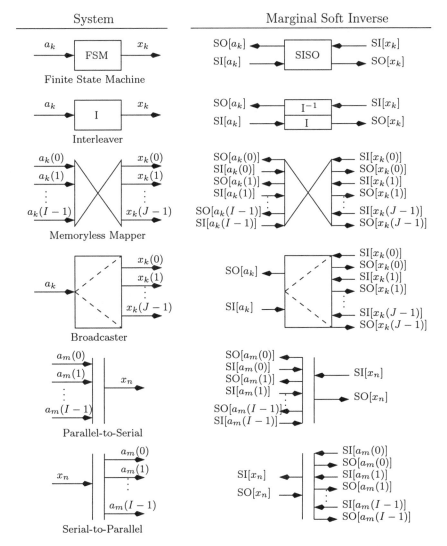

Figure 2.7. Several common systems and the associated marginal soft inverses in implicit index block diagrams.

2.8(b). The *soft output demodulator (SODEM)* block represents the translation from the observation z_n to the soft information $\mathrm{SI}[x_n]$. For example, in the metric domain this is $\mathrm{MI}[x_n] = (z_n - x_n)^2/N_0$. In some cases we will implicitly assume this block by showing the soft-in on the noise-free channel signals as the input to the detector. We also introduce the convention of showing soft-out information ports that need not be computed as terminating into a circle.

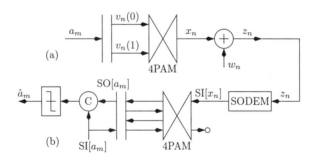

Figure 2.8. (a) The mapping of a bit sequence to a 4-PAM sequence and (b) the optimal detector using the blocks in Fig-2.7. Note that for a given bit-pair, the mapper is defined as in Fig-1.4

For the special case of $n \in \{0, 1\}$ only, the processing captures all of the cases discussed in the Examples 1.3-1.6. In the notation of Fig-2.8(b), however, it is implicit that n may range over a much larger set. Due to the finite memory characteristic of the 4-PAM mapper, this optimal detector is the concatenated detector which basically repeats the "one-shot" detection described in Examples 1.3-1.6 for every two input bits (*i.e.*, similar to the description in Example 2.1).

The soft inverse of the 4-PAM bit labeling is determined by the MAP detector for the system in isolation. For example, interpreting $v_n(0)$ and $v_n(1)$ as a_0 and a_1 in the "one-shot" problem considered in the examples of Chapter 1, it follows that

$$p(x_n = -A | z_n) = p(v_n(0) = 0)p(v_n(1) = 1)p(z_n | x_n = -A) \quad (2.7a)$$

$$\equiv p(v_n(0) = 0)p(v_n(1) = 1)e^{\frac{-(z_n + A)^2}{N_0}} \quad (2.7b)$$

which implies that the sum-product marginal soft-inverse of the mapper produces $PO[x_n = -A] = PI[v_n(0) = 0]PI[v_n(1) = 1]$. Similarly, for example,

$$\frac{p(v_n(0) = 0 | z_n)}{p(v_n(0) = 0)} = p(v_n(1) = 0)p(z_n | x_n = -3A) \quad (2.8a)$$

$$+ p(v_n(1) = 1)p(z_n | x_n = +3A) \quad (2.8b)$$

$$\equiv p(v_n(1) = 0)e^{\frac{-(z_n + 3A)^2}{N_0}} + p(v_n(1) = 1)e^{\frac{-(z_n - 3A)^2}{N_0}} \quad (2.8c)$$

so that the soft inverse produces

$$PO[v_n(0) = 0] = PI[v_n(1) = 0]PI[x_n = -3A]$$
$$+ PI[v_n(1) = 1]PI[x_n = +3A] \quad (2.9)$$

───────────────────────────────── *End Example*

The *broadcaster* simply reproduces the input a_m at each of the output ports – *i.e.*, $x_k(j) = a_k$ for $j = 0, 1, \ldots J - 1$. This important block is often shown implicitly in block diagrams by simply splitting a signal line to provide input to several parallel systems. For example, the parallel concatenation of Fig-2.3(b) has an implicit broadcaster block where a_m is sent to each of the two systems. The soft inverse of the broadcaster, provided by the special case of (2.5), in the sum-product semi-ring is

$$P[\mathbf{x}(a_k), a_k] = \prod_{j=0}^{J-1} \text{PI}[x_k(j) = a_k] \times \text{PI}[a_k] \qquad (2.10a)$$

$$\text{PO}[a_k] = P[\mathbf{a}, \mathbf{x}(\mathbf{a})]/\text{PI}[a_k] \qquad (2.10b)$$

$$\text{PO}[x_k(j)] = P[\mathbf{x}(a_k), a_k]/\text{PI}[x_k(j)] \qquad (2.10c)$$

which can be simplified further as

$$\text{PO}[a_k] = \prod_{j=0}^{J-1} \text{PI}[x_k(j) = a_k] \qquad (2.11a)$$

$$\text{PO}[x_k(j)] = \prod_{i \neq j} \text{PI}[x_k(i) = x_k(j)] \times \text{PI}[a_k = x_k(j)] \qquad (2.11b)$$

Thus, the soft-output for a given input or output is simply the combination of soft-in information associated with other variables. For a concrete example, if $J = 2$, and $a_k \in \{0, 1\}$, then $\text{PO}[a_k = 0] = \text{PI}[x_k(0) = 0]\text{PI}[x_k(1) = 0]$, $\text{PO}[x_k(0) = 0] = \text{PI}[x_k(1) = 0]\text{PI}[a_k = 0]$, and $\text{PO}[x_k(1) = 0] = \text{PI}[x_k(0) = 0]\text{PI}[a_k = 0]$. Note that if the soft-in information on a_k is uniform in the $J = 2$ case, then $\text{PO}[x_k(0)] \equiv \text{PI}[x_k(1)]$ and $\text{PO}[x_k(1)] \equiv \text{PI}[x_k(0)]$ so that on the right-hand side, the soft-information is just switched.

The marginal soft-inverse of an FSM is an important special case since many encoders and channels are modeled as FSMs. The most celebrated method for carrying out the equivalent of the general soft inversion in (2.5) is via the forward-backward algorithm. The algorithm described in Section 1.3.2.2 need only be modified to provide soft information on the FSM output as well as the FSM input, and to convert this soft information into extrinsic form. Specifically, for the MSM version, the soft outputs are

$$\text{MO}[a_k] = \text{MSM}_0^{K-1}[a_k] - \text{MI}[a_k] \qquad (2.12a)$$

$$\text{MO}[x_k] = \text{MSM}_0^{K-1}[x_k] - \text{MI}[x_k] \qquad (2.12b)$$

$$= \min_{t_k : x_k} \text{MSM}_0^{K-1}[t_k] - \text{MI}[x_k] \qquad (2.12c)$$

where, the MSM of the transition $\text{MSM}_0^{K-1}[t_k]$ may be expressed in terms of the forward and backward state MSMs, and the transition metric $\text{M}_k[t_k]$ as in (1.67). In fact, since $\text{M}_k[t_k] = \text{MI}[x_k] + \text{MI}[a_k]$, we may convert to extrinsic information by modifying the completion operation in (1.67) in the spirit of (2.6)

$$\text{MO}[a_k] = \min_{t_k:a_k} \left[\text{MSM}_0^{k-1}[s_k] + \text{MI}[x_k(t_k)] + \text{MSM}_{k+1}^{K-1}[s_{k+1}] \right] \quad (2.13a)$$

$$\text{MO}[x_k] = \min_{t_k:x_k} \left[\text{MSM}_0^{k-1}[s_k] + \text{MI}[a_k] + \text{MSM}_{k+1}^{K-1}[s_{k+1}] \right] \quad (2.13b)$$

Example 2.5. —————————————————————————————
Returning to the numerical example considered in Example 1.11, we have that $\text{MSM}_0^{11}[a_3 = -1] = 20.4$ and $\text{MSM}_0^{11}[a_3 = +1] = 22.4$. Recall that the two input values have different a-priori probabilities, so that, in this case $\text{MI}[a_3 = -1] = 0.357$ and $\text{MI}[a_3 = +1] = 1.20$. In this example, then, $\text{MO}[a_3 = -1] = 20.4 - 0.357 = 20.0$ and $\text{MO}[a_3 = -1] = 22.4 - 1.20 = 21.2$. Notice that the non-uniform a-priori information on $\{a_k(\zeta)\}_{k \neq 3}$ has been included in $\text{MO}[a_3]$ – *i.e.*, this can be viewed as a nuisance parameter with $\text{MO}[\cdot]$ representing the associated p-generalized likelihood.

It is also stated in Example 1.11 that $\text{MSM}_0^{11}[x_7 = 0.707] = 22.9$. Since $z_7 = -0.729$, the associated soft-in metric for $x_7 = 0.707$ follows from (1.57b)

$$\text{MI}[x_7 = 0.707] = \frac{(-0.729 - 0.707)^2}{N_0} + \ln(\sqrt{\pi N_0}) = 1.95 \quad (2.14)$$

It follows that the extrinsic output metric is $\text{MO}[x_7 = 0.707] = 23.1 - 1.95 = 21.15$.
————————————————————————————— *End Example*

Example 2.6. —————————————————————————————
With the block diagram conventions established, we may express the optimal concatenated detectors for several of the previous examples in this form. Specifically, for the serial concatenation in Example 2.1, the optimal processing shown in Fig-2.4 may be expressed as shown in Fig-2.9. The analogous parallel concatenation of a mapper with finite block memory and an FSM has significantly different characteristics. This is considered in Problem 2.3.
————————————————————————————— *End Example*

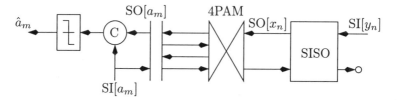

Figure 2.9. The optimal concatenated detector for the system of Example 2.1 and Fig-2.4 expressed in terms of the standard marginal soft inverse blocks.

2.2.1.1 Explicit Index Marginal Soft Inverse Blocks

Many of the standard blocks for the implicit index block diagrams in Fig-2.7 are not meaningful for the explicit index convention. For example, the parallel to serial conversion is meaningless since all of the variables are denoted explicitly. The corresponding explicit index diagram for the parallel to serial converter is shown in Fig-2.10. In the

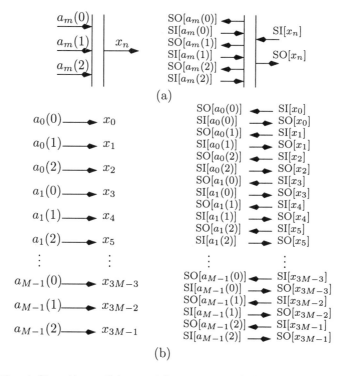

Figure 2.10. A (3 to 1) parallel to serial converter and the corresponding marginal soft inverse in (a) implicit and (b) explicit index block diagrams.

explicit index convention, the simplicity of this system and its soft inverse becomes apparent – *i.e.*, the system simply renames the variables.

This shows the relative advantages of the two approaches: the implicit index convention is more compact, but the explicit index convention often reveals the system structure more clearly.

Only two of the modules in Fig-2.7 have useful counterparts in the explicit index versions. These are the mapper and the broadcaster. In other words, since all variables are denoted explicitly the only "systems" worth denoting are those in which some variables are processed (mappers) or broadcast. The convention that we adopt, therefore is shown in Fig-2.11 for the example of the parallel concatenation. Thus, we con-

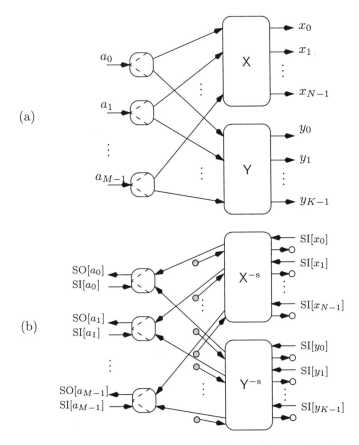

Figure 2.11. (a) The parallel concatenation of Fig. 2.3(b) and (b) the suboptimal concatenated detector implied by this model, both shown in the implicit index convention.

tinue to use round-edged blocks for single-index systems or "nodes" and only denote the broadcaster node with the special block similar to the symbol used for the implicit index version. In Fig-2.11 we also introduce the convention of representing uniform soft-in information (*e.g.*, zero metrics) by shaded circles.

2.3   Iterative Detection Conventions

A concatenated detector was determined to be optimal for several of the examples in Sections 2.1 and 2.2. However, for many systems of practical interest, a concatenated detector based on marginal soft inverses is not optimal. This may be the case, for example, if there is a subsystem that has memory equal to the input symbol length. In this case, the concatenated detector implied by the vector-mapping system diagrams may be optimal, but still prohibitively complex. For example, consider inserting an interleaver in the simple serial and parallel concatenations shown in Fig-2.3, as shown in Fig-2.12. For the serial concatenation, an

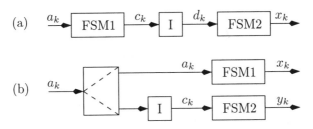

Figure 2.12.   (a) Serial and (b) parallel concatenation of two FSMs with interleaving.

SOA for FSM2 would need to pass soft information equivalent to $p(\mathbf{z}|\mathbf{c})$ back to the SOA for FSM1 in order to maintain sufficiency. Similarly, for the parallel concatenated system, an optimal concatenated detector would need to pass soft information of the form $p(\mathbf{z}|\mathbf{a})$ and $p(\mathbf{z}|\mathbf{c})$ from the SOA associated with FSM1 and FSM2, respectively. For reasonably sized interleavers, this is prohibitively complex. Thus, the interleaver breaks up the structure of the relatively simple "super-trellis" thus disallowing the relatively simple optimal processing.

As mentioned in Example 2.2, for the non-interleaved serial concatenation, a reasonable suboptimal detector could be based on passing marginal soft-information on the intermediate symbols (*i.e.*, x_n in Fig-2.3(a)). This could still be performed for the interleaved system in Fig-2.12(a). Specifically, a SISO could produce marginal soft information on d_k based on the soft-in information on x_k and the structure of FSM2, which could then be deinterleaved and used as the soft-in on c_k for a second SISO (or hard-out detector such as the Viterbi algorithm) that exploits the structure of FSM1. In fact, this is typically an effective method since it allows the outer detector, which exploits the structure of FSM1, to use some reliability information based on the structure of

FSM2. This reliability information is insufficient, however, so that the process is suboptimal.

This motivates the iterative detector shown in Fig-2.13(a) for the interleaved, serially concatenated system shown in Fig-2.12(a) where each SISO is the marginal soft-inverse. In this receiver, the processing

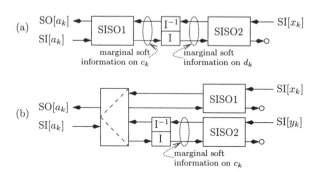

Figure 2.13. The iterative detectors associated with the parallel and serial concatenated system in Fig-2.12.

begins just as described above with SISO2 being activated to produce marginal soft information on d_k. This is then deinterleaved and used as soft-in on c_k. Next, SISO1 is activated, producing marginal soft-out information on both the input a_k and c_k. Combining $SI[a_k]$ and $SO[a_k]$, a final decision may be made. However, the soft-out information on c_k directly implies some beliefs on d_k via the interleaver permutation. This suggests executing SISO2 again, but with the $SI[d_k]$ set by the interleaved versions of the soft-out on c_k from SISO1. After this second activation of SISO2, a new soft-out on d_k is available, which can then be deinterleaved and used as soft-in on c_k for a second activation of SISO1. This process can be continued with soft information on c_k and d_k continually refined. After the iteration process is terminated by some stopping criterion, a final decision may be made on a_k by combining $SI[a_k]$ and $SO[a_k]$ from SISO1. We refer to the sequence of activations of SISO2, the interleaver/deinterleaver, and SISO1, and interleaver/deinterleaver as one iteration. For the first iteration, all internal soft-in information is set to uniform (*e.g.*, zero in the metric domain). That is, on the first iteration, the $SI[d_k]$ for SISO2 is uniform. Note that the soft-out information on x_k need not be computed in this example, nor does the soft-out on a_k except for the final iteration.

Note that the iterative detector in Fig-2.13(a) corresponds to the system block diagram in Fig-2.12(a) with each subsystem replaced by the corresponding marginal soft-inverse. This is the standard convention for constructing iterative (or "turbo" detectors). For example, the iterative detector associated with the interleaved, parallel concatenation is shown in Fig-2.13(b) and is constructed according to this convention. The *activation schedule* for this iterative decoder is not as obvious as that for the iterative detector in Fig-2.13(b). For example, SISO1 and SISO2 can be activated in parallel, followed by the interleaver/deinterleaver, then the soft inverse of the broadcaster; thus defining a single iteration. An alternate activation schedule is SISO1 → the soft broadcaster → the interleaver/deinterleaver → SISO2 → the soft broadcaster; defining an iteration. While the first schedule may seem more reasonable at first glance, in practice, it may be more complex and perform virtually the same as the second schedule. To illustrate this more clearly, consider the case where the a-priori information on a_k is uniform. Then, since the broadcaster only has two outputs and the soft-out on a_k is not required until the final iteration, the soft inverse of the broadcaster can be replaced by an exchange with a combining operation performed at the end of the final iteration. This notion is illustrated in Fig-2.14.

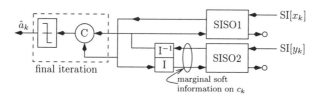

Figure 2.14. The iterative detector in Fig-2.13(b) shown for uniform a-priori information on a_k.

The two activation schedules suggested are illustrated in Fig-2.15 with the interleaver/deinterleaver and soft broadcaster activations shown symbolically as an exchange between SISOs. In the serial schedule

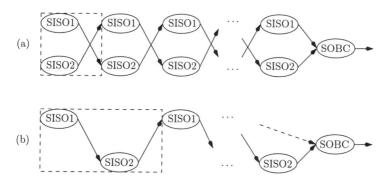

Figure 2.15. Two reasonable activation schedules for the iterative detector in Fig. 2.13(b): (a) parallel activation, and (b) serial activation started by SISO1. One iteration is shown in by the dashed box.

shown, SISO1 begins the iteration and the final decision is made by combining the most recent soft-out information on a_k from the two SISOs. Alternatively, this can be accomplished by combining the soft-in and soft-out information on a_k after the final activation of the interleaver/deinterleaver as shown in Fig-2.14. There is of course another serial schedule started by SISO2. Note that the parallel activation schedule may be viewed as running these two serial activation schedules using two separate iterative detectors which do not interact until the final iteration, after which the most recent soft-out on a_k from each detector is combined to make the final decision. The effectiveness of these schedules depends on a number of factors including the sensitivity of the iterative detectors to the initial conditions, the convergence rate, the structure of the two FSMs, the stopping criterion, etc. The main point is that, even after drawing the block diagram of the iterative detector, a number of other parameters must be specified to fully define the algorithm. This is summarized in the next section.

2.3.1   Summary of a General Iterative Detector

In this section we summarize the standard iterative detection technique as follows:

- Given a system comprising a concatenated network of subsystems, construct the marginal soft inverse of each subsystem. The marginal soft inverse is found by considering the subsystem in isolation with independent inputs and a memoryless channel. Specifically, adopting either the MAP-SyD or MAP-SqD criterion and working in either the probability or metric domain, the marginalizing and combining operators are defined according to Table 1.1. Using these operators, specify an algorithm to compute the extrinsic soft outputs for the system inputs and outputs as defined in (2.5).
- Construct the block diagram of the iterative detector by replacing each subsystem by the corresponding marginal soft inverse and connecting these soft inverses accordingly. Specifically, each connection between subsystems in the system block diagram is replaced by a corresponding pair of connections in the iterative detector block diagram so that the soft-out port of each is connected to the soft-in port of the other.
- Specify an activation schedule that begins by activating the soft inverses corresponding to some subsystems providing the global outputs and ends with activation of some soft inverses corresponding to subsystems with global inputs.
- Specify a stopping criterion.

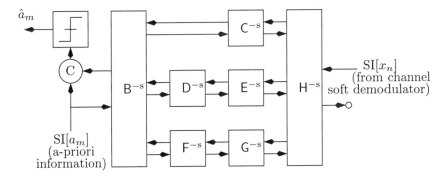

Figure 2.16.   The iterative detector implied by the Fig-2.1.

■ Take the soft-inputs on global output symbols as the channel likelihoods (metrics) obtained by appropriate soft-demodulation. The
soft-inputs for the global inputs are the a-priori probabilities (metrics) which are typically uniform.

■ At the activation of each subsystem soft inverse, take as the soft-in on
the digital inputs/outputs the soft-outputs from connected subsystem
soft inverses. If no soft information is available at the soft-in port,
take this to be uniform soft-in information (*i.e.*, this applies to the
first activation of soft inverses that have inputs or outputs that are
internal or hidden variables).

A common stopping criterion is that a fixed number of iterations
are to be performed with this number determined by computer simulation. For example, while formal proofs of convergence for complicated
iterative detectors do not exist, in most cases of practical interest, the
performance improvement from iteration reaches a point of diminishing
returns. Thus, one can select as the number of iterations the smallest
number that achieves the "full iteration gain." Alternatively, performance may be sacrificed to reduce the number of iterations. It is also
possible to define a stopping rule that results in variable number of iterations. For example, if there is a test to see that further iterations will
not alter the final decisions, this may be used as a stopping criterion.

In most cases of practical interest, there is either a natural activation schedule, or different activation schedules produce similar results.
Thus, for the most part, the iterative detector is specified once the block
diagram is given and the subsystem soft inverses are determined. For
example, the iterative detector for the general concatenated system in
Fig-2.1 is shown in Fig-2.16.

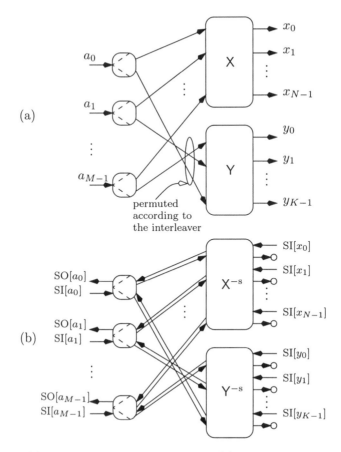

Figure 2.17. (a) The system block diagram and (b) the iterative detector for the interleaved, parallel concatenation shown in Fig. 2.12(b) using the explicit index convention.

2.3.2   Explicit Index Block Diagrams

While the emphasis thus far has been placed on implicit index block diagrams and their associated iterative detectors, the same guidelines apply to the explicit index version. For example, the explicit index block diagram for the interleaved parallel concatenated system in Fig-2.12(b) is shown in Fig-2.17 along with the associated iterative detector. Note that the soft inverse of each of the FSMs can be implicitly computed in an efficient manner using the forward-backward algorithm. In fact, the explicit index convention can shed considerable light on this process. We return to this topic and explicit index models in Section 2.6.

2.4   Iterative Detection Examples

Based on the principle described in Section 2.3, many interesting and powerful applications of iterative detection can be demonstrated. In the following subsections, we focus on several applications in detail. In some of these examples, our focus is on the details of the algorithms, while in others it is on the performance improvements and trade-offs. Since all of the examples considered in this section are based on AWGN channels, we first summarize some simplifications that can be made in MSM-based iterative detectors for this special case.

2.4.1   Normalization Methods and Knowledge of the AWGN Noise Variance

Examples 1.11-1.12 demonstrates that normalization may be desired for numerical stability. There are a number of reasonable normalization strategies. Specifically, recall that, in the probability domain, multiplication by any positive constant produces an equivalent soft measure. One reasonable choice is to normalize so that the soft-information sums to one (*i.e.*, we refer to this as sum-to-unity normalization).

$$P'[u_k] = \frac{P[u_k]}{\sum_{\tilde{u}_k} P[\tilde{u}_k]} \tag{2.15}$$

This may be applied to any algorithm operating in the probability domain (*e.g.*, APP, GAP, etc.). Furthermore, it may be applied to any intermediate soft-information. For example u_k could be s_k, x_k, a_k, etc. Software or digital hardware implementations, however, are best implemented in the metric domain. Implementation in the probability domain requires frequent normalization of all soft-information quantities for numerical stability. It is possible to select the normalization convention to reduce the storage requirements by one. For example, if the normalization in (2.15) is used, then there is no need to store $P'[u_k = 0] = 1 - \sum_{u_k \neq 0} P'[u_k]$.

In the metric domain, one may subtract any finite constant from the $|\mathcal{U}|$ values of $M[u_k]$ and still maintain an equivalent soft measure. A common scheme is to subtract the smallest of the $|\mathcal{U}|$ metrics from all metrics. This ensures that the absolute value of the metrics is minimized. If the metric of a particular conditional value is subtracted from each value, then the storage requirements are again reduced by one. Specifically, these two schemes are

$$M'[u_k] = M[u_k] - \min_{\tilde{u}_k} M[\tilde{u}_k] \tag{2.16a}$$

$$M'[u_k] = M[u_k] - M[u_k = 0] \tag{2.16b}$$

Under the convention of (2.16b), $M'[u_k = 0] = 0$. Again, these conventions may be mixed and matched in a particular iterative detector. For example, in a metric-based iterative detector, the forward and backward metrics in a SISO could be normalized every several recursions by subtracting off the smallest metric (*i.e.*, as in (2.16a)) and the $MO[a_k]$ values could be normalized for each value of k via the convention in (2.16b). Note that, for binary u_k, the convention in (2.16b) requires only one number to be stored for the soft information on u_k (*i.e.*, akin to the negative log-likelihood).

For MSM-based algorithms, one can work with an isomorphic soft-measure that often provides an important simplification. Specifically, the soft information measure $M^{(C)}[u_k] = CM[u_k]$, where C is a fixed positive constant can be used instead of $M[u_k]$ without affecting the combining and marginalizing operations. This is because multiplication by a positive constant commutes with the $\min(\cdot)$ (or $\max(\cdot)$) operation. An important application of this is to iterative detection on AWGN channels. For this case, if the global system inputs $a_k(\zeta)$ are uniformly distributed over $\mathcal{A}$, then knowledge of the noise power is not required if one works with $M^{(N_0)}[\cdot]$. This is not the case for sum-product (APP) or min*-sum (M*SM) algorithms (*e.g.*, see Problem 2.8).

To demonstrate this, recall that for a system with independent inputs $\{a_m(\zeta)\}$ and outputs $\{x_n(\mathbf{a}(\zeta))\}$, the joint soft information for an AWGN channel is

$$P[\mathbf{a}, \mathbf{x}(\mathbf{a})] \equiv \exp\left(-\frac{\|\mathbf{z} - \mathbf{x}(\mathbf{a})\|^2}{N_0}\right) p(\mathbf{a}) \qquad (2.17a)$$

$$M[\mathbf{a}, \mathbf{x}(\mathbf{a})] \equiv \frac{1}{N_0}\|\mathbf{z} - \mathbf{x}(\mathbf{a})\|^2 + [-\ln p(\mathbf{a})] \qquad (2.17b)$$

where one constant has been absorbed already. Considering the isomorphic joint metrics given by $M^{(N_0)}[\mathbf{a}, \mathbf{x}(\mathbf{a})] = N_0 M[\mathbf{a}, \mathbf{x}(\mathbf{a})]$, the key is that the min-sum processing yields equivalent marginal soft information with the same correspondence. More precisely, consider $MSM^{(N_0)}[a_n]$ as the marginal soft-information obtained by min-marginalization of $M^{(N_0)}[\mathbf{a}, \mathbf{x}(\mathbf{a})]$. Then it follows that

$$MSM^{(N_0)}[a_n] \triangleq \min_{\mathbf{a}:a_n} M^{(N_0)}[\mathbf{a}, \mathbf{x}(\mathbf{a})] \qquad (2.18a)$$

$$= \min_{\mathbf{a}:a_n} N_0 M[\mathbf{a}, \mathbf{x}(\mathbf{a})] \qquad (2.18b)$$

$$= N_0 \min_{\mathbf{a}:a_n} M[\mathbf{a}, \mathbf{x}(\mathbf{a})] \qquad (2.18c)$$

$$= N_0 MSM[a_n] \qquad (2.18d)$$

As a result, a min-sum algorithm can be run using metrics $M^{(N_0)}[\cdot]$ and the final marginal soft information measures are isomorphic to the standard MSMs.

The key to this relation is that the mapping defining the isomorphism commutes with the marginalizing operator. It is simple to verify that the above argument does not hold for $\min^*(\cdot)$ marginalization, or equivalently, in general

$$\left[\sum_{\mathbf{a}:a_n} (P[\mathbf{a}, \mathbf{x}(\mathbf{a})])^{N_0}\right]^{1/N_0} \neq \sum_{\mathbf{a}:a_n} P[\mathbf{a}, \mathbf{x}(\mathbf{a})] \tag{2.19}$$

In summary, an iterative detector or soft-out algorithm using min-sum processing can multiply *all* metrics considered by a *single* finite, positive constant C and the output soft information is equivalent to that produced by the standard MSM version multiplied by C. For an AWGN channel one can use $MI[x_n(\mathbf{a})] = \|z_n - x_n(\mathbf{a})\|^2$ and $MI[a_m] = -N_0 \ln p(a_m)$ with min-sum marginalizing and combining. For the case of uniform global system inputs, $MI[a_m] \equiv 0$, the value of N_0 is not required. Note that, since the marginal metric information on the global system input is typically only used to obtain final decisions via thresholding, the distinction between $M[\cdot]$ and $M^{(N_0)}[\cdot]$ is irrelevant in practice. Thus, in the following, we do not distinguish between these two cases and use multiplication of the metrics in min-sum algorithms by a single positive constant without explicit notation.

It is worth emphasizing two details of the above development. First, if $p(a_n)$ is not uniform for the global system inputs, then the value of N_0 is required for min-sum algorithms. Second, for the case of uniform global inputs, input metrics on subsystem inputs are implicitly normalized. In other words, if u_k is the input to a subsystem but not a global system input, then during the iteration process $MO[u_k]$ from one soft inverse will be used as $MI[u_k]$ for another soft inverse. This soft information is passed without modification – i.e., do not multiply $MI[u_k]$ by N_0 as part of the soft inversion process. These details are illustrated in the following example.

Example 2.7. ───
Consider an isolated ISI-AWGN system with independent, BPSK modulated input having $p(a_k = -1) = p$ and $p(a_k = +1) = 1 - p$ for all k. The channel is the 5-tap ISI channel $(\sqrt{3}/6, 1/2, 1/\sqrt{3}, 1/2, \sqrt{3}/6)$. The MAP-SyD and MAP-SqD receivers are run using the estimate $\hat{N}_0$ for the noise power. The impact of estimation error on the MAP-SyD and MAP-SqD receivers is shown in Fig-2.18. Note that with the relatively

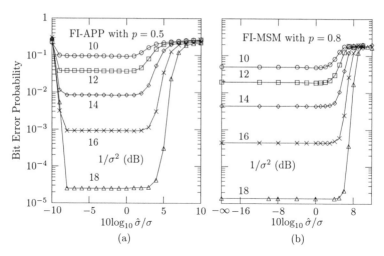

Figure 2.18. The impact of estimation error on the (a) MAP-SyD receiver (APP-forward-backward) with $p = 0.5$, and (b) the MAP-SqD (MSM-forward-backward) with $p = 0.8$. Note that $\sigma^2 = N_0/2$ and $E_b/N_0 = 2/\sigma^2$.

large a-priori bias ($p = 0.8$), the performance of the MSM algorithm degrades quickly with overestimation of the noise power. However, underestimation of N_0 has virtually no impact on the performance. Thus, in practice one may assume $N_0 \rightarrow 0$, or equivalently $p = 0.5$ without significantly degrading the performance. The MAP-SyD receiver is sensitive to both over and under estimation of the noise level. However, the performance is fairly robust over an asymmetric interval around the true value. Specifically, the APP algorithm is more robust to underestimation of the noise variance than to overestimation. Also, this tolerance interval is relatively constant for different SNRs.

An iterative example is shown in Fig-2.19. This is the performance of the iterative row-column detector for the two-dimensional ISI-AWGN channel as described in Section 5.4. The qualitative results in Fig-2.19, however, are common to other iterative detection applications.

Comparing the results in Fig-2.18 and Fig-2.19, the impact of estimation error on iterative detectors is similar but more severe. For the APP case, there is a tolerance interval which is smaller than the corresponding interval in Fig-2.18(a) and this interval varies with SNR. Moreover, inside of this interval, the performance varies more than in the non-iterative application. This is also true for the MSM case.

End Example

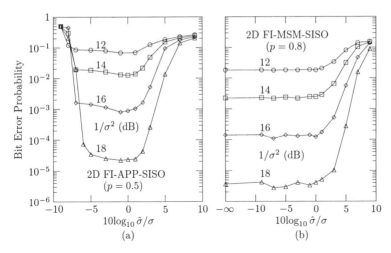

Figure 2.19. The impact of noise-power estimation error on the performance of an iterative detector using forward-backward SISOs: (a) APP-based SISOs with $p = 0.5$ and (b) MSM-based SISOs with $p = 0.8$. All curves are obtained after 5 iterations with no significant further gain achieved with further iteration. The noise variance estimate is used without modification for all iterations.

2.4.2 Joint "Equalization" and Decoding

Coding and equalization[4] are indispensable in TDMA cellular mobile systems operating in frequency selective fadingfading channels. In addition, although this is not optimal from an information theoretic point of view, interleaving is almost always used as a method to obtain time diversity with reasonable complexity and delay in those slowly fading channels. Due to the presence of the interleaver, the traditional receiver consisting of a Viterbi equalizer/deinterleaver/hard-decision Viterbi decoder, performs poorly, the reason being that such a segregated receiver does not jointly combine the time diversity of the code with the frequency diversity of the channel [AnCh97]. In this section we describe how iterative detection can be utilized in trellis-coded/interleaved systems transmitted over ISI static and frequency selective fading channels.

The communication system under consideration is shown in Fig-2.20, with the transmitter consisting of a trellis coded modulation (TCM) encoder and a symbol interleaver. The interleaved symbols are formatted in bursts and output to the inner ISI channel. Assuming perfect channel state information (CSI) available at the receiver, the optimal receiver front-end consists of a filter matched to the overall response of

---

[4]We emphasize that the term "equalization" is used here to denote ISI-mitigation, and does not imply linear or decision feedback equalization.

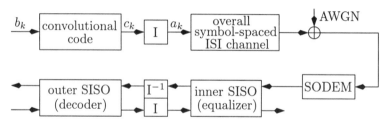

Figure 2.20.   TCM in ISI channel and the associated iterative detection network for perfect-CSI

the channel and pulse shape, followed by symbol-spaced sampling and whitening [Fo72]. The equivalent discrete-time observation model can be written as

$$z_k = \sqrt{E_s} \sum_{n=0}^{L-1} f_n^{(k)} a_{k-n} + w_k \qquad (2.20)$$

where $f_n^{(k)}$ is the nth tap of the overall channel at time k, a_k is the coded and interleaved symbol and w_k is a white complex Gaussian noise with $\mathbb{E}\{|w_k|^2\} = N_0$. The code symbols and the channel taps are both normalized to unit energy.

By realizing that the transmission system described above consists of a serial concatenation of inner (ISI channel) and outer (encoder) FSMs through an interleaver, the iterative receiver suggested in [PiDiGl97, AnCh97] and shown in Fig-2.13 can be utilized for joint equalization and decoding.

Example 2.8. ————————————————————

The communication system described above was simulated using the following parameters. A binary input stream of length 1710 bits is encoded by a rate $R = 1/2$, 16-state convolutional code with generator matrix $G = [1 + D^3 + D^4 \ \ 1 + D + D^2 + D^4]$. Two interleaver structures were examined: i) a block interleaver of size $57 \times 30 = 1710$ and ii) a pseudo-random interleaver of length 1710. In both cases, the interleaved symbols are mapped to a QPSK constellation using Gray coding and transmitted to the channel in bursts of 57 symbols. A 3-tap channel was considered, and the following four scenarios were simulated:

a. the worst-case static channel for uncoded systems with $\mathbf{f}_k = [0.5 \ \ \sqrt{0.5} \ \ 0.5]^{\mathrm{T}}$,

b. an equal-power complex Gaussian fading channel with independent taps, constant over the 1710 symbol frame and frame-to-frame independent,

c. an equal-power complex Gaussian fading channel with independent taps, constant over each burst and burst-to-burst independent, and

d. an equal-power fading channel with independent taps, burst-to-burst independent, and autocorrelation function given by the Clarke spectrum [Cl68]

$$R_f(m) = J_0(2\pi\nu_d m) \qquad (2.21)$$

where $J_0(\cdot)$ is the zero-order Bessel function of the first kind and ν_d is the normalized Doppler spread ($\nu_d = 0.005$ is used in this example). This last model corresponds to the wide sense stationary uncorrelated scattering (WSSUS) assumption [Be63].

The SISO blocks at the receiver are both fixed interval algorithms exchanging APP-type extrinsic information.

Results are presented for the static channel (a) and for the first five iterations in Fig-2.21. It is clear from this figure that iterative detec-

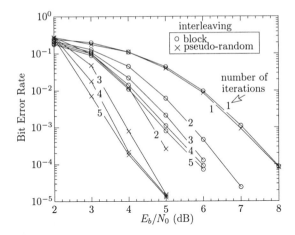

Figure 2.21.   Iterative receiver performance with block and pseudo-random interleaving for the worst 3-tap static channel (curves for the first five iterations are shown)

tion provides a significant gain over the first iteration; a fact that was also pointed out in [PiDiGl97]. Contrary to the results presented in [PiDiGl97] though, a slightly better performance is achieved here for the pseudo-random interleaved system at each iteration. As an example, a BER of 10^{-3} is achieved at 7 dB and 3.5 dB at the first and fifth iteration with pseudo-random interleaving, while the corresponding numbers from [PiDiGl97] are 7.5 dB and 4.3 dB respectively. This small differences can be attributed to the different SISO algorithm utilized herein.

Furthermore, performance depends on the type of interleaver used: with block interleaving, a 2 dB iteration gain is achieved, while pseudo-random interleaving provides an additional 1.5 dB gain at the fifth iteration. At least two possible reasons can be offered to explain this behavior. Assuming that the entire system of encoder/interleaver/channel is viewed as an overall code, it is conceivable that the code resulting from the pseudo-random interleaving is more powerful than the one resulting from block interleaving. In addition, it is possible that the iterative detection algorithm more accurately approximates the optimal detector (*i.e.*, MAP-SqD, or MAP-SyD) in the case of pseudo-random interleaving. The two above mentioned mechanisms are not mutually exclusive. Nevertheless, the fact that the performance enhancement is observed only after the second iteration, lends credence towards the latter explanation.

Performance curves are presented in Fig-2.22 for different channel models for the two interleaving schemes and the first and fifth iteration. The BER performance for case (b) with either interleaver is significantly

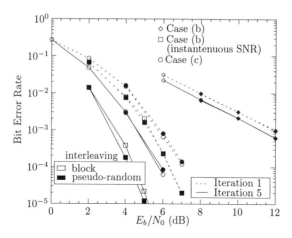

Figure 2.22. Iterative receiver performance with block and pseudo-random interleaving for different fading channel models (curves for the first and fifth iterations are shown)

worse than all other cases. This was expected since in this situation, there is no time diversity built into the system due to the interleaver and consequently the presence of the code can only slightly – if at all – increase the performance over an uncoded system. An additional interpretation of this result stems from the fact that the performance represented by case (b) is the result of averaging the performance of static channels (similar to case (a)) over an ensemble of channel shapes and powers under an average power constraint $\mathbb{E}\{\sum_{n=0}^{L-1}|f_n^{(k)}(\zeta)|^2\} = 1$.

In order to separate the effect of power averaging from that of shape averaging, the same curves were also plotted versus the instantaneous E_b/N_0. Comparing these curves with the corresponding ones for the static channel (a), it is evident that shape averaging is not detrimental. In fact, the shape-averaged performance is slightly better since the static channel represents the worst case performance (at least for an uncoded system).

To further investigate the effect of channel shape in the overall performance, statistics on the relative difference of errors due to the static channel (a) and the different channel shapes (b) were collected. In particular, the quantity $\frac{\sum_i N_a^i - N_b^i}{\sum_i N_a^i} \times 100$ was evaluated where N_a^i and N_b^i represent the number of errors in the ith experiment for the case (a) and (b) (channel power is always normalized to unity) respectively. Results are presented in Table 2.1 for $E_b/N_0 = 5\text{dB}$ and for block and random interleaving up to the fifth iteration. It is evident from these data, that the

		iteration			
	1	2	3	4	5
block	96.8	96.8	96.8	94.4	95.7
random	95.9	95.9	46.7	34.7	36.0

Table 2.1.   Percentage of channel shapes better than $[0.5 \ \sqrt{0.5} \ 0.5]^T$

static channel (a) is almost always worse than any other channel shape, when block interleaving is present, regardless of the iteration. This explains the big difference in performance between cases (a) and (b) (when presented versus the instantaneous SNR) for block interleaving (*i.e.*, 1.5 dB at 10^{-4}). On the other hand, when random interleaving is employed, after a couple of iterations, only a small percentage of channels introduce less errors than channel (a). In fact, only 36% more errors are due to channel (a) in the fifth iteration, which explains the agreement between the corresponding BER curves.

Performance curves for the quasi-independent channel model (c) are also shown in Fig-2.22. The time diversity provided by the burst-to-burst independence is evident from the slope of the curves. Nevertheless, iterative detection provides a much smaller gain compared to that obtained with the static channel – *i.e.*, only 1.2 dB at 10^{-4} as compared to up to 4 dB for the static channel. Moreover, the choice of interleaving does not affect the performance. In addition, simulation results not presented here, showed that the performance curves for the quasi-independent/fast fading channel model (d) are almost identical with those for channel model (c), which means that almost all the time diversity of the system is provided through the independent fading of the bursts. Note that the similarity between cases (c) and (d) is only true when the receiver has perfect CSI (refer to Chapter 4 for a treatment of the problem when perfect CSI is not available at the receiver).

The results presented suggest that there is no significant difference between the *average* performance obtained with pseudo-random and block interleaving for fading ISI channels. Furthermore, iteration provides relatively small improvements in this average performance over fading channels (e.g., approximately 1-2 dB for 5 iterations). However, significant gains in performance for a given channel realization are obtained by using iterative detection and choosing a pseudo-random interleaver over a block interleaver. If one were to average the performance over not only the short-term fading statistics, but the user mobility profile (mobility, shadowing, path-loss, etc) as well, significant improvement would be expected. This may be reflected in, for example, a lower outage probability, especially for low-mobility users.

———————————————————————— *End Example*

Example 2.9. ————————————————————————————

Iterative detection can also be used effectively when no interleaver is present between the two subsystems. Consider the case of the 16-state, rate 1/2, convolutional code with Gray-mapped QPSK modulation and a 3-tap, $\mathbf{f}^{\mathrm{T}} = [1\ 2\ 1]$ ISI channel. This is a special case of the serial concatenation considered in Example 2.2 so that the optimal detector can be implemented on the joint, or super-trellis. According to the development in Example 2.2, the optimal processing can be carried out on a trellis with no more than $|\mathcal{S}^{(\mathrm{CC})}| \times |\mathcal{S}^{(\mathrm{ISI})}| = 2^4 \times 4^2 = 256$ states. However, since both FSMs are simple, the joint trellis can be defined based on the states $\mathbf{b}_{k-6}^{k-1}$. More precisely, a transition in the super-trellis corresponds to $\mathbf{b}_{k-6}^{k}$ which uniquely determines $\mathbf{a}_{k-2}^{k}$ and thus, the output of the ISI channel $x_k(\mathbf{a}_{k-2}^{k})$.

Alternatively, an iterative detector can be used which operates the same way as above, just omitting the interleaver/deinterleaver pair. Specifically, the ISI and convolutional code SISOs operate on the respective FSM trellises and exchange soft information on the QPSK symbols. Fig-2.23 shows the performance of several receivers for this serial concatenation without interleaving. The performance of the MAP sequence detector for the global system is shown as is that of the min-sum iterative receiver. The curve label HID is for a Viterbi detector for the ISI channel and Hamming distance decoding of the convolutional code. Comparing

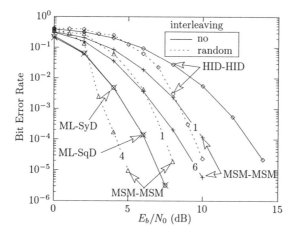

Figure 2.23. Joint equalization and decoding with and without an interleaver.

to the similar system with interleaving, we note that the iteration gain is smaller and the difference between APP and MSM processing is larger in this example. Also shown are the corresponding performance curves when a pseudo-random interleaver is present. Note that it is not easy to make quantitative comparisons between the two systems since the interleaver changes the system structure (*i.e.*, it may be thought of as modifying the code, or as modifying the ISI channel). It is interesting, however, to see that the performance gaps for each system is similar, especially in light of the fact that the optimal detector for the interleaved system cannot be simulated.

——— End Example

2.4.3 Turbo Codes

2.4.3.1 Parallel Concatenated Convolutional Codes

The application that fueled the wide spread emergence of iterative detection techniques is so-called "turbo coding" [BeGlTh93, BeGl96, BeMo96, HeWi98, VuYu00]. These are codes comprising a parallel concatenation of convolutional codes separated by an interleaver. These are also called *Parallel Concatenated Convolutional Codes (PCCC)*.[5] In this section we consider a particular PCCC which is similar to the original turbo code described by Berrou, Glavieux, and Thitmajshima [BeGlTh93, BeGl96]. We also give a fairly detailed presentation

---

[5]Many use the term "turbo codes" to describe various members of a family of turbo-like codes that have similar distance spectrum properties and may be decoded using iterative detection principles. Thus, we use the more descriptive term PCCC.

of a specific MSM-based iterative decoder. This decoder is the subject of the case study implementation in Chapter 6.

The encoder, which is shown in Fig-2.24, consists of two 4-state constituent encoders, a pseudo-random interleaver, and a puncture mapper. The constituent encoders are *recursive systematic convolutional codes*

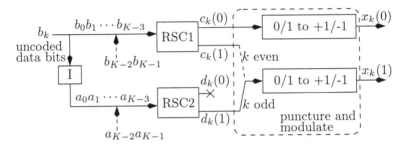

Figure 2.24.   The PCCC encoder considered.

(RSCs). Thus, one of the code outputs is the uncoded input and the other is a parity bit based on a feedback encoding procedure. The specific constituent encoders considered are 4-state RSCs, with the parity bit generated by the generator polynomial

$$g(D) = \frac{1 + D^2}{1 + D + D^2} \tag{2.22}$$

where D is the unit delay operator. A realization of this encoder is shown in Fig-2.25, where all variables are in $\{0, 1\}$ and the summers are modulo two (other structures are also possible – *e.g.*, see Problem 2.10). The state of the encoder is defined as $s_k = (p_{k-1}, q_{k-1})$, which are the

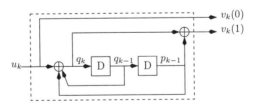

Figure 2.25.   The RSC encoder for the constituent codes of the PCCC example.

outputs of the two delay operators. Using the notation in Fig-2.25, the output and next-state are computed via

$$s_{k+1} = (p_k, q_k) = (q_{k-1}, p_{k-1} \oplus q_{k-1} \oplus u_k) \tag{2.23a}$$
$$v_k(0) = u_k \tag{2.23b}$$
$$v_k(1) = q_k \oplus p_{k-1} = q_{k-1} \oplus u_k \tag{2.23c}$$

$s_k = (p_{k-1}\ q_{k-1})$	$u_k = 0$			$u_k = 1$		
	$v_k(0)$	$v_k(1)$	$s_{k+1} = (p_k\ q_k)$	$v_k(0)$	$v_k(1)$	$s_{k+1} = (p_k\ q_k)$
$0 = (0\ 0)$	0	0	$0 = (0\ 0)$	1	1	$1 = (0\ 1)$
$1 = (0\ 1)$	0	1	$3 = (1\ 1)$	1	0	$2 = (1\ 0)$
$2 = (1\ 0)$	0	0	$1 = (0\ 1)$	1	1	$0 = (0\ 0)$
$3 = (1\ 1)$	0	1	$2 = (1\ 0)$	1	0	$3 = (1\ 1)$

Table 2.2. The next-state and output tables for the constituent codes of the PCCC example.

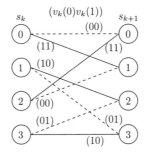

Figure 2.26. The trellis for the constituent codes of the PCCC example. Transitions corresponding to zero inputs are shown dashed.

From these relations, the next-state and output tables are as shown in Table 2.2 and the trellis structure is as shown in Fig-2.26.

Both constituent codes are started in the zero state and two tail bits are used to terminate each RSC encoder into the zero state. More precisely, $K - 2$ input bits $\mathbf{b}_0^{K-3}$ are accepted from the source. The first encoder encodes these bits along with two tail bits b_{K-2}, b_{K-1} to produce the two output sequences $c_k(0)$ (systematic) and $c_k(1)$ (parity) for $k = 0, 1 \ldots (K - 1)$. The bit sequence $\mathbf{b}_0^{K-3}$ is permuted to obtain $\mathbf{a}_0^{K-3}$ according to a known interleaver rule – i.e., $a_k = b_{I(k)}$ for $k = 0, 1, \ldots (K - 3)$. Two tail bits a_{K-2}, a_{K-1} are added to this sequence to terminate the trellis into the zero state. Since the RSC is a recursive FSM, the tail bits required to terminate into the zero state differ according to the value of s_{K-3}, but are easily computed using the information in Table 2.2. This is shown in Fig-2.27 where the effect of the initial state being set to zero is shown. Note that the first and last two transitions are restricted due to the initial state information and the tail bits, respectively.

The systematic bit of RSC1 is always sent and the systematic bit of RSC2 is never sent. The parity bit streams of each RSC are punctured

s_{K-2}	u_{K-2}	u_{K-1}
0	0	0
1	1	1
2	1	0
3	0	1

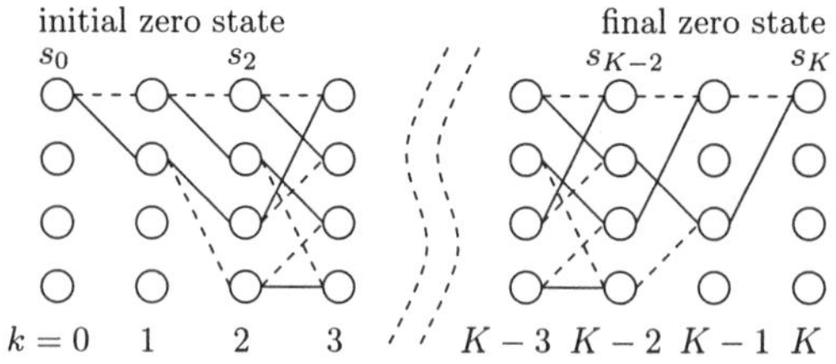

Figure 2.27. Termination of the trellis for the constituent codes of the PCCC example. Values shown for the constituent encoder inputs u_k are those required to drive the encoder to $s_K = 0$ for a given state at time $K - 2$.

and multiplexed so that the overall code parity bit is $c_k(1)$ (the parity from RSC1) and $d_k(1)$ (the parity from RSC2) for even and odd k, respectively. Thus, the rate of the overall code is approximately $1/2$ (*i.e.*, accounting for the tail bits, it is $\frac{1}{2} - \frac{1}{K}$). The two code output bits are modulated using a BPSK scheme – *i.e.*, $x_k = (-1)^{v_k}$ where $v_k \in \{0, 1\}$ is a coded bit.

An AWGN channel is assumed so that the observation is

$$z_k(i) = \sqrt{E_s} x_k(i) + w_k(i) \quad k = 0, 1 \ldots (K-1), \quad i = 0, 1 \quad (2.24)$$

where $w_k(0)$ and $w_k(1)$ are realizations of independent real-valued AWGN sequences, each with zero mean and variance $N_0/2$. The energy per coded symbol E_s is set to $E_b/2$ under the assumption of fixed transmit power and throughput (*i.e.*, a bandwidth expansion of two). It follows that the pdf of the received samples is

$$p_{z_k(\zeta;i)|x_k(\zeta;i)}(z_k(i)|x_k(i)) = \frac{1}{\sqrt{\pi N_0}} \exp\left(\frac{-[z_k(i) - \sqrt{E_s} x_k(i)]^2}{N_0}\right) \quad (2.25)$$

A metric for $x_k(i)$ is therefore $\mathrm{M}[x_k(i)] = [z_k(i) - \sqrt{E_s} x_k(i)]^2/N_0$ for $x_k(i) = \pm 1$.

Since all variables considered are binary-valued, it is reasonable to store only the difference between metric values. We adopt the normalization convention in (2.16b) so that the metric of $v_k = 0$, for all binary variables v_k, is zero. This may be accomplished using

$$\mathrm{M}[x_k(i)] = ([z_k(i) - \sqrt{E_s} x_k(i)]^2 - [z_k(i) - \sqrt{E_s}]^2)/N_0 \quad (2.26a)$$

$$\mathrm{M}[x_k(i) = +1] = 0 \quad (2.26b)$$

$$\mathrm{M}[x_k(i) = -1] = 4\sqrt{E_s} z_k(i)/N_0 \quad (2.26c)$$

The metric in (2.26) is valid for either min-sum or min*-sum processing. In the following we focus on min-sum processing and multiple all metrics by $N_0/(4\sqrt{E_s})$ to obtain[6]

$$\text{M}[x_k(i) = +1] = 0 \qquad\qquad \text{M}[x_k(i) = -1] = z_k(i) \qquad (2.27)$$

The iterative decoder is shown in Fig-2.28 where the SISOs are the marginal soft inverses of the RSCs implemented using the forward-backward algorithm. The soft inverse of the modulation and puncture map-

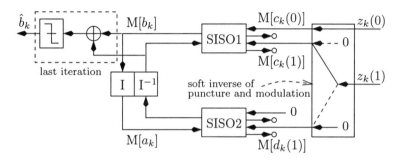

Figure 2.28.   The MSM-based iterative decoder for the PCCC in Fig-2.24.

ping is a simple distribution of the channel metrics in (2.27) according to the puncturing scheme. More precisely, the following input metric information is assigned

$$\text{MI}[c_k(0) = 1] = z_k(0) \qquad (2.28a)$$

$$\text{MI}[c_k(1) = 1] = \begin{cases} z_k(1) & k \text{ even} \\ 0 & k \text{ odd} \end{cases} \qquad (2.28b)$$

$$\text{MI}[d_k(0) = 1] = 0 \qquad (2.28c)$$

$$\text{MI}[d_k(1) = 1] = \begin{cases} 0 & k \text{ odd} \\ z_k(1) & k \text{ even} \end{cases} \qquad (2.28d)$$

Due to the structure of the soft inverse of the modulator and puncture mapper, the assignment in (2.28) need only be done once (*i.e.*, subsequent activation of the soft inverse of the puncture mapping does nothing and is therefore not required). Final decisions on b_k can be made by thresholding the $\text{MI}[b_k] + \text{MO}[b_k]$ after the final activation of the interleaver/deinterleaver. Note that this corresponds to activation of a soft inverse broadcaster as described in Section 2.3.

---

[6]This convention is adopted since the implementation study in Chapter 6 uses min-sum processing. To run an M*SM-based algorithm, one should replace $z_k(i)$ by $4\sqrt{E_s}z_k(i)/N_0$ everywhere in the following.

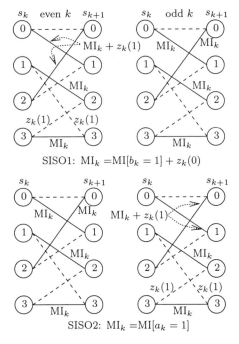

Figure 2.29. Transition metrics for the two SISOs in the PCCC example considered.

Notice that the output of RSC1 is $(c_k(0), c_k(1))$ so that the metrics for each of these quantities should be added for each transition metric. Many of the transition metrics in each of the SISOs are zero due to the normalization scheme. Specifically, by the normalization in (2.26)-(2.27) the input metric of $c_k(i) = 0$ and $d_k(i) = 0$ are always zero. Furthermore, the puncturing and similar normalization of the metrics on a_k and b_k imply that several other metrics will be zero. Specifically, we normalize so that $\mathrm{MI}[b_k = 0]$, $\mathrm{MO}[b_k = 0]$, $\mathrm{MI}[a_k = 0]$, and $\mathrm{MO}[a_k = 0]$ are all zero. With this assumption, the trellis for each of the two SISOs is shown in Fig-2.29.

In order to ensure that the normalization convention that the metric of a zero conditional value is zero, it must be enforced at each completion operation. Specifically, let $F_{k-1}[s_k]$ and $B_{k+1}[s_{k+1}]$ be the forward and backward state metrics in the forward-backward algorithm, then the completion operation in (2.13) can be specialized to incorporate the normalization convention. Specifically, for SISO1 and even k, this yields

$$\mathrm{MO}[b_k = 1] = \min_{t_k:b_k=1} (F_{k-1}[s_k] + \mathrm{MI}[c_k(1)] + B_{k+1}[s_{k+1}])$$

$$- \min_{t_k:b_k=0} (F_{k-1}[s_k] + \mathrm{MI}[c_k(1)] + B_{k+1}[s_{k+1}]) + z_k(0) \quad (2.29)$$

For SISO1 and odd k this simplifies to

$$\text{MO}[b_k = 1] = \min_{t_k : b_k = 1} (\text{F}_{k-1}[s_k] + \text{B}_{k+1}[s_{k+1}])$$
$$- \min_{t_k : b_k = 0} (\text{F}_{k-1}[s_k] + \text{B}_{k+1}[s_{k+1}]) + z_k(0) \quad (2.30)$$

Similarly, for SISO2 and even k, the completion is

$$\text{MO}[a_k = 1] = \min_{t_k : a_k = 1} (\text{F}_{k-1}[s_k] + \text{B}_{k+1}[s_{k+1}])$$
$$- \min_{t_k : a_k = 0} (\text{F}_{k-1}[s_k] + \text{B}_{k+1}[s_{k+1}]) \quad (2.31)$$

and for odd k

$$\text{MO}[a_k = 1] = \min_{t_k : a_k = 1} (\text{F}_{k-1}[s_k] + \text{MI}[d_k(1)] + \text{B}_{k+1}[s_{k+1}])$$
$$- \min_{t_k : a_k = 0} (\text{F}_{k-1}[s_k] + \text{MI}[d_k(1)] + \text{B}_{k+1}[s_{k+1}]) \quad (2.32)$$

The edge information should be used when initializing the forward state metrics. Specifically, for each SISO and for every iteration $\text{F}_{-1}[s_0 = 0] = 0$ and $\text{F}_{-1}[s_0 \neq 0] = \infty$. As discussed in Examples 1.10 and 1.11, the tail bits should be accounted for by setting $\text{MI}[b_k]$ and $\text{MI}[a_k]$ for $k = K - 2, K - 1$ accordingly. In this case where the final state is known and output metric information need be computed for the tail bits, this may be accomplished by proper initialization of $\text{B}_K[s_K]$. Specifically, in each SISO for every iteration $\text{B}_K[s_K = 0] = 0$ and $\text{B}_K[s_K \neq 0] = \infty$. Initializing to infinity may not be practical in many implementations. This is discussed in Section 6.2. Alternatively, one can collapse the first two transitions in either direction according to Fig-2.27 and initialize $\text{F}_1[s_2]$ and $\text{B}_{K-2}[s_{K-2}]$ directly.

Simulation results for this example PCCC system are shown in Fig-2.30 and Fig-2.31. The activation used is that shown in Fig-2.15(b). So, one iteration is SISO1 $\rightarrow$ I/I^{-1} $\rightarrow$ SISO2 $\rightarrow$ I/I^{-1}. The min-sum and min*-sum (using the metrics in (2.26)) iterative decoding algorithms are compared for $K = 1024$. The improvement associated with APP-based processing over MSM-based processing is greatest at low SNR and is less than 0.5 dB over the SNR range of interest.

The effect of interleaver size K is illustrated in Fig-2.31. Also shown is the performance of the rate $1/2$, 128-state convolutional code with largest minimum distance [LiCo83] using soft-in decoding, and a baseline uncoded system which uses half of the channel bandwidth. Comparing the approaches at a BER of 10^{-4}, a PCCC with $K = 512$ provides approximately one additional dB of coding gain relative to the traditional convolutional code. With a $K = 16384$ PCCC, this additional

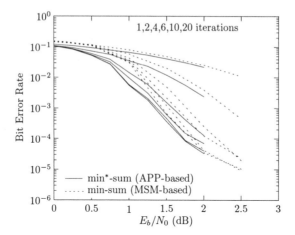

Figure 2.30. The performance of min-sum and min*-sum iterative decoding of the example PCCC with $K = 1024$.

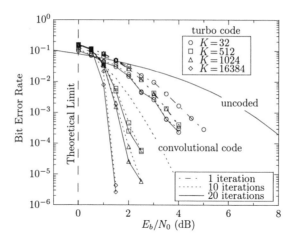

Figure 2.31. The performance of min-sum iterative decoding of the example PCCC for various K.

gain is approximately 2.2 dB. This is only approximately 1 dB from the constrained channel capacity for the rate 1/2, BPSK-AWGN channel.

All interleavers used were selected at random and held fixed for all simulations. Some improvement at low error rates can be obtained by using a so-called *semi-random* (S-random) interleaver (*e.g.*, [DoDi95, FrWe99, DaMo99]). This alleviates the so-called "knee" in the PCCC performance curve. While a detail explanation of why turbo codes are good low-SNR codes is not the focus of this book, this knee effect is evidence of the intuitive reason. Specifically, traditional codes are designed to maximize the minimum distance of the code without regard to the number of minimum distance neighbors. As the bound development in Chapter 1 suggests, this is a good design metric for sufficiently high SNR, but may not be the most effective at low SNR. At low SNR, the coefficients multiplying the Q-functions in the bounds become more important than the minimum distance. Therefore, at low SNR, the *distance spectrum* properties of the code are more important than the minimum distance [PeSeCo96]. In fact, at sufficiently high E_b/N_0, the traditional convolutional code will outperform the PCCC because it has a larger minimum distance. The transition region between these two regimes is manifested by the knee in the PCCC curve. The S-random interleaver is designed to alleviate this effect by carefully considering the effects of the input sequences that produce minimum distance pairwise errors for the PCCC (*i.e.*, these are weight two input sequences).

There are a number of turbo-like codes that have similar distance spectrum properties, impressive performance, and may be decoded using iterative detection principles. These include serially concatenated convolutional codes [BeDiMoPo98b], low-density parity check (LDPC) codes [Ga62, Ga63, Ma99], repeat-accumulate (RA) codes [Mc99], and self-concatenated codes [Di97, DiPo97]. LDPCs, in particular, are an attractive alternative to PCCCs for a number of reasons as described in Section 2.6.3.

2.4.3.2   Serially Concatenated Convolutional Codes

Serially Concatenated Convolutional Codes (SCCCs) were introduced in [BeDiMoPo98b] as an alternative to the original turbo codes, which were PCCCs [BeGlTh93]. As shown in Fig-2.32, in an SCCC the se-

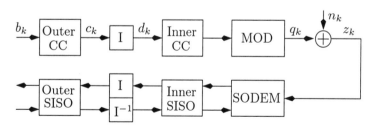

Figure 2.32.   Serial concatenation of CCs and iterative detection network for perfect CSI.

quence of source bits b_k is partitioned into blocks and convolutionally encoded using a rate R_o outer CC, producing K coded symbols c_k. These symbols are fed to an inner CC of rate R_i through a pseudo-random

symbol interleaver[7] of length K. The output symbols are mapped to
the constellation complex symbols x_k, resulting in an overall code rate
of $R = R_o R_i \log_2 |\mathcal{Q}|$ (bits per channel use). The complex symbols x_k
are transmitted through an AWGN channel, resulting in the complex
baseband post-correlator model

$$z_k = \sqrt{E_s} q_k + n_k \qquad (2.33)$$

where n_k is white complex circular Gaussian noise with $\mathbb{E}\{|n_k|^2\} = N_0$,
E_s is the symbol energy, and the symbols q_k are normalized to unit
energy. The structure of a SCCC is one of a serial concatenation of
two FSMs through an interleaver and therefore it permits the iterative
receiver shown in Fig-2.32.

Example 2.10. ————————————————————————————————
The SCCC system under consideration consists of an outer 4-state, rate
1/2 RSC connected through a length $K = 16384$ symbol pseudo-random
interleaver to an inner 4-state, rate 2/3 RSC. The corresponding generator matrices are given by

$$G_o(D) = \begin{bmatrix} 1 & \frac{1+D^2}{1+D+D^2} \end{bmatrix} \quad G_i(D) = \begin{bmatrix} 1 & 0 & \frac{1+D^2}{1+D+D^2} \\ 0 & 1 & \frac{1+D}{1+D+D^2} \end{bmatrix}$$

The output symbols are mapped to an 8PSK constellation with Gray encoding, resulting in an overall code rate $R = 1/2 \times 2/3 \times \log_2 8 = 1$(bits
per transmitted symbol). The performance of the iterative receiver utilizing APP- and MSM-type SISOs is shown in Fig-2.33. In these simulation results fixed interval forward-backward SISOs are utilized in the
iterative receiver. We observe that an E_b/N_0 loss of 0.35 dB is experi-
enced by the MSM-type SISO compared to the APP-type SISO, in the
10th iteration and at a BER level of 10^{-5}, which might be crucial in this
application.
———————————————————————————————— *End Example*

2.4.4   Multiuser Detection

In this application, we consider the detection of multiple data sequences that interfere with each other on a common channel. In particular, we consider a system of the form in Fig-2.34(a), where each user
data stream is coded, interleaved, and sent through a common channel. The soft inverse of the multiuser channel may be computed using

---

[7]In [BeDiMoPo98] it was shown that bit interleaving yields better performance with a slightly
more complicated decoder structure.

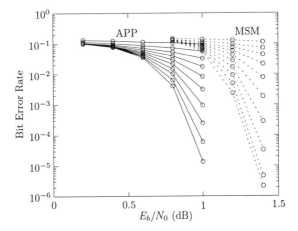

Figure 2.33.   Performance of iterative detection receiver for the SCCC.

the forward-backward algorithm based on a generalization of the Unger-boeck metrics derived in Example 1.13. Thus, first we concentrate on the "inner channel" by temporarily ignoring the code structure in order to derive the soft inverse of the multiuser channel.

Consider the case of M users, with the noise-free signal of user m described as

$$y_m(t; \{a_k(\zeta; m)\}_{k=0}^{K-1}) = \sum_{i=0}^{K-1} a_i(\zeta; m) h_m(t - iT) \qquad (2.34)$$

where $a_i(m)$ is an independent input sequence taking values in $\mathcal{A}(m)$ and $h_m(t)$ is the overall channel impulse response for user m. Thus, $h_m(t)$ captures the effects of all noise-free channel distortions and pulse-shaping. In the multiuser detection literature, $h_m(t)$ is often referred to as the *signature waveform* for user m. The total received signal is the sum of these M signals corrupted by AWGN, specifically,

$$r(\zeta, t) = y(t; \{a_k(\zeta; m)\}) + n(\zeta, t) \quad t \in \mathcal{T} \qquad (2.35a)$$

$$y(t; \{a_k(\zeta; m)\}) = \sum_{m=0}^{M-1} y_m(t; \{a_k(\zeta; m)\}_{k=0}^{K-1}) = \sum_{i=0}^{K-1} \mathbf{a}_i^{\mathrm{T}} \mathbf{h}(t - iT)$$

$$(2.35b)$$

where $\mathbf{a}_i$ and $\mathbf{h}(t)$ are the $(M \times 1)$ vectors with components $\{a_i(m)\}_{m=0}^{M-1}$ and $\{h_m(t)\}_{m=0}^{M-1}$, respectively. This is the inner channel in Fig-2.34(a).

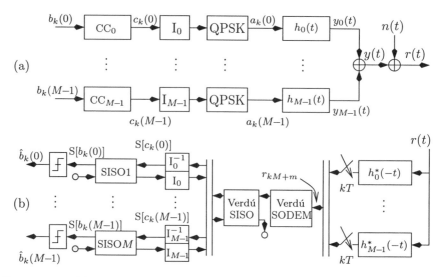

Figure 2.34. (a) The multiuser system and (b) iterative detector considered.

The negative-log likelihood functional for all users data based on the observation waveform may be expressed in terms of

$$\int_T r(t)y^*(t; \{\mathbf{a}_k(\zeta; m)\})dt = \sum_{i=0}^{K-1} \mathbf{a}_i^H \mathbf{r}_i \qquad (2.36a)$$

$$\mathbf{r}_i = \int_T r(t)\mathbf{h}^*(t - iT)dt \qquad (2.36b)$$

$$\int_T |y(t; \{\mathbf{a}_k(\zeta; m)\})|^2 dt = \sum_{i=0}^{K-1}\sum_{j=0}^{K-1} \mathbf{a}_j^H \mathbf{R}(j - i)\mathbf{a}_i \qquad (2.36c)$$

$$\mathbf{R}(j - i) = \int_T \mathbf{h}^*(t - jT)\mathbf{h}^T(t - iT)dt \qquad (2.36d)$$

The expressions in (2.36) are simply the extensions of (1.75)-(1.76) to the $(M \times 1)$ vector symbol/channel model in (2.35). In particular, $\mathbf{r}_i$ is the $(M \times 1)$ output of a bank of matched filters to the individual user signature waveforms with $r_i(m)$ denoting the i-th output sample of the matched filter for user m.[8] Similarly, assuming that $h_m(t)$ is nonzero only on $[0, (L+1)T)$, we obtain the Ungerboeck metric recursion

---

[8]In the case of time-varying signature waveforms, such as in a long-code spread-spectrum system, the correlations should be implemented directly in place of the matched filtering. The following development may be easily generalized to this case. Also note that any user delay information has been absorbed into the definition of $h_m(t)$.

analogous to that in (1.77)

$$
M_0^k[\{\mathbf{a}_i\}_{i=0}^k] \triangleq \frac{1}{N_0} \left[\sum_{i=0}^k \sum_{j=0}^k \mathbf{a}_j^H \mathbf{R}(j-i)\mathbf{a}_i - 2\Re \left\{ \sum_{i=0}^k \mathbf{a}_i^H \mathbf{r}_i \right\} \right]
$$

$$
+ \sum_{i=0}^{K-1} -\ln p(\mathbf{a}_i) \tag{2.37a}
$$

$$
= M_0^{k-1}[\mathbf{a}_0^{k-1}] + M_k[\mathbf{a}_k, \{\mathbf{a}_{k-m}\}_{m=1}^L] \tag{2.37b}
$$

$$
M_k[\mathbf{a}_k, \{\mathbf{a}_{k-m}\}_{m=1}^L] = \frac{2}{N_0} \Re \left\{ \mathbf{a}_k^H \left[\frac{1}{2} \mathbf{R}(0)\mathbf{a}_k - \mathbf{r}_k + \sum_{i=1}^L \mathbf{R}(i)\mathbf{a}_{k-i} \right] \right\}
$$

$$
- \ln p(\mathbf{a}_k) \tag{2.37c}
$$

where $\mathbf{R}(i) = \mathbf{0}$ for $|i| > L$, which follows from the FIR assumption on $h(t)$, has been used.

The expression in (2.37c) means that the MAP receiver for detection of $\mathbf{a}_k$, $\{\mathbf{a}_k\}_{k=0}^{K-1}$, or $a_k(m)$ can be implemented by processing on a trellis with $(|\mathcal{A}|^M)^L = |\mathcal{A}|^{LM}$ states[9] – i.e., the state is $\{\mathbf{a}_{k-m}\}_{m=1}^L$. The most celebrated example of this is the case where each $h_m(t)$ has duration T, but where each user has been delayed by a different, known amount τ_m – i.e., $h_m(t)$ has support in $[\tau_m, T + \tau_m)$. This is often referred to as the *asynchronous multiuser channel*. In this case, the processing may be carried out on a trellis with $|\mathcal{A}|^M$ states, with each transition, corresponding to the k-th symbol of all M users, having metric defined in (2.37c). However, in this case, ordering the users such that $0 \le \tau_0 < \tau_1 < \cdots \tau_{M-1} < T$, the same results can be achieved using a trellis with $|\mathcal{A}|^{M-1}$ states. To show this, we note two methods for computing the quadratic form. First, one can sum down columns first via

$$
\mathbf{v}^H \mathbf{Q} \mathbf{w} = \sum_{i=0}^{N-1} \sum_{j=0}^{N-1} v_i^* q_{i,j} w_j = \sum_{i=0}^{N-1} v_i^* \left[\sum_{j=0}^{N-1} q_{i,j} w_j \right] \tag{2.38}
$$

where $q_{i,j}$ is element of $\mathbf{Q}$ at row i and column j. Second, one can sum the n-th row and column together

$$
\mathbf{v}^H \mathbf{Q} \mathbf{w} = \sum_{n=0}^{N-1} \left[v_n^* w_n q_{n,n} + \sum_{i=0}^{n-1} v_i^* q_{i,n} w_n + \sum_{j=0}^{n-1} v_n^* q_{n,j} w_j \right] \tag{2.39}
$$

---

[9] For simplicity, we assume that all users employ the same modulation format so that $|\mathcal{A}(m)| = |\mathcal{A}|$. The sets $\mathcal{A}(m)$ may differ, however, to account for different received signal energies.

Applying (2.39) to the first two terms inside the real-part operator in (2.37c) and (2.38) to the third, yields

$$M_k[\mathbf{a}_k, \mathbf{a}_{k-1}] = \sum_{m=0}^{M-1} M_{k,m}[\{a_k(i)\}_{i=0}^{m}, \{a_{k-1}(i)\}_{i=m+1}^{M-1}] \tag{2.40}$$

with

$$
\begin{aligned}
M_{k,m}[\{a_k(i)\}_{i=0}^{m}, \{a_{k-1}(i)\}_{i=m+1}^{M-1}] = {}& MI[a_k(m)] \\
& + \frac{2}{N_0} \Re\left\{ a_k^*(m) \left[\tfrac{1}{2} a_k(m) r_{m,m}(0) - r_k(m) \right.\right. \\
& \left.\left. + \textstyle\sum_{j=0}^{m-1} r_{m,j}(0) a_k(j) + \sum_{j=m+1}^{M-1} r_{m,j}(1) a_{k-1}(j) \right] \right\}
\end{aligned} \tag{2.41}
$$

The recursion in (2.41) uses the fact that $r_{i,m}(1) = 0$ for $i \geq m$, which follows from the assumption on the order of the delays τ_k (*i.e.*, see Problem 2.15). Thus, by exploiting knowledge of the relative user delays, the transition metrics $M_k[\cdot]$ can be computed in M steps. Furthermore, the processing can be performed on a $|\mathcal{A}|^{M-1}$-state trellis with state $s_{k,m} = \{a_k(i)\}_{i=0}^{m-1} \cup \{a_{k-1}(i)\}_{i=m+1}^{M-1}$, where m cycles through $0, 1, \ldots (M-1)$ for each value of k. The notation $a_{kM+m} = a_k(m)$ may be used to denote this convention which is often explained in terms of modulo M indexing for a_i (*e.g.*, see [Ve98]). In this form, each transition in the trellis corresponds to the transmission of one symbol of one user and incorporates one sample from one user's matched filter. We refer to this as the Verdú trellis with Verdú metric $M_{k,m}[a_k(m), s_k(m)]$ after the original description – *i.e.*, Verdú showed that MAP-SyD and MAP-SqD could be performed using this trellis [Ve84, VePo84, Ve86]. The relation between these two trellises is illustrated in Fig-2.35.

The above development provides the definition of a multiuser channel SISO. We consider the case of $L = 1$ with the assumption of known delays ordered as described above. We therefore use a Verdú SISO based on the $|\mathcal{A}|^{M-1}$-state trellis and the above Verdú metric. An iid binary data sequence was generated for each user and encoded using the rate 1/2, 16-state convolutional code described in Section 2.4.2. A different interleaver of size 1710 was selected at random for each user and used to interleave the 4-ary output of the convolutional code (*i.e.*, symbol interleaving). The output of each interleaver was then mapped to a QPSK modulation format using the Gray map. The initial and final states of the code trellis are known to the receiver and no edge information was assumed for the multiuser trellis. The associated iterative detector is illustrated in Fig-2.34(b). Note that the Verdú soft demodulator computes the metrics $M_{k,m}[\cdot]$ from the matched filter outputs and the

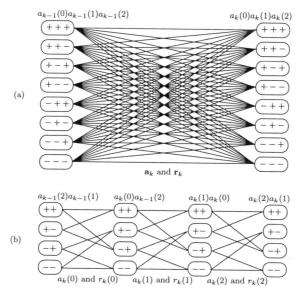

Figure 2.35. The direct trellis and the simplified Verdú trellis for $M = 3$ and $\mathcal{A}(m) = \{+1, -1\}$.

multiuser SISO runs the forward-backward algorithm on the associated Verdú trellis. All processing was done using min-sum soft inverses.

Three scenarios are considered for the selection of the signature waveforms $\{h_m(t)\}$. The first example scenario is taken from [MoGu98], where $M = 3$ highly correlated users with equal power share the channel. Specifically, for normalized signature sequences (*i.e.*, $\int |h_m(t)|^2 dt = 1$), the correlations are

$$\mathbf{R}(0) = \begin{bmatrix} 1 & \frac{2}{3} & \frac{1}{3} \\ \frac{2}{3} & 1 & \frac{2}{3} \\ \frac{1}{3} & \frac{2}{3} & 1 \end{bmatrix} \qquad \mathbf{R}(1) = \begin{bmatrix} 0 & \frac{1}{3} & \frac{2}{3} \\ 0 & 0 & \frac{1}{3} \\ 0 & 0 & 0 \end{bmatrix} \qquad (2.42)$$

This may represent a severe like-signal interference channel where the individual user waveforms have not been designed for good multiple access interference rejection. The results of this simulation are shown in Fig-2.36. The multiuser SISO has $4^{3-1} = 16$ states.

Several baseline receivers are also shown for comparison. The *conventional* detector is a single user matched-filter detector which completely ignores the multiple access interference (MAI). The conventional detector can be used for either soft-in decoding (SID) or hard-in decoding (HID) of the convolutional codes. In the latter, the matched filter outputs are thresholded to make decisions on $a_k(m)$ which are deinterleaved and used for Hamming distance Viterbi decoding. In the SID case, the

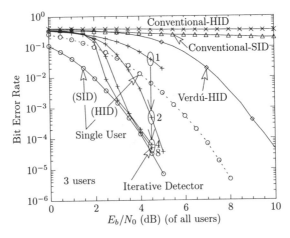

Figure 2.36. Performance of various receivers for a channel with 3 highly correlated users.

matched-filter outputs are deinterleaved and used to form soft-in metrics for Euclidean distance decoding of the convolutional codes. A hard-out Verdú multiuser detector with hard-in decoding of the convolutional codes is also considered. Finally, the performance of the corresponding single-user channel (*i.e.*, with no MAI) with HID and SID of the convolutional code is shown.

The conventional detector fails over the SNR region of interest. The Verdú multiuser detector with HID is approximately 3 dB (in SNR) worse than the corresponding single user case with HID and approximate 5.5 dB worse than the soft-decision decoded, single-user channel. The iterative receiver outperforms the Verdú-HID by approximately 2.5 dB on the first iteration. Thus, as in the single-user channel, there is approximately a 2.5 dB gain associated with SID over hard-decision decoding. With iteration, however, the joint structure of the multiuser channel and the convolutional codes is exploited. After four iterations the performance of the iterative receiver is close to that of the single-user channel with SID. Note that there appears to be a threshold effect for this high-correlation case where the iterative detector fails to provide substantial gains at low SNR. Finally, the receiver in [MoGu98] is similar to that of Fig-2.34(b) with the multiuser SISO replaced by an approximation. This approximation, which is directly analogous to the processor discussed in Example 2.15, yields a degradation of approximately 1.5 dB in the threshold relative to the iterative processing described here [GoCh00].

For a second example, we consider the system simulated in [HaSt97] which is a four-user system with relatively low cross correlation and equal received power for each user. This may represent a heavily loaded code-division multiple access (CDMA) system with power control. Specifically, for normalized signature waveforms, the correlation is

$$
\mathbf{R}(0) = \begin{bmatrix} 1 & \frac{-2}{7} & \frac{1}{7} & 0 \\ \frac{-2}{7} & 1 & \frac{-2}{7} & \frac{1}{7} \\ \frac{1}{7} & \frac{-2}{7} & 1 & \frac{-2}{7} \\ 0 & \frac{1}{7} & \frac{-2}{7} & 1 \end{bmatrix} \qquad \mathbf{R}(1) = \begin{bmatrix} 0 & \frac{1}{7} & \frac{2}{7} & \frac{-1}{7} \\ 0 & 0 & \frac{1}{7} & \frac{2}{7} \\ 0 & 0 & 0 & \frac{1}{7} \\ 0 & 0 & 0 & 0 \end{bmatrix} \qquad (2.43)
$$

In this case, a slight variation on the system of Fig-2.34(b) was used based on bit-interleaving. Specifically, the two coded bits from each convolutional encoder were serialized, interleaved, and modulated using BPSK modulation. Thus, the multiuser SISO has $2^{4-1} = 8$ states. Simulation results are shown in Fig-2.37. Since the multiuser channel

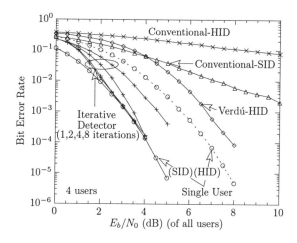

Figure 2.37.   Performance of various receivers for 4-user channel.

is less severe in this example than in the 3-user example, the performance degradation of all receivers relative to the single-user channel is less severe. This includes a less severe threshold effect for the iterative receiver. Note that the performance difference between the Verdú-HID and the first iteration of the iterative receiver is similar to that of the HID-SID for the single-user channel.

As a final example, we consider a *synchronous* multiuser channel with severe MAI. It was pointed out in [Ve98] that this is a particularly severe channel for which the Verdú detector for an uncoded system fails. Specifically, we consider the 2-user system with $\mathbf{R}_1 = \mathbf{0}$ (synchronous) and $r_{i,j}(0) = 1$ for $i, j \in \{0, 1\}$. In this degenerate case, one user's signal can completely cancel the other user's signal. Simulation results for this system are shown in Fig-2.38. As expected, hard-decision decoding of

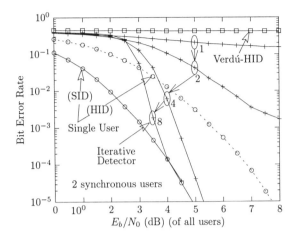

Figure 2.38.   The severe 2-user synchronous multiuser system.

the codes is not possible when the Verdú detector fails. However, using the Verdú SISO and multiple iterations, the code structure can be exploited to mitigate the MAI. In the synchronous case, the multiuser SISO degenerates to a soft inverse mapper. This may be the reason that iteration gain is observed for many more iterations than in the asynchronous cases considered. Specifically, because the synchronous multiuser channel is a memoryless mapping of $\mathbf{a}_k$, its soft inverse is memoryless on the associated soft-information, so that it takes more iterations for the information to propagate between the code SISOs.

In summary, for the multiuser channel, iterative detection may be used to incorporate global system structure (*e.g.*, error correction coding) in mitigating MAI. As a result, significant performance gains are observed relative to a receiver that considers the MAI and code structure separately. In some severe cases, these gains can effectively enable a system to operate when segregated processing fails. Additional recent research in this area can be found in [Mo97b, Mo98, MoGu98, VaWo98, ReScAlAs98, AlReAsSc99, WaPo99, GoCh00].

2.5   Finite State Machines SISOs

The SISO for an FSM is such an important component of many iterative detectors, that we describe several variations in detail in this section. The baseline SISO module is the standard, fixed-interval (FI) forward-backward algorithm described in Section 1.3.2.2 and modified to provide extrinsic soft-out information on the FSM inputs and outputs in Section 2.2.1.

We distinguish between two types of variations from the baseline forward-backward SISO. First, there are variations on the architecture of the algorithm which do not change the soft information produced. More precisely, different algorithms are discussed that, as the forward-backward algorithms does, carry out the equivalent of the general marginal soft-inversion operation in (2.5). These variations generally take advantage of a different form of algorithm scheduling, some special aspect of the underlying FSM, or even possibly a dramatically different method carrying out the soft inversion.

Second, we consider SISO algorithms based on different *combining windows*. For example, the FI-SISO is constructed based on an algorithm that computes, for example, the $\mathrm{MSM}_0^{K-1}[u_k]$ for the isolated FSM under the assumption of independent inputs and a memoryless channel. In some cases, it may be advantageous to use a combining window which is a subset of the full interval $\{0, 1, \ldots (K-1)\}$. When the entire interval is not used, the marginal soft inverse has not been computed, rather, the SISO represents an approximation.

An important component of this development is the clear identification of the desired soft information to be computed for the isolated system. Once this is specified, there may be more than one algorithm or architecture that can compute this soft information. Unless specified otherwise, algorithms discussed in this section are semi-ring algorithms so that the algorithmic duality discussed in Section 1.2.1.2 applies. We prefer to develop the results in this section using the MSM soft output which provides good intuition and using the APP version to verify some details by equations.

We consider the three basic combining windows shown in Fig-2.39. In computing the soft-out on a quantity u_k associated with the transition t_k, the *fixed-interval (FI)*, *fixed-lag (FL)*, and *sliding-window (SW)* SISOs, use soft-in information over combining windows $\{0, 1, \ldots (K-1)\}$, $\{0, 1, \ldots (k+D)\}$, and $\{(k-D), \ldots k \ldots (k+D)\}$.[10] It is implicitly assumed that, for a FL or SW algorithm, when one of the edges of the boundary exceeds the FI range, the value is replaced by the final edge. For example, this would be explicitly denoted for the right edge of the FL combining window as $\min(k+D, K-1)$. Variations on these combining windows are discussed in Section 2.5.6.

Thus, we develop a given SISO by specifying an algorithm to compute the soft information $\mathrm{MSM}_{k_1}^{k_2}[u_k]$ for appropriately defined k_1 and k_2,

---

[10]The term sliding window is used by some authors to describe what we refer to as fixed-lag algorithms. We follow the terminology that is standard in the estimation literature (*e.g.*, [Me95]).

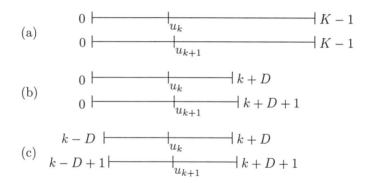

Figure 2.39.　Three combining windows considered in detail.

which implies the marginal soft inverse as described in Section 2.2. A critical component of this development is to carefully identify the soft output being produced so that algorithm equivalences can be identified.

2.5.1　The Forward-Backward Fixed-Interval SISO

As shown in Section 1.3.2.2, $\text{MSM}_0^{K-1}[u_k]$ can be computed by running forward and backward add-compare-select (ACS) recursions. The extrinsic soft-outputs for u_k can then be obtained by the completion operation in (2.12) or (2.13). Note that the forward and backward recursions may be performed in parallel or sequentially as shown in Fig-2.40. In either case, however, the transition metrics must be stored for

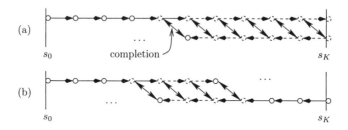

Figure 2.40.　Two different schedules for executing the FI-SISO via the forward-backward algorithm: (a) the serial forward backward schedule and (b) the parallel forward and backward recursions. The dashed lines represent locations for which the completion step has been performed and the transition metrics can be released from memory.

K time locations. The practical difference between the two schedules in Fig-2.40(a) and Fig-2.40(b) is the associated latency and area requirements of the implied architectures. Specifically, the parallel schedule has half the latency, but requires twice as many ACS processors as the serial schedule. Note that each ACS processor node in Fig-2.40 represents a set or *bank* of $|S|$ ACS units, one for each state. This is the most primitive processing unit we consider. Architectures for carrying out the processing of an ACS bank in parallel or serial are discussed in Chapter 6.

Thus, in this section we roughly characterize the area, latency, and computational complexity of a specific algorithm/architecture by counting in units of ACS banks. The FI-SISO can be implemented with $\mathcal{O}(K)$ (*i.e.*, order K) latency and $\mathcal{O}(K)$ computational complexity. The area associated with computational (*e.g.*, ACS) units is $\mathcal{O}(1)$. Specifically, executing parallel forward and backward recursions requires 2 ACS banks, $2N$ ACS recursions, and N ACS-clock cycles.

2.5.2 Fixed-Lag SISOs

2.5.2.1 Forward-Backward FL-SISO

A fixed-lag SISO is based on the computation of $\text{MSM}_0^{k+D}[u_k]$. This can be computed in a straightforward manner using the forward-backward algorithm operating on the interval $\{0, 1, \ldots k, \ldots (k + D)\}$. This may be seen by inspection of Fig-1.13. If the right edge of the trellis corresponds to s_{k+D+1}, then the desired MSM is computed when the backward state metrics are initialized to a constant (*e.g.*, zero). Specifically, the forward ACS recursion is the standard one in (1.59b) and the backward ACS is

$$\text{MSM}_{k+d}^{k+D}[s_{k+d}] = \min_{t_{k+d}:s_{k+d}} \text{MSM}_{k+d+1}^{k+D}[s_{k+d}] + \text{M}_{k+d}[t_{k+d}] \qquad (2.44)$$

This backward recrusion is executed for $d = D - 1, \ldots 0$, initialized by $\text{MSM}_{k+D+1}^{k+D}[s_{k+D+1}] \equiv 0$.

Thus, the FL-SISO computes a forward recursion identical to that of the FI-SISO. A new backward recursion is started after each forward ACS step. This is illustrated in Fig-2.41. Fixed-lag algorithms are sometimes referred to as *continuous decoding algorithms* since they can release the information on u_k before observing z_i for $i \geq (k + D)$. This is primarily of interest in non-iterative algorithms since it may dictate the latency of the system. For example, a *fixed-lag Viterbi algorithm* performs a traceback of D steps after each forward ACS step. As a result, this FL-VA produces a hard decision that is consistent with thresholding $\text{MSM}_0^{k+D}[u_k]$. Thus, one may view the backward recursion of the FL-MSM algorithm just described as replacing the traceback operation of the Viterbi algorithm, thus eliminating the need to store survivors, but at the cost of significantly more complexity. This execution of the forward-backward FL-SISO is illustrated in Fig-2.41.

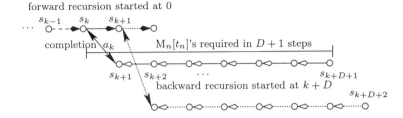

Figure 2.41. The forward-backward or bi-directional fixed-lag algorithm. Solid lines represent operations performed to compute the soft outputs at time k (*i.e.*, MO[a_k] and/or MO[x_k]). The finely-dashed lines correspond to the computations occurring to compute the soft-outputs at time $k + 1$. The coarsely-dashed lines correspond to information from previous computations. This convention is also used on future figures in this section.

In fact, the computational complexity of the FL-SISO is dominated by the repetition of the backward recursion. Neglecting edge effects, for each forward ACS step, D backward ACS steps are taken. Thus, the FL-SISO is approximately D times more computationally complex than the corresponding FI-SISO algorithm. The potential advantage is that the transition metrics can be released from memory after being processed by the forward recursion. For many practical iterative decoder implementations, the circuit area associated with memory is a major concern. Thus, as discussed in Chapter 6, FL-SISOs are attractive for systems with large block sizes (*i.e.*, large K).

The issue of initialization of the backward recursion has been the source of some confusion in the literature (*e.g.*, see Problem 2.20). From the shortest path intuition associated with the MSM soft output and Fig-1.13, it is clear that the backward state metrics should be initialized uniformly in order to compute the metrics of shortest paths from any state s_0 to any state s_{k+D+1} based on the observation $\mathbf{z}_0^{k+D}$. This is most easily seen formally using the APP version. In particular, the backward recursion for $p(\mathbf{z}_{k+d}^{k+D}|s_{k+d})$ is as given in (1.70b). Note that, by definition this is initialized by setting $p(\mathbf{z}_{k+D}^{k+D}|s_{k+D})$, which can be computed based on the probability of transition t_{k+D} as

$$p(z_{k+D}|s_{k+D}) = \sum_{t_{k+D}:s_{k+D}} p(\mathbf{z}_{k+D}|t_{k+D})p(a_{k+D}) \qquad (2.45)$$

Finally, note that (2.45) corresponds to performing the sum-product backward step in (1.70b) for $p(z_{k+D}|s_{k+D})$ with the convention that $p(\mathbf{z}_{k+D+1}^{k+D}|s_{k+D+1}) = 1$. That is, in the probability domain, the backward state recursion parameters (*i.e.*, equivalent to APPs) should be uniform. Note that the above development basically constitutes a re-derivation of the forward-backward algorithm in the sum-product semiring.

The fixed-lag version of the forward-backward algorithm was suggested independently by a number of authors [BeMoDiPo96, ChCh98, Vi98, KwKa98] with the initialization of the backward recursion sometimes suggested incorrectly (*e.g.*, see Problem 2.20). Alternatively, it was suggested that by selecting sufficiently large D, the initialization of the backward recursion was not important (*i.e.*, essentially taking advantage of the merging of the implicit backward survivor sequences). Uniform initialization was justified formally in [ChCh98] and independently in [MoAu99b].

2.5.3 Forward-Only (L^2VS) FL-SISO

For a single conditional value of $u_k(\zeta) = u_k$, the desired soft-output can be computed using only a forward recursion. Specifically, a standard forward ACS recursion can be run with the transition metrics for transitions not consistent with $u_k(\zeta) = u_k$ set to infinity. With this modification, the forward metric terminating into state s_{k+D+1} is $\mathrm{MSM}_0^{k+D}[u_k, s_{k+D+1}]$. This process is shown in Fig-2.42(b). Marginalizing out s_{k+D+1}, the desired soft outputs can be obtained.

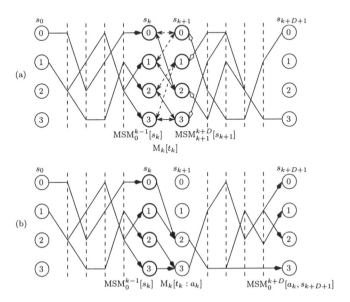

Figure 2.42. The equivalence of (a) a forward and backward recursion and (b) a constrained forward ACS recursion. Shown for the standard 4-state simple-FSM trellis and $u_k = a_k = +1$.

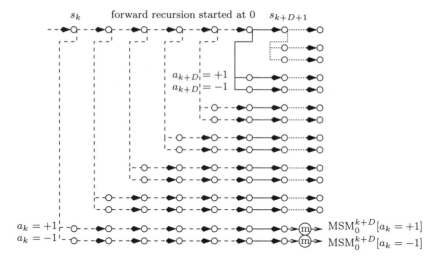

Figure 2.43. The L²VS fixed-lag algorithm. (Line style convention as described in Fig-2.41.)

Note that this requires a *distinct* constrained forward ACS recursion for each conditional value of u_k. Furthermore, these $|\mathcal{U}|$ constrained forward ACS recursions may be started from a given unconstrained forward ACS. More precisely, the constrained forward ACS is defined as

$$\text{MSM}_{k_1}^{k+d}[u_k, s_{k+d+1}] = \min_{t_{k+d}:s_{k+d+1}, u_k} \text{MSM}_{k_1}^{k+d-1}[u_k, s_{k+d}] + \text{M}_{k+d}[t_{k+d}]$$

(2.46)

In order to use this approach in an FL mode, a single unconstrained forward ACS is run only for the purpose of spawning constrained forward ACS recursions. Along with each unconstrained ACS update, a set of $|\mathcal{U}|$ constrained forward ACS recursions is started. Furthermore, D other sets of constrained forward ACS recursion previously spawned should be updated via (2.46). Finally, the "oldest" of these constrained ACSs is mature with state metrics $\text{MSM}_0^{k+D}[u_k, s_{k+D+1}]$. These can be marginalized to obtain the desired soft output metrics

$$\text{MSM}_0^{k+D}[u_k] = \min_{s_{k+D+1}:u_k} \text{MSM}_0^{k+D}[u_k, s_{k+D+1}]$$

(2.47)

Note that, for sufficiently large D, the condition $s_{k+D+1} : u_k$ reduces to minimization over all s_{k+D+1}. The execution of this algorithm is summarized in Fig-2.43

In order to verify mathematically that the desired soft output is produced, the APP version can be used. Specifically, the L²VS algorithm computes

$$p(\mathbf{z}_0^{k+d}, u_k, s_{k+d+1}) = \sum_{s_{k+d}:(u_k, s_{k+d+1})} p(\mathbf{z}_0^{k+d}, u_k, s_{k+d+1}, s_{k+d}) \quad (2.48a)$$

$$= \sum_{s_{k+d}:(u_k, s_{k+d+1})} p(\mathbf{z}_0^{k+d-1}, u_k, s_{k+d}) p(z_{k+d}, s_{k+d+1} | s_{k+d}) \quad (2.48b)$$

$$= \sum_{s_{k+d}:(u_k, s_{k+d+1})} p(\mathbf{z}_0^{k+d-1}, u_k, s_{k+d}) p(z_{k+d} | t_{k+d}) p(a_{k+d}) \quad (2.48c)$$

which is the constrained sum-product recursion. Note that the desired soft-out information $p(\mathbf{z}_0^{k+D}, u_k)$ can be obtained by summing over s_{k+D+1} after the recursion has reached $d = D$. Finally, the constrained recursion is started by an unconstrained recursion since $p(\mathbf{z}_0^{k-1}, u_k, s_k) = p(\mathbf{z}_0^{k-1} s_k) p(u_k)$ – the past observations and current state are independent of $u_k(\zeta)$.

This algorithm was originally suggested by Lee [Le74] and rediscovered by Li, Vucetic, and Sato [LiVuSa95] so we refer to it as the L²VS algorithm.

2.5.3.1 Comparison of FL Architectures

It is important to realize that the FL forward-backward algorithm and the L²VS algorithm produce *exactly* the same soft information when the same marginalizing and combining operators and normalization methods are used. This is clear, because the soft output information produced by each algorithm was carefully identified and found to be the same.

The equivalence of the L²VS and the forward-backward (or "bidirectional") FL algorithms was shown in [ChCh98], where it was also noted that the L²VS algorithm has greater computational complexity than the forward-backward FL algorithm by approximately a factor of $|\mathcal{U}|$. Comparing Figs. 2.41 and 2.43, it seems that the computational complexity is offset by a higher degree of parallelism in the L²VS version. However, the backward recursion in the forward-backward FL-SISO can also be executed in parallel as shown in Fig-2.44. This is done by starting each backward recursion $D + 1$ steps before it is required for completion, and updating each of these staggered backward recursions in parallel.

Thus, both architectures allow one soft-output computation per ACS clock with a sufficient number of ACS units. A careful accounting of the storage requirements and computational complexity of the various FL-SISOs is given in Table 2.3. For reasonable values of D (*e.g.*, 5 to 7 times $\log_{|\mathcal{A}|} |\mathcal{S}|$), the computational complexity of the L²VS architecture is approximately $|\mathcal{U}|$ times greater than that of the equivalent FL bidirectional structures. The storage requirements of the L²VS approach

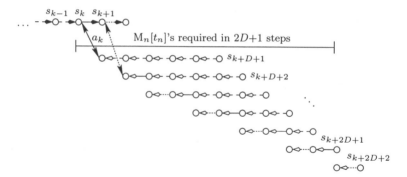

Figure 2.44. The forward-backward FL-SISO with pipelined backward recursion. (Line style convention as described in Fig-2.41.)

Algorithm	# ACS ops.	Total Storage
L²VS	$|\mathcal{U}|D + 1$	$|\mathcal{S}|(|\mathcal{U}|D + |\mathcal{A}| + 1)$
fwd-bwd (serial)	$D + 1$	$|\mathcal{S}|(D|\mathcal{A}| + 1)$
fwd-bwd (parallel)	$D + 1$	$|\mathcal{S}|(D(2|\mathcal{A}| + 1) - |\mathcal{A}| + 1)$

Table 2.3. Complexity of various FL-SISO architectures. Values listed are required per soft-output produced – *e.g.*, $\mathrm{MSM}_0^{k+D}[u_k]$ for all possible values of u_k. In addition to those operations listed, each algorithm requires $|\mathcal{U}|$ completion operations (see (2.12)).

are roughly $|\mathcal{U}|/|\mathcal{A}|$ times greater than that of the corresponding serial bi-directional FL algorithm shown in Fig-2.41 and $|\mathcal{U}|/(2|\mathcal{A}| + 1)$ times that of the parallel bi-directional architecture in Fig-2.44. These measures do not account for various tricks for storage and computation such as exploiting soft-output normalization to reduce storage requirements slightly (*e.g.*, see Section 2.5.6). The area requirements for the L²VS approach are also approximately $|\mathcal{U}|$ times that of the parallel forward-backward architecture.

In summary, the forward-backward FL algorithm is preferred over the L²VS algorithm by virtually every measure. Furthermore, the backward recursion for forward-backward version may be partially pipelined so that the bi-directional version has a greater amount of flexibility in trading latency and area. However, the L²VS has an intuitive structure which may be helpful in conceptualizing algorithms or architectures that may latter be translated into a more efficient bi-directional form (*e.g.*, see Section 4.2.3).

2.5.4   Sliding Window SISOs

From the development of the fixed-interval and fixed-lag algorithms, it is clear that the sliding window SISO can be implemented by carrying out a forward and backward recursion around each transition. Specifically, $\mathrm{MSM}_{k-D}^{k-1}[s_k]$ and $\mathrm{MSM}_{k+1}^{k+D}[s_{k+1}]$ can be computed by forward and backward ACS recursions, respectively, starting with uniform initialization.[11] This process is computationally inefficient, performing approximately $2D$ times as much computation as the corresponding FI-SISO. However, these computations can be carried out in parallel if one allocates sufficient parallel hardware resources. This concept is illustrated in Fig-2.45 where the maturation process of the soft information is shown against ACS clocks in a fully parallel architecture.

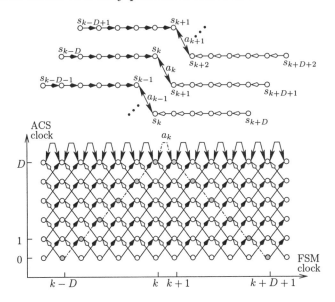

Figure 2.45.   The SW-SISO using forward-backward recursions.

Example 2.11. —————————————————————————————

The PCCC system considered in Section 2.4.3.1 was simulated with each fixed interval SISO replaced by a sliding window SISO with the results shown in Fig-2.46. The min-sum combining rules were used with various values of D.

Note that for $D \geq 16$ there is little degradation from the fixed-interval SISO. This roughly follows the rules of thumb established for the traceback depth in the Viterbi algorithm. Specifically, if the traceback depth is at least 5 to 7 times $\log_2 |\mathcal{S}|$ for an FSM with binary inputs, then the performance degradation relative to that of a fixed-interval Viterbi

---

[11]Note that for $k - D$ close to the left edge, some a-priori state information should be accounted for (see Problem 2.13).

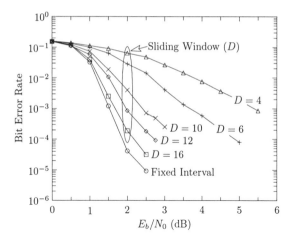

Figure 2.46. Performance of a turbo decoder for the PCCC of Section 2.4.3.1 with min-sum SW-SISOS.

algorithm in negligible [HeJa71]. Thus, one might expect that selecting D larger than 10 to 14 in the SW-SISO would yield acceptable results.

—— *End Example*

2.5.5   A Tree-Structured SISO

The forward and backward recursions used to develop the SISOs are based entirely on the decoupling property of state conditioning in (1.85). Namely, the shortest path problem (*i.e.*, MSM computations) based on the observation interval $\{0, 1, \ldots K - 1\}$ can be decoupled into two shortest path problems when conditioning on s_k – *i.e.*, one for the interval $\{0, 1, \ldots k - 1\}$ and another for $\{k, k+1, \ldots K - 1\}$. The forward ACS operation combines the solution to these two problems by marginalizing out the state s_k (*e.g.*, see (1.59b)). This is the basic dynamic programming principle behind the Viterbi algorithm. This decoupling property can be exploited more aggressively by fusing together conditional shortest-path solutions based on larger observation intervals. In this manner, algorithms for soft-inversion of the FSM (SISOs) can be developed that are significantly different from the forward-backward algorithm. In this section we show how the SISO computations can be carried out with low-latency based on a tree structure.

The low-latency architecture is derived by formulating the SISO computations in terms of a combination of a prefix and suffix operations. To obtain this formulation, define $C[s_k, s_m]$, for $m > k$, as the MSM of state pairs s_k and s_m based on the soft-inputs between them – *i.e.*, $C[s_k, s_m] \triangleq \mathrm{MSM}_k^{m-1}[s_k, s_m]$. Note that, generally, there are $|\mathcal{S}|^2$ values of $C[s_k, s_m]$ corresponding to all combinations of left and right edge states. The forward and backward state MSMs can be obtained from these values via

$$\text{MSM}_0^{k-1}[s_k] = \min_{s_0} C[s_0, s_k] \tag{2.49a}$$

$$\text{MSM}_k^{K-1}[s_k] = \min_{s_K} C[s_k, s_K] \tag{2.49b}$$

Thus, if one can obtain $\{C[s_0, s_k]\}_{k=1}^{K-1}$ and $\{C[s_k, s_K]\}_{k=1}^{K-1}$ then the forward and backward state MSMs are directly available via the marginalization operations in (2.49).

The MSMs $C[s_{k_0}, s_m]$ and $C[s_m, s_{k_1}]$ for $k_0 \le m \le k_1$, can be fused together in the following sense

$$C[s_{k_0}, s_{k_1}] = C[s_{k_0}, s_m] \otimes_C C[s_m, s_{k_1}] \triangleq \min_{s_m} \left[C[s_{k_0}, s_m] + C[s_m, s_{k_1}] \right]. \tag{2.50}$$

which follows directly from the definition of $C[\cdot]$ and implicitly defines the *C-fusion operator* $\otimes_C$. This process, which is much like an ACS operating on multiple steps through the trellis, is illustrated in Fig-2.47.

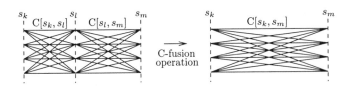

Figure 2.47. The C-fusion operator.

The semi-ring properties of the marginalization and combining operators imply that the associated fusion operator $\otimes_C$ is also *associative*. Note that the required quantities in (2.49) can be obtained by

$$C[s_0, s_k] = C[s_0, s_1] \otimes_C C[s_1, s_2] \otimes_C \cdots C[s_{k-1}, s_k] \tag{2.51a}$$

$$C[s_k, s_K] = C[s_k, s_{k+1}] \otimes_C \cdots C[s_{K-2}, s_{K-1}] \otimes_C C[s_{K-1}, s_K] \tag{2.51b}$$

Thus, the computation of $\{C[s_0, s_k]\}_{k=1}^{K-1}$ and $\{C[s_k, s_K]\}_{k=1}^{K-1}$ may be viewed as *prefix* and *suffix* operations, respectively. Many fast algorithms exist for solving prefix problems in parallel, with a common application being the design of fast adder circuits. These algorithms are based on tree structures and typically have latency that is logarithmic in K.

One example tree-structured SISO algorithm is illustrated in Fig-2.48. The outputs may be marginalized as in (2.49) to provide the desired forward and backward state MSMs which can then be used to perform

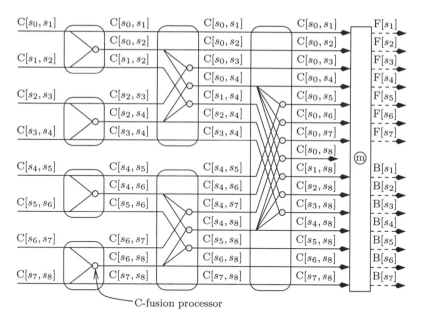

C-fusion processor

Figure 2.48. A tree-SISO shown for $K = 8$. Dashed lines represent only forward or backward information while solid lines represent bi-directional information.

completion via (2.12). The inputs to this tree-SISO are the transition metrics $M_k[t_k] = C[s_k, s_{k+1}]$. Note that, at each stage in the tree, all C-fusion operations could be performed in parallel if sufficient hardware resources are available. It is straightforward to generalize this example to an tree-SISO which has $\log_2 K$ stages. Thus, given sufficient hardware resources, one could complete the equivalent of the forward-backward recursions in $\log_2 K$ C-fusion clock cycles.

Note that, in general, the C-fusion operation in (2.50) has complexity that is significantly larger than the standard ACS operations on a sparsely connected trellis. In fact, for sufficient separation between the end-points of the intervals in (2.50) (*i.e.*, so that all states are connected), the complexity of the C-fusion operation is that of $|\mathcal{S}|^2$, $|\mathcal{S}|$-way ACS operations. In contrast, for an FSM with inputs a_k, the standard one-step ACS operation involves $|\mathcal{S}|$, $|\mathcal{A}|$-way ACS operations. Thus, if the C-fusion process is to be performed in the same amount of time as a standard ACS operation, the C-fusion processor will have significantly larger area.

Some simplification can be obtained by converting to either forward or backward information at the earliest point possible in the tree computation. This saves computation (and storage) since the "FC" and "BC" fusion operations require combining and marginalization over fewer pos-

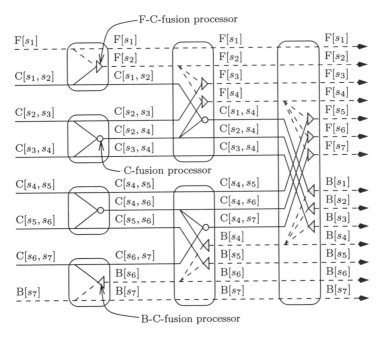

Figure 2.49. The tree-SISO in Fig-2.48 with simplification in the edge computations at the earliest possible stage. Dashed lines represent only forward or backward information while solid lines represent bi-directional information.

sibilities. More specifically, these modules perform

$$\text{MSM}_0^{k-1}[s_k] = \min_{s_m} \left[\text{MSM}_0^m[s_m] + \text{C}[s_m, s_k] \right] \quad (2.52a)$$

$$\text{MSM}_k^{K-1}[s_k] = \min_{s_m} \left[\text{C}[s_k, s_m] + \text{MSM}_m^{K-1}[s_m] \right] \quad (2.52b)$$

This is shown in Fig-2.49 for the example in Fig-2.48. Notice that of the 17 C-fusion operations shown in Fig-2.48, all but 4 can be computed using the simpler FC or BC fusion processors.

This architecture has been generalized in [BeCh00] where a careful accounting of the computational complexity has been performed. It was found that, for reasonably small values of $|S|$, this tree-SISO architecture has $\mathcal{O}(\log_2 K)$ latency and $\mathcal{O}(K \log_2 K)$ computational complexity. Thus, the exponential speed-up relative to the standard forward-backward algorithm comes at a cost of an asymptotic increase in computational complexity that is $\log_2 K$. The impact on the circuit area, however, may be the most important aspect since the tree-SISO architecture is highly parallel. The use of several tree-SISOs operating on subintervals is also described in [BeCh00]. This can further reduce the latency of a SISO operation as discussed in Section 2.5.6.

2.5.6   Variations on Completion and Combining Windows

As described in Example 1.11, for a non-recursive FSM the completion operation can be performed on the state metric instead of the transition metric. If soft-out information is not required for the FSM output, this fact may be used to alleviate SISO computation and/or storage requirements. For example, the SISO for the inner ISI channel in the joint equalization and decoding application has this characteristic. In such cases, the extrinsic output metric for the FSM input a_k is $\text{MO}_{k_1}^{k_2}[a_k] = \text{MSM}_{k_1}^{k_2}[a_k] - \text{MI}[a_k]$ and the MSM term may be computed by marginalizing $\text{MSM}_{k_1}^{k_2}[s_{k+m}]$ for any value of $m \in \{1, \ldots L\}$. Here we have assumed that the combining window is such that $k - k_1 > L$ and $k_2 - k > L$. We refer to this process as *L-early completion*. In this special case, for example, the fixed-interval forward-backward SISO need only store forward and backward state metrics every L time steps.

Fixed-lag and sliding-window SISOs may be modified to take advantage of *L*-early completion. For example, in the fixed-lag algorithm, completion can be done on L input symbols based on a single backward recursion. Specifically, the D-step backward recursion yields $\text{MSM}_0^{k+D}[s_{k+1}]$ which can then be marginalized to produce the soft-out information $\text{MO}_0^{k+D}[a_{k-m}]$ for $m \in \{0, \ldots L-1\}$. Thus, a *minimum-lag* SISO algorithm can be implemented which performs one backward recursion every L forward steps. For a given minimum lag D, it is expected that the minimum-lag algorithm will perform at least as well as the fixed-lag counterpart and is also approximately L times less complex. If $D < L$, no backward recursion is required for this special case [KoBa90] (see Problem 2.21).

Minimum-lag algorithms can be implemented using a similar idea even for arbitrary FSMs. In the general case, a single backward recursion of $(D + H - 1)$ steps is used to perform H completion operations. This yields a minimum-lag algorithm with lag D. More precisely, this yields $\text{MO}_0^{k+D+H-1}[u_{k+m}]$ for $m = 0, 1, \ldots (H-1)$. Note that u_k can be either the FSM input or output in this case. This requires storage of the forward state metrics for H consecutive times, as opposed to one time for the FL version, and reduces the computational complexity by approximately a factor of H relative to the FL-SISO since a backward recursion is performed only once for every H forward steps.

This same concept can be applied to both the forward and backward recursions in the SW-SISO, yielding a *minimum half-window* SISO. This concept may be used to trade-off parallelism for area by tiling an observation interval with subwindows and applying a minimum half-window

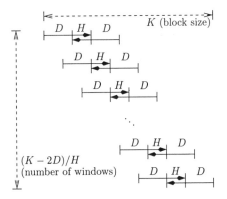

Figure 2.50.   Minimum half window SISOs used on a tiled observation interval.

SISO on each subinterval. This concept is illustrated in Fig-2.50. When $H = 1$, this corresponds to the parallel SW-SISOs illustrated in Fig-2.45.

2.5.7   Soft-Output Viterbi Algorithms

The term Soft-Output Viterbi algorithm (SOVA) is used in the literature to describe a variety of algorithms based on augmenting the Viterbi algorithm to provide soft information on the FSM inputs (*e.g.*, [Ba87, HaHo89, BeAdAnFa93, Vu97]). The SOVA described in [Ba87] computes the fixed-lag MSM for binary input FSMs – *i.e.*, $\text{MSM}_0^{k+D}[a_k]$ using a method similar to that of the L^2VS structure. Instead of spawning two constrained forward ACS recursions for each conditional value of a_k as in the L^2VS structure, the SOVA of [Ba87] runs one unconstrained forward ACS and, for each state s_k, stores the survivor sequence, the state metric, and reliability information for the survivor sequence. More precisely, the reliability information associated with the survivor sequence entering state s_k is $\Delta\text{MSM}_0^{k-1}[\breve{a}_{k-d}, s_k]$ for $d = 1, 2 \ldots D$, where $\breve{a}_{k-d}$ is the symbol value implied by the survivor and $\Delta\text{MSM}_0^{k-1}[\breve{a}_{k-d}, s_k]$ is the difference in $\text{MSM}_0^{k-1}[a_{k-d}, s_k]$ for $a_{k-d} = \breve{a}_{k-d}$ and its compliment. After a standard ACS and survivor path update step, $\Delta\text{MSM}_0^{k}[\breve{a}_{k-d}, s_{k+1}]$ can be computed by tracing back on the two survivors compared in the ACS for s_{k+1}. It is assumed that the states have been defined in a convention such that the two competing paths entering state s_{k+1} correspond to different values of a_k. This allows one to initialize the reliability information for $d = 1$ by the difference in the path metrics in the associated ACS step. The update the reliability information for $d > 1$ is accomplished by tracing back on both of the survivor paths associated with the ACS competitors. The updating process performed during this traceback operation is roughly equivalent to D ACS steps, although the computational complexity depends on the number disagreements between the two paths being traced.[12]

Intuitively, therefore, the two constrained forward ACS recursions associated with the L^2VS structure may be replaced using a traceback and update operation which, using MSM differences, requires roughly half the computation. This correspondence is shown in [FoBuLiHa98], where the SOVA of [Ba87] is referred to as the improved-SOVA to distinguish it from the SOVA in [HaHo89], which does not compute MSMs. The SOVA in [Ba87, FoBuLiHa98] has been extended to non-binary FSM inputs in [CoXiXi99] which is a rediscovery of the L^2VS structure for a FL-MSM algorithm.

The SOVA of [Ba87, FoBuLiHa98] has computational complexity less than half that of the L^2VS while producing the same output. The fixed-lag forward-backward algorithm, however, has half the complexity of the L^2VS without any traceback operation and also produces an equivalent output. Furthermore, unlike all of the SISO algorithms discussed in this section, SOVA algorithms are non-semi-ring algorithms because they utilize a traceback operation (*i.e.*, there is no APP-version of the SOVAs in [Ba87, HaHo89]). Also, these SOVAs do not naturally produce soft-out information on the FSM output. In many cases, therefore, the FL forward-backward algorithm is preferred because of its relative simplicity and generality. However, SOVAs may be reasonable choices for the SISOs associated with a PCCC iterative decoder with BPSK or QPSK modulation since soft-out information is required only for the FSM inputs (*i.e.*, see Problem 2.12). Furthermore, the degradation in performance relative to an FL-MSM SISO associated with the simple SOVA in [HaHo89] is smallest when the constituent codes have a small number of states.

2.6   Message Passing on Graphical Models

Using the implicit index block diagram convention with the notion of marginal soft inverses, iterative detectors for several practical applications were presented in Section 2.4. A great deal more insight into the iterative detection process can be obtained, however, by considering explicit index block diagrams and the corresponding processing. Consider, for example, the parallel concatenated system in Fig-2.17. If X and Y are FSMs, then their soft inverses can be implemented using the forward-backward algorithm. With explicit index block diagrams, it will be made

---

[12]The SOVA in [HaHo89] provides further complexity reduction.

apparent that the forward-backward algorithm is nothing more than an iterative detection algorithm (or concatenated detector) on a subsystem decomposition of the FSM. In other words, there is a concatenated system model for the FSM such that application of the iterative detection principles yields the optimal detector for the isolated FSM (*i.e.*, the forward-backward algorithm). Thus, there is really nothing to distinguish the soft information passed internally in the forward-backward algorithm and that passed between the soft broadcasters and X^{-s} and Y^{-s} in Fig-2.17.

In fact, as illustrated by Example 2.9, the drawing of boundaries between subsystems is somewhat arbitrary. How one partitions the system block diagram into subsystems does not affect the overall system input output relation, but it does affect the associated concatenated/iterative detector. This brings us back to the discussion at the beginning of this chapter about the conditions for optimality of a concatenated detector. Specifically, recall that a concatenated detector is optimal if the soft information passed from the channel observations is a sufficient statistic for detection of the desired input variable. With explicit index block diagrams, the condition for this statistical sufficiency becomes clear – *i.e.*, the explicit index block diagram cannot have "cycles." This will be explained in detail in the following subsections.

Implicit index block diagrams are *graphical system models* and the corresponding iterative detection processing, described in Section 2.3.2, is a message passing algorithm on the associated graph. Thus, in the following, we use the terms "graphical model" and "explicit index block diagram" interchangeably. Interpretation of iterative or turbo detection as message passing on graphs has been widely celebrated in the communications and error correction coding community [McMaCh98, KsFr98, AjMc00]. This is in part because the technique has been well known in computer science for some time [Pe86, Pe88, GoVa89, JeLaOl90, ShSh90, LaSp88, Je96]. The distinction being that, in the most interesting applications considered in communications, message passing (iterative detection) is done on graphs with cycles and is therefore suboptimal. Given the motivation of iterative detection in Section 2.2 and the various applications explored in Section 2.4 it may not seem surprising that iterative detection is useful, even when it is suboptimal. However, the effectiveness of message passing on graphs with cycles was not widely appreciated in the computer science literature. Thus, the suggestion of iterative detection by Gallager [Ga63] and independently by

Berrou, Glavieux, and Thitmajshima [BeGlTh93, BeGl96] represented a significant practical breakthrough.[13]

While we use explicit index block diagrams as a simple and natural graphical model, several other conventions are more common. The correspondence between these models and explicit index block diagrams is given in Section 2.6.4. At this point, however, we note that the explicit index block diagrams are *directed* graphs (*i.e.*, we think of inputs to systems producing outputs) with implicit nodes for the system inputs and outputs. Furthermore, except for the trivial case of renaming variables (*e.g.*, see Fig-2.10), variables are not connected without a system node separating them. Thus, our graphical models are essentially *bipartite* graphs meaning that there are types of nodes which can only be connected to each other.

2.6.1   Optimality Conditions for Message Passing

Reconsider the serial concatenated system shown in Fig-2.4 and the associated concatenated detectors in Fig-2.5 with the SOAs in Fig-2.5 interpreted as the associated soft inverses. The optimality of the equivalent detectors in Fig-2.5(a) and Fig-2.5(b) is far more apparent by the structure of the explicit index diagram in the latter than by the implicit index block diagram of the former. Specifically, if the soft information passed on x_1 is equivalent to $p(\mathbf{z}|x_1)$ then MAP-SyD decision for a_2 and a_3 is obtained using a sum-product soft inverse of the system X_1. Similarly, the MAP-SqD decision for a_1 can be obtained using a max-product soft inverse of X_1 if the message passed on x_1 is equivalent to the generalized likelihood $g(\mathbf{z}|x_1)$ with nuisance parameter set $\{a_m(\zeta)\}_{m\neq 1}$.

For an arbitrary system with global inputs $\{a_m\}$, the optimal decision on a_m is obtained via message passing under the same condition. Specifically, if a_m is the input to a subsystem Q, then the soft information passed by the soft inverse of Q in a concatenated detector should be a sufficient statistic for detection of a_m. Recall that, in the absence of any feedback in the system block diagram, it was reasoned that this implies that the message passed inward from the channel must be a sufficient statistic for detection of a_m based on the observation sequence.

In graph terminology, our assumption of no feedback loops means that there are no *directed cycles* in the graphical model. Specifically,

---

[13]In [Pe88, Exercise 4.7], Pearl suggested that belief propagation on a graph with cycles could be used as a reduced-complexity approximation to the optimal inference algorithm. Also, in [MuWeJo99, We00] Pearl is attributed with a personal communication suggesting this approach. In view of the effectiveness of iterative detection in communication applications (*e.g.*, turbo codes), probability propagation or message passing on graphs with cycles has become an active area of research in computer science [We00, MuWeJo99, Fr98, FrMc98].

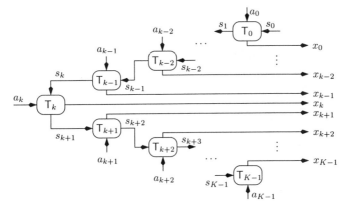

Figure 2.51. The system block diagram for an arbitrary FSM drawn to emphasize the view of a_k as the input to a concatenated system with nuisance parameters.

a directed cycle exists when a directed path through the graph from a given vertex returns to that vertex. Thus, by the development at the beginning of the chapter, we expect that any system represented by an explicit index block diagram with no directed loops should have a MAP detector based on the associated concatenated detector (*i.e.*, message passing). An example of this notion is illustrated in Fig-2.51 where an arbitrary FSM has been decomposed into a concatenation of "transition subsystems." This graphical model for the FSM has the property that the outputs are obtained sequentially since the output x_k cannot be obtained until the value of s_k is available. When a MAP-SyD or MAP-SqD decision for a_k is desired, a_k should be viewed as the single input to the system with $\{a_i\}_{i \neq k}$ viewed as nuisance parameters. The associated optimal concatenated detector for a_k is shown in Fig-2.52. This detector is obtained by replacing all systems by the corresponding soft inverses[14] and passing the appropriate messages. Specifically, with the soft-in information on x_i being the marginal likelihoods (metrics) from the channel and the soft-in information on the inputs a_i being the a-priori probabilities (metrics), the MAP decision on a_k is obtained by passing message from right to left in Fig-2.52. This proceeds by first activating $\mathsf{T}_0^{-\mathrm{s}}$, then $\mathsf{T}_1^{-\mathrm{s}}$, etc. on the upper portion of the concatenated detector and $\mathsf{T}_{K-1}^{-\mathrm{s}}$, then $\mathsf{T}_{K-2}^{-\mathrm{s}}$, etc. in the lower half of the concatenated detector.

---

[14]Notice that regardless of whether a_i is considered as an input or a nuisance parameter for the system T_i, the soft inversion remains the same. This foreshadows the development of marginal soft inverses for systems with channel nuisance parameters contained in Chapter 4.

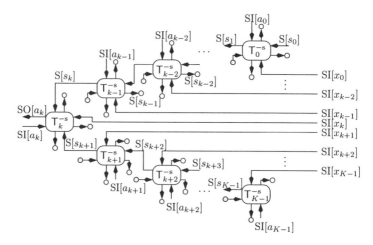

Figure 2.52. The optimal concatenated (message passing) detector for a_k based on the model in Fig-2.51.

The message passing on the upper and lower parts of the detector in Fig-2.52 can be scheduled independently, but the soft output on a_k cannot be computed until a message has been received by T_k^{-s} from both parts, after which SO$[a_k]$ is produced, combined with SI$[a_k]$, and thresholded. Careful inspection of the processing in Fig-2.52 reveals that the message passing in the upper and lower parts corresponds to the standard forward and backward recursions of the forward-backward algorithm. Specifically, for sum-product marginalizing and combining, the message passed from T_i^{-s} to T_{i+1}^{-s} is P$[s_{i+1}] \equiv p(\mathbf{z}_0^i, s_{i+1})$ and that passed from T_{j+1}^{-s} to T_j^{-s} in the lower portion is P$[s_{j+1}] \equiv p(\mathbf{z}_{j+1}^{K-1}|s_{j+1})$. Note that these are, in fact, sufficient statistics for detection of a_k from $\mathbf{z}_0^i$ and $\mathbf{z}_{j+1}^{K-1}$, respectively.

It follows that, by connecting the ports of the transition soft inverses, the entire FSM soft inverse can be computed by passing messages. This is illustrated in Fig-2.53. The system in Fig-2.53(a) is identical to that in Fig-2.51, just redrawn to emphasize the desire to compute the extrinsic soft outputs for all a_i. Similarly, the message passing or "iterative" algorithm in Fig-2.53(b) is the concatenated detector in Fig-2.52 with messages passed in both directions. The soft inverse of a transition subsystem is shown in Fig-2.54. Specifically, activation of T_k^{-s} does one update of the forward and backward state recursions to incorporate z_k and one completion for each of the system input (a_k) and output (x_k) variables (see Problem 2.24). Therefore, the forward-backward SISO may be viewed as the application of the iterative detection principles defined in Section 2.3 to a subsystem decomposition of the FSM.

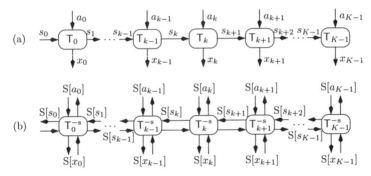

Figure 2.53. (a) An explicit index block diagram for an arbitrary FSM, and (b) the associated concatenated detector.

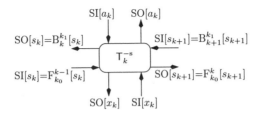

Figure 2.54. The soft inverse of of the transition subsystem. Activation is equivalent to one update of the backward and forward state metric recursions, and completion for both the FSM input and output.

It is somewhat odd to think of the processing illustrated in Fig-2.53(b) as iterative detection since it is perhaps more accurately described as concurrently processing K concatenated detectors, one for each value of k, of the form shown in Fig-2.52. Consider two ways of viewing the connection between the processing in Fig-2.52 for each value of k and the bi-directional message passing in Fig-2.53(b). First, one could realize that messages passed from T_0^{-s} through the upper part of the concatenated detector in Fig-2.52 to T_k^{-s} could be updated and passed along to T_{k+1}^{-s} providing the corresponding message for the concatenated detector for a_{k+1}. Similarly, the message from the lower portion of the concatenated detector in Fig-2.52 could be updated and passed along to provide the analogous information for the concatenated detector associated with a_{k-1}. This leads to the standard forward and backward recursions in the fixed-interval forward-backward SISO, each executed serially. Specifically, for the forward recursion, messages are passed in the processing of Fig-2.53(b) by successive, serial activation of T_0^{-s}, T_1^{-s}, etc. The backward recursion is similarly viewed as activation of T_{K-1}^{-s}, T_{K-2}^{-s}, etc. Since there is no "cross-talk" between the SI/SO ports in Fig-2.54 corresponding to the forward and backward recursions, these two serial activations proceed independently and their individual scheduling need not be coordinated. This fact was illustrated in Fig-2.40. The execution of serial forward and backward message passes in parallel, is referred to as the *fully serial* or *two-way schedule* in more general graphs as discussed in Section 2.6.1.2.

The other interpretation is that the concatenated detector of Fig-2.52 is basically replicated for each location k. Messages for each can be passed toward the particular input variable of interest as was described for a_k in the context of Fig-2.52. This can also be accomplished by activating all soft inverse nodes in Fig-2.53(b) in parallel K times. This notion has also previously been described. In particular, the soft out information for the sliding window SISO is the result of the first D such parallel activations of the processing nodes in Fig-2.53(b). Conversely, Fig-2.45 provides a good visualization process of the maturation of the soft-outputs for each T_k^{-s}. This is referred to as the *fully parallel* or *flooding schedule* and generalizes other graphs as discussed in Section 2.6.1.2.

The conclusions drawn above generalize to more complicated graphs. Specifically, any explicit index block diagram implies an optimal detector or marginal soft inverse of the corresponding system provided that there are no *undirected cycles* in the graphical model. An undirected cycle is a loop in the graph that can be traced when the direction of the edges in neglected (*i.e.*, a directed cycle is an undirected cycle). We use the term *cycle-free graph* to refer to a graph (directed or undirected) that has no undirected cycles. Conversely, we refer to a graph with undirected cycle as a *loopy* graph or graph with cycles. Before formalizing this optimality claim, it is helpful to consider an example of graphical models with cycles and cycle-free graphical models with different structure than the simple graph in Fig-2.53(a).

Example 2.12. ───

Consider the PCCC in the implicit index block diagram in Fig-2.24. The corresponding explicit index diagram is shown in Fig-2.55 with the FSM graph from Fig-2.53(a) adopted for each constituent code. Note that puncturing and state termination are shown. This graph has no directed cycles (feedback), but does have undirected loops. For example, one can go from $\mathsf{T}_1^{(1)}$ to $\mathsf{T}_k^{(1)}$ along the trellis of RSC1, then through the broadcaster and interleavers to $\mathsf{T}_1^{(2)}$. Going from $\mathsf{T}_1^{(2)}$ to $\mathsf{T}_0^{(2)}$ along the trellis from RSC2, then back through the interleaver and broadcaster, one arrives back at the node $\mathsf{T}_1^{(1)}$. Therefore, based on the above claim regarding the optimality of message passing on cycle-free graphs, one

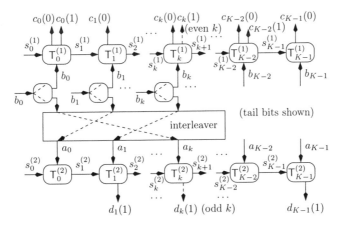

Figure 2.55. The explicit index block diagram of the PCCC shown in Fig-2.24.

cannot conclude that message passing on the graphical model of Fig-2.55 is optimal.[15]

The decoder used in Section 2.4.3.1 is message passing on the graphical model in Fig-2.55 with a specific activation schedule. More precisely, if only the structure of the subgraph corresponding to RSC1 is considered, the processing of SISO1 is accomplished by message passing on this subgraph as described above. Next, activation of all the soft inverse broadcaster nodes is performed. Message passing on the subgraph corresponding to RSC2 is then performed to carry out the soft inversion of RSC2. The soft inverses of the broadcasters are executed again, completing a single iteration of the detector in Fig-2.28.

This process is suboptimal, in general, because one cannot ensure the soft information associated with $c_k(i)$, $d_k(i)$ and b_k are weighted equally for each value of k. Intuitively, for example, because of the cycles, the soft-in information for $c_{10}(0)$ maybe "counted" more times than that for $c_{631}(0)$.

It is interesting to generalize the concept of the SW-SISO to this more general graphical model. For example, the connections leading into and out of the broadcaster associated with a specific input b_k can be considered. In Fig-2.56 we have shown this subgraph for the actual $K = 1024$ interleaver used in the simulations of Section 2.4.3.1 with the input bit of interest being b_{500}. Notice that this subgraph has no cycles, so that, by the above claim, *optimal* detection of b_{500} based on the associated subset of observations can be accomplished by message

---

[15]This is not to say that it is always suboptimal. Specifically, the absence of cycles is a sufficient, but not necessary condition for the optimality claim.

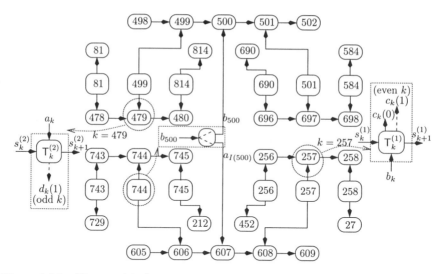

Figure 2.56. The graphical system model for bit b_{500} for the PCCC of Fig-2.55. This represents a subgraph without loops for a region around the variable b_{500}. The associated message passing algorithm provides optimal detection of $b_{500}(\zeta)$ based on the associated subset of observations. (Based on the actual $K = 1024$ interleaver used in the simulations of Section 2.4.3.1.)

passing on this graph. This diagram is analogous to cutting the upper and lower sections of the graph in Fig-2.51 after T_{k-1} and T_{k+2}. The associated concatenated detector for b_{500} would pass messages "inward" toward the node corresponding to the soft inverse of the broadcaster for $k = 500$. The exact scheduling of this will be discussed in the following. The important point is that, assuming that the above claim is true, one can obtain the MAP-SyD or MAP-SqD decision for b_{500} based on the observations $z_n(0)$ and $z_m(1)$ for $n \in \{$ 81, 256, 257, 258, 498, 499, 500, 501, 502, 584, 690, 743, 744, 745, 814 $\}$ and $m \in \{$ 256, 258, 498, 500, 502, 584, 690, 744, 814 $\} \cup \{$ 27, 479, 605, 607, 609, 697,729 $\}$. Note that, for example, $z_{452}(1)$ is not used because of the puncturing (*i.e.*, the only soft-in information propagated from the node 27 in the lower-right is MI[a_{27}]).

This subgraph provides an example of a cycle-free graph with interesting structure. Furthermore, it provides intuition as to why turbo decoding is effective. Based on intuition gained by the simulations of the SW-SISO in Example 2.11 and the rule-of thumb regarding traceback depth in the Viterbi algorithm, it is reasonable to expect that if the radius of the subgraph in Fig-2.56 is increased the resulting decision for b_{500} will approach the optimal MAP decision based on the entire observation sequence. This is because, presumably, one could create a cycle-free subgraph akin to that in Fig-2.56, but with larger radius. Op-

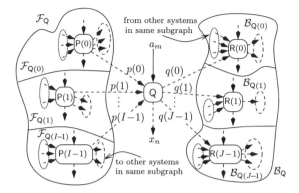

Figure 2.57. The general case of message passing on a graph with no directed or undirected cycles.

timal detection on this subgraph, obtained via message passing, should capture most of the strongly relevant observation information. In other words, the soft-in information from observations that are connected to b_{500} only through a very long sequence of nodes will have little effect on the MAP detection of b_{500}. Thus, if the graphical model of a system only has very long cycles, then it may be expected that message passing will be effective. In fact, as we will see, message passing on graphs with short cycles is generally less effective than it is on graphs with long cycles.

——————————————————————————————— *End Example*

2.6.1.1 Optimal Message Passing on Cycle-Free Graphs

The optimality claim for message passing on cycle-free graphs is verified by considering the system shown in Fig-2.57. Specifically, assume that a system has been represented in explicit index form and that this graph has no cycles. Each global input a_i is the input to only one system node (possibly a broadcaster) and each global output x_j is the output of one system node. System nodes may or may not have global system inputs and/or outputs. Marginal soft-in information for each global system input and output variable is available. Consider solving the general problem stated in (1.37) – *i.e.*, computing the marginal soft-output for a_m based on all soft-in information and the global system structure. Because there are no cycles, we can isolate the system Q that receives a_m as an input as illustrated in Fig-2.57. Specifically, no output of Q can return as an input to a system node that provides an input to Q or a directed cycle would occur. Similarly, an input of Q cannot be an input to a system in $\mathcal{B}_Q$ or an undirected cycle will exist.

We can conclude that if $S_{\mathcal{F}_Q}[p(0), p(1), \ldots p(I-1)]$ and $S_{\mathcal{B}_Q}[q(0), q(1), \ldots q(J-1)]$ are available then these can be combined with $\text{SI}[x_n]$, $\text{SI}[a_m]$

and marginalized over the structure of Q to provide the desired soft output on a_m. Here, for example, $S_{\mathcal{F}_Q}[\mathbf{p}]$ represents the effect of combining all soft-in information for global system inputs and outputs contained in the subgraph that describes the system structure on the subset of indices $\mathcal{F}_Q$ and marginalizing over the configurations of that subgraph consistent with $\mathbf{p}$.

Furthermore, the cycle-free assumption implies that no system in $\mathcal{F}_Q$ can have outputs driving two systems $P(i)$ and $P(i')$ for $i \neq i'$, otherwise there would be an undirected cycle occurring through the connection to Q. By similar logic, the outputs of $P(i)$ and $P(i')$ cannot drive any common system other than Q. Thus, a partition of $\mathcal{F}_Q$ into disjoint subgraphs describing the structure on $\{\mathcal{F}_{Q(i)}\}$ is obtained by the lack of cycles. As a consequence of this partition, the soft information on $\mathbf{p}$ must factor into the combination of $S_{\mathcal{F}_{Q(i)}}[p_i]$ over $i = 0, 1, \ldots I - 1$.

Similar reasoning leads to the conclusion that $\mathcal{B}_Q$ is partitioned as shown in Fig-2.57 and that the soft information on $S_{\mathcal{B}_Q}[\mathbf{q}]$ factors into the combination of $S_{\mathcal{B}_Q(j)}[q(j)]$. Together this leads to

$$S[a_m] = \underset{\mathbf{p}:a_m}{\textcircled{m}} \left(S_{\mathcal{F}_Q}[\mathbf{p}] \,\textcircled{c}\, S_{\mathcal{B}_Q}[\mathbf{q}(\mathbf{p}, a_m)] \,\textcircled{c}\, SI[x_n] \,\textcircled{c}\, SI[a_m] \right) \qquad (2.53a)$$

$$S_{\mathcal{F}_Q}[\mathbf{p}] = \overset{I-1}{\underset{i=0}{\textcircled{c}}} S_{\mathcal{F}_{Q(i)}}[p_i] \qquad\qquad S_{\mathcal{B}_Q}[\mathbf{q}] = \overset{J-1}{\underset{j=0}{\textcircled{c}}} S_{\mathcal{B}_{Q(j)}}[q_j] \qquad (2.53b)$$

This argument can then be applied inductively to the nodes $\{P(i)\}$ and $\{R(j)\}$ to show that $S[a_m]$ based on the entire graph structure and all soft-in information can be obtained by message passing and updating using the corresponding soft inverses. More precisely, the soft-in messages from all a_i and x_j need to propagate to the node Q.

The above argument is only a slight generalization of the discussion regarding the optimality of message passing in Fig-2.52 and Fig-2.53(b). Specifically, the message passing inward to Q^{-s} on $\mathcal{F}_Q$ and $\mathcal{B}_Q$ corresponds to the forward and backward recursions in the forward-backward algorithm, respectively. It follows that this reasoning can be applied to each input variable a_i and/or each output variable x_j. As in the FSM example, it is possible to perform this operation for all input and output variables by message passing with a specific schedule. This is discussed further in the next section.

2.6.1.2   Scheduling and Convergence for Cycle-Free Graphs

Scheduling defines the propagating pattern of messages in the graph and is not unique. Roughly speaking, for any node Q^{-s} in the message passing algorithm corresponding to the subsystem node Q in a cycle-free graphical system model, if a message has propagated from every other processing node, the soft output information has converged (*e.g.*, see [AjMc00, KsFrLo00]). There are two extreme activation schedules, the fully serial schedule and fully parallel schedule.

In the fully parallel schedule, every node is activated simultaneously. With this schedule, the messages will converge (or stabilize or mature) after at most diam(M) iterations where diam(M) is the diameter of the cycle-free graph M (*i.e.*, the maximum number of subsystem nodes separating any two variables). This schedule is also called the flooding schedule [KsFr98].

In fully serial or two-way [KsFr98] schedule, messages begin at the *leaf nodes*, which correspond to the global system inputs and outputs, and propagate *inwards* by activating each node one-by-one. A node is activated when a message has been received from all connected nodes except one. When the node is activated, a message is passed out only on that one node (*i.e.*, this is called the *all-but-one rule* in [KsFrLo00]). For example, in the FSM graph in Fig-2.53, the T_{K-1}^{-s} and T_0^{-s} are activated initially and pass information inward only (*i.e.*, backward and forward recursion, respectively). Once a node has passed an inward message, it waits to receive a message back from that direction and is configured to send a message *outward*. After some period ($\leq$ diam(M)), there will be some nodes that have received messages from all of their neighbors (*i.e.*, the nodes in the "center" of the graph). Then, the messages will begin to propagate outward and each node will receive messages from all neighbors. For example, in the FSM case of Fig-2.53, the inward passes correspond to running the forward and backward recursions inward to the center of the graph. When these two recursions meet (or pass each other) this corresponds to the switch from inward to outward message passes. Also, at the time that a message has been received from all the neighbors of a particular node, the soft-outputs on the global system inputs and outputs have converged and can be output. This schedule is also referred to as the generalized forward-backward schedule [KsFrLo00] and the inward-outward schedule [AjMc00]. With this schedule, each node is activated twice. In addition, it is possible to construct hybrid schedules from these two extreme cases.

The reason we may describe scheduling somewhat imprecisely is that once the soft-out information has converged, further activation of the soft inverse nodes has no effect. For example, consider the FSM model and message passing algorithm in Fig-2.53. Assume that, by some schedule, the algorithm has converged. The forward and backward state messages on the ports of the soft inverse in Fig-2.54 then correspond to $k_0 = 0$ and $k_1 = (K - 1)$ and the soft-out information on a_k and x_k are based on the entire observation interval. If the soft inverse of T_k is activated again (*i.e.*, after convergence), none of the soft-out information changes. Thus, in the above description of scheduling options, when we describe "inward-only" message propagation, it is only for computational efficiency. In fact, at each activation, the soft inverse may produce soft-out messages on all of its ports. For example, in a flooding schedule for the system of Fig-2.53, if all soft-out ports produce soft information for each activation, then the soft outputs on a_k and x_k represent "preliminary" versions of the final soft-outputs. In fact, after D activations using flooding, the soft-outputs are exactly those produced by the sliding window SISO. Thus, the exact soft inverse of a system represented by a cycle-free graph can be computed using message passing with a significant flexibility in scheduling between the fully parallel and serial extremes.

Example 2.13. —————————————————————————————————
Consider the two-way schedule applied to the cycle-free graph in Fig-2.56. The node corresponding to b_{500} is the "center" node in this graph so, if soft-out information for only b_{500} is sought, only the inward message passes are required. The start of this schedule is the parallel activation of the soft inverse nodes of (81, 498, 814, 690, 502, 584, 729, 212, 452, 27, 605, 609) with inward-only message passing. The eight soft broadcaster nodes connected to the nodes above are activated next, but only to pass the inward messages through (*i.e.*, no soft-out on b_k is produced). Next, in the upper left quadrant, for example, the nodes 478 and 480 are activated, while node 499 waits for a message from the connected soft broadcaster. Next, node 479 is activated. The 499 soft broadcaster is activated next, followed by the activation of the 499 node in the trellis of RSC1. By this time, node 500 in the RSC1 trellis has received a message from both the left and right, so it is activated next to pass inward to the soft broadcaster for b_{500}. The symmetric process has taken place on the lower half of the graph, so the soft broadcaster would switch to an outward message mode. More importantly the soft output on b_{500} is mature and can be output. If soft outputs on b_i for other values of i are desired, the outward messages should be passed until they reach the soft broadcaster for b_i.
——————————————————————————————— *End Example*

2.6.2   Revisiting the Iterative Detection Conventions

Given the interpretation of message passing on graphs described above, we should reconsider the iterative detection principles stated in Section 2.3. As stated in Section 2.3.2, the iterative detection conventions may be viewed as message passing on graphs which may have cycles. With loops in the graphical models, however, convergence cannot be guaranteed in general. Thus, in general there is no optimal schedule or stopping criterion when the graph has cycles. For example, it is difficult to define what "inward" and "outward" relative to a specific node when there are cycles in the graph. The real challenge, it seems, is determining good graphical models for systems and subsystems. Once a system has been represented by a graphical model, such as an explicit index block diagram, the iterative detector is basically fixed except for the scheduling and the stopping criterion.

It is worth reemphasizing that the conventions described in Section 2.3 and message passing on graphs with cycles (*i.e.*, also called Belief Propagation on loopy graphs as described in Section 2.6.4) are one in the same. Thus, all of the applications considered in Section 2.4 are valid examples of message passing on graphs. The message passing interpretation provides another view and illustrates that the implicit index methods used in Section 2.4 are only one of many possible activation schedules for message passing on an underlying loopy graphical model. In the implicit index block diagram approach, there are essentially two types of messages passed. First, the system is represented as a concatenated network of subsystems, with each subsystem represented by a cycle-free subgraph. Messages are passed *on* these subsystem graphs and *between* these subsystem graphs. The messages passed on the subsystem graphs are used to compute the subsystem marginal soft inverse exactly. Even with the demarcation of the subgraphs according to subsystems (*i.e.*, super-nodes in the graph), the explicit index block diagrams for the examples in Section 2.4 have loops. For example, the graph in Fig-2.11 has cycles. The development of this section has made it clear that the distinction of message passes "in" and "between" subsystem soft inverses is somewhat artificial.[16]

The use of a fixed-lag or sliding window SISO, to approximate the subsystem soft inverse, however, does change the algorithm. This is because the message passing paradigm requires that the soft inverse of every system be used in the associated iterative detector. As stated in Section 2.5, changing the combining window fundamentally changes the soft-out information produced. On the other hand, using the tree-SISO of Section 2.5.5 in place of a forward-backward SISO is perfectly within the principles of iterative detection described in Section 2.3 and the message passing conventions described above. Specifically, it is just a

---

[16]The "between" messages in the explicit index approach usually correspond to physically tangible variables in the system model.

computation architecture for computing the exact marginal soft inverse of an FSM. This is expounded upon in the following example.

Example 2.14. ───────────────────────────────────

Consider the decoding of the PCCC using SW-SISOs as described in Example 2.11. By the above reasoning this is *not* in keeping with the iterative detection principles and is therefore not the standard message passing on the graph of Fig-2.55. It is closely related to a particular activation schedule for message passing on this graph. Specifically, consider the *incomplete flooding* schedule on the subgraph of RSC1 which passes messages D times. This is exactly equivalent to the SW-SISO process as suggested by Fig-2.45. Messages are then passed to the nodes in the subgraph corresponding to RSC2 (*i.e.*, activation of the soft broadcaster nodes). Again, D message passes are performed on the subgraph of RSC2 which is equivalent to firing a SW-SISO for SISO2. These messages are then passed back through the soft broadcasters and another incomplete flooding schedule on the trellis of RSC1 is executed. It is on this second round of message passing on the RSC1 subgraph that the SW-SISO iterative detector and the message passing algorithm begin to differ. Specifically, the node $[\mathsf{T}_k^{(1)}]^{-\mathrm{s}}$ has active forward and backward state messages at its ports. In contrast, the SW-SISO is activated with uniform information on the forward and backward state metrics each time it is activated.

The message passing algorithm described above can be viewed as the iterative detector with SW-SISOs, but with a modified state metric initialization convention for the SW-SISOs. Consider Fig-2.45 to be a description of the processing associated with the SW-SISO for RSC1. The above message passing algorithm corresponds to continuing the upward progression of the forward-backward recursions from one activation to the next. For each activation, however, the FSM transition metrics change according to the information on the uncoded input bits after each round of message passing on the RSC2 subgraph. In practice this means that the SW-SISO can be used for a standard message passing algorithm, but this requires the additional storage of the latest forward and backward state metrics between activations of the SW-SISO. While this represents considerable overhead, it will provide significantly better performance for small values of D.

─── *End Example*

The graphical modeling view also makes it clear that one can use a model for a given system that has cycles in it and apply message passing to approximate the optimal detector. In other words, while most of the examples we have considered thus far have a natural decomposition (*i.e.*, they were constructed by concatenating subsystems), one may try to artificially decompose a system into a loopy concatenation of subsystems. The primary potential advantage of performing message passing on such a loopy graph, as opposed to optimal detection on the global system structure, is complexity reduction. A toy example of this process is considered in the next example.

Example 2.15. ———————————————————————
Consider the special case of an ISI-AWGN channel as described in (1.62) with $L = 2$, $f_0 = f_2 = 1$, $f_1 = 2$, and BPSK modulation (*i.e.*, $\mathcal{A} = \{+1, -1\}$). The data sequence is independent and uniformly distributed over $\mathcal{A}$ and the system is normalized such that $\mathbb{E}\{a_k^2(\zeta)\|\mathbf{f}\|^2\}/\mathbb{E}\{w_k^2(\zeta)\} = 2E_b/N_0$.

An explicit index block diagram for this system and the corresponding iterative detector are shown in Fig-2.58 and Fig-2.59, respectively.

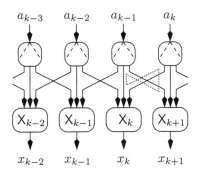

Figure 2.58. A loopy system diagram for a three-tap ISI channel. An (undirected) cycle of length four is shown.

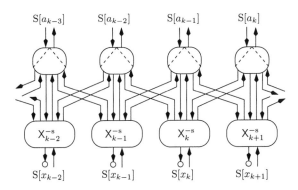

Figure 2.59. The iterative detector associated with the graph in Fig-2.58.

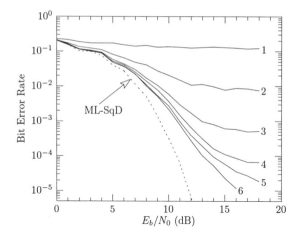

Figure 2.60.   The performance of iterative detector in Fig-2.59 compared against the optimal decision rule.

Each of the soft inverse nodes associated with the channel outputs can be activated in parallel, followed by a parallel activation of all broadcaster nodes. These two parallel activations define one iteration for the following discussion.

The performance of the MSM-based iterative detector is shown in Fig-2.60 along with that of the optimal ML-SqD. Significant improvements in performance were not observed after six iterations. The minimum cycle length in the system diagram is 4. This is measured in terms of edges between soft inverse nodes. For example, a cycle of length four in Fig-2.58 is: broadcaster for a_k, X_{k+1}, broadcaster for a_{k-1}, X_k.

Thus, iterative detection in this system is not particularly effective – *i.e.*, the approximate 3 dB degradation in SNR relative to the optimal processing is larger than that observed with systems having long cycles. Nonetheless, this simple example shows how one may reformulate the system model (*i.e.*, adding cycles) and use iterative detection as an effective suboptimal detector. One iteration of this algorithm is substantially less complex than the corresponding Viterbi algorithm or MSM-SISO. However, six iterations of this algorithm represents substantially more computation than the optimal detector (see Problem 2.25).

This example is continued in Chapter 3 to illustrate variations on the iterative detection conventions that may yield complexity reductions and/or improved performance.

——————————————————————————— *End Example*

Although no complexity reduction was achieved in Example 2.15, if the ISI channel were sparse, a significant complexity reduction would be achieved. Furthermore, a sparse ISI channel would have long cycles in the graphical model so that the iterative algorithm is expected to perform near optimally. This application is considered in detail in Chapter 3.

2.6.3 Valid Configuration Checks

The MAP data detection problem, or more generally, the marginal soft inversion of a system is stated in (1.37). The structure of the system is enforced only through the mapping of the input sequence $\mathbf{a}$ to the outputs $x_n(\mathbf{a})$. Notice that marginal soft information on $\{x_n\}$ can be combined to form $S[\mathbf{x}]$ without regard to whether the sequence $\mathbf{x}$ is actually consistent with some input sequence. This must be accounted for in the marginalization operation by only marginalizing over valid input output pairs. This can be done by defining a *valid configuration check* function $V(\mathbf{a}, \mathbf{x})$. This is an indicator function for a valid configuration so that $V(\mathbf{a}, \mathbf{x}) = 1$ for valid input output pairs and $V(\mathbf{a}, \mathbf{x}) = 0$ for invalid pairs. With this definition, it can be verified that the soft-inverse of a system can be computed by replacing (2.5) by

$$S[\mathbf{a}] = \overset{M-1}{\underset{m=0}{\text{\textcircled{c}}}} SI[a_m] \qquad S[\mathbf{x}] = \overset{N-1}{\underset{n=0}{\text{\textcircled{c}}}} SI[x_n] \qquad (2.54\text{a})$$

$$SO[a_m] = (\ \underset{\mathbf{a}:a_m,\mathbf{x}}{\text{\textcircled{m}}}\ S[\mathbf{x}] \,\text{\textcircled{c}}\, S[\mathbf{a}] \,\text{\textcircled{c}}\, S[V(\mathbf{a}, \mathbf{x})]\,)\,\text{\textcircled{c}}^{-1} SI[a_m] \qquad (2.54\text{b})$$

$$SO[x_n] = (\ \underset{\mathbf{a},\mathbf{x}:x_n}{\text{\textcircled{m}}}\ S[\mathbf{x}] \,\text{\textcircled{c}}\, S[\mathbf{a}] \,\text{\textcircled{c}}\, S[V(\mathbf{a}, \mathbf{x})]\,)\,\text{\textcircled{c}}^{-1} SI[x_n] \qquad (2.54\text{c})$$

where, for a validity check variable v, $S[v = 1] = I_c$ and $S[v = 0] = I_m$. For example, in the probability domain, the soft information for a $v = 1$ (valid) is 1, and $v = 0$ (invalid) is 0, while it is zero and infinity, respectively, in the metric domain. Thus, invalid configurations do not contribute to the soft-out information because they do not affect the marginalization.

Comparing (2.5) and (2.54) yields an interesting interpretation. Although the $V(\mathbf{a}, \mathbf{x})$ was added just based on maintaining the equivalent soft inversion, (2.54) may be viewed as the soft inversion problem stated for a system with inputs $\{a_m\}$ and $\{x_n\}$, selected freely from the digital alphabets $\mathcal{A}$ and $\mathcal{X}$, respectively, and output $V(\mathbf{a}, \mathbf{x})$. For some systems it is simpler to describe and exploit the structure through validity checks than by the direct input output relations. This is illustrated in the following simple example.

Example 2.16. ───────────────────────────────

Consider a binary block code. This code has uncoded input bits b_0, $b_1, \ldots b_{K-1}$ and coded outputs $c_0, c_1, \ldots, c_{N-1}$, each in $\{0,1\}$. The structure of the code can be described by either the $(N \times K)$ *generator matrix* $\mathbf{G}$ or the $((N-K) \times N)$ *parity check matrix* $\mathbf{P}$. The code is represented directly using the generator matrix via $\mathbf{c}(\mathbf{b}) = \mathbf{Gb}$ where all arithmetic is modulo two. The code may also be represented by a validity check of the form $V(\mathbf{b}, \mathbf{c})$ using the fact that $\mathbf{Pc}(\mathbf{b}) = \mathbf{0}$. Thus, the validity indicator is one if $\mathbf{Pc}(\mathbf{b}) = \mathbf{0}$ and zero otherwise (*i.e.*, the configuration check is only a function of $\mathbf{c}$).

The code can be decoded directly using the generator matrix structure or using the parity check structure. For example, MSM-based soft-out information on b_k using the generator matrix structure is obtained using

$$\text{MO}[b_k] = \min_{\mathbf{b}:b_k} \sum_{i \neq k} \text{MI}[b_i] + \sum_{n=0}^{N-1} [z_n - \text{q}(c_n(\mathbf{b}))]^2 \qquad (2.55)$$

where $c_n(\mathbf{b})$ is the n-th element of $\mathbf{c}(\mathbf{b}) = \mathbf{Gb}$ and $\text{q}(\cdot)$ is the mapping from $\{0,1\}$ to the channel modulation (*e.g.*, for BPSK $\text{q}(0) = +1$ and $\text{q}(1) = -1$).

The equivalent processing using the parity check structure is obtained using

$$\text{MO}[b_k] = \min_{\mathbf{b}:b_k, \mathbf{c}} \sum_{i \neq k} \text{MI}[b_i] + \sum_{n=0}^{N-1} [z_n - \text{q}(c_n)]^2 + \text{M}[V(\mathbf{b}, \mathbf{c})] \qquad (2.56)$$

This still requires checking to verify that $\mathbf{b}$ and $\mathbf{c}$ correspond. If, however, one is interested on soft-out information on c_n, then the parity check structure yields

$$\text{MO}[c_n] = \min_{\mathbf{b}, \mathbf{c}:c_n} \sum_{k=0}^{K-1} \text{MI}[b_k] + \sum_{j \neq n} [z_j - \text{q}(c_j)]^2 + \text{M}[V(\mathbf{b}, \mathbf{c})] \qquad (2.57)$$

If the independent input variables are uniformly distributed, then this may be simplified to

$$\text{MO}[c_n] = \min_{\mathbf{c}:c_n} \sum_{j \neq n} [z_j - \text{q}(c_j)]^2 + \text{M}[V(\mathbf{c})] \qquad (2.58)$$

where $V(\mathbf{c})$ indicates a zero parity check.

─────────────────────────────────── *End Example*

The key to using validity checks efficiently is to obtaining an efficient structure for $V(\mathbf{a}, \mathbf{x})$ (*i.e.*, a factorization). In other words, one can find a graphical representation of the validity check and use message passing on this graph to perform the soft inversion process in (2.54). If the graphical model is cycle free, then this provides the exact soft inverse (*i.e.*, optimal detection) and if it has loops, then message passing is a suboptimal method of approximating the soft inverse (*i.e.*, suboptimal detection). A simple example of the former is illustrated in the next example.

Example 2.17. —————————————————————————————
Consider the *single parity check (block) code* having $K = N - 1$ and $\mathbf{P} = \mathbf{p}^T = [\, 1\ 1\ \cdots\ 1\,]$ which simply enforces even parity. The single parity check is $\mathbf{p}^T \mathbf{c} = \sum_{i=0}^{N} p_i c_i$, where the arithmetic is modulo two. This check has the trellis representation as illustrated in Fig-2.61. Specifically,

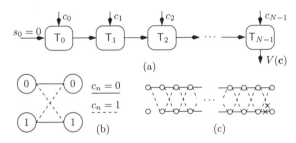

(a)

$c_n = 0$

$c_n = 1$

(b)

(c)

Figure 2.61. A trellis representation of a single parity check: (a) the validity check as the output of the final transition, (b) the individual transition structure, and (c) the parity check implemented by removing transitions.

consider the state to be $s_n = \sum_{i=0}^{n-1} p_i c_i$ and each transition has metric $\mathrm{MI}[c_n]$. Starting with $s_0 = 0$, each state represents a partial parity check sum. On the final transition, the validity check output $V(\mathbf{c}) = V(s_{N-1}, c_{N-1})$ is produced. Note that the only effect of the validity check is to disallow all transitions into the state $s_N = 1$. Thus, we can delete those transitions by forcing termination of the trellis into the zero state (*i.e.*, $s_N = 0$) and implicitly implement the validity check.

Soft inversion or decoding of this simple code can then be done by running the forward-backward algorithm on this two-state trellis. Specifically, in the min-sum semi-ring, when the input bits b_k are uniformly distributed on $\{0, 1\}$, running the forward-backward algorithm on the pruned trellis of Fig-2.61(c) with $\mathrm{MI}[c_n] = (z_n - \mathsf{q}(c_n))^2$ performs the computation in (2.58). Notice that if $\mathrm{MO}[b_k]$ is sought, the minimization must be done only over all $\mathbf{c}$ consistent with $b_k(\zeta) = b_k$. This may or may not be necessary depending on whether the goal is soft inversion or decoding of the isolated single-parity check. In the former, soft-out information on b_k is typically not needed. In the latter, this may be simply implemented in the case of systematic codes by removing all transitions t_n in the trellis of Fig-2.61 inconsistent with $c_n = b_n$.

———————————————————————————————————— *End Example*

The previous example illustrates that the FSM trellis may be viewed as a validity check for the FSM structure. In fact, we have used this interpretation implicitly throughout Chapters 1 and 2 by writing the marginalization as $\mathbf{t}_0^{K-1}$: u_k where it is assumed that the metric of any invalid transition in the trellis is infinite. This may also be viewed as breaking a global configuration check (*e.g.*, $V(\mathbf{t})$) into a set of local validity checks (*e.g.*, $\{V_k(t_k)\}$). In general, the single validity check $V(\mathbf{a}, \mathbf{x})$ in (2.54c) may be replaced by a set of validity checks $\{V_i(\mathbf{a}, \mathbf{x})\}_i$. Typically, $V_i(\cdot)$ checks only the local validity of configuration of the system and therefore depends on $\mathbf{a}$ and $\mathbf{x}$ only locally. Using the conventional

$$S[V(\mathbf{a}, \mathbf{x})] = \bigodot_i S[V_i(\mathbf{a}, \mathbf{x})] \qquad (2.59)$$

a single, global validity check can be decomposed into a set of local checks without affecting the computation in (2.54). More specifically, each local check has veto power; if one of these local checks fail, then the global check fails. An example of this is illustrated next.

Example 2.18. ————————————————————————————————————
The parity check of a block code can be represented as $N - K$ single parity checks. Thus, a given block code can be decomposed into $N - K$ single-parity check codes. This concept is illustrated in Fig-2.62(a). The associated graph has cycles since some subsets $\{c_n\}$ will be involved in more than one parity check. For iid, uniformly distributed input bits b_k, the message passing algorithm is then run as shown in Fig-2.62(b). More specifically, this algorithm does not accept or produce soft information on b_k. The soft inverse of each validity check does not produce soft-out information on the V_n and the soft-in for all validity check variables is always the same. In the following, we may drop the SI/SO ports associated with the validity check variables in the diagrams for compactness (*i.e.*, they are implicit). After running the suboptimal iterative parity check decoder, the soft-out information on c_n can be thresholded to determine the MAP estimates $\hat{c}_n$ (*i.e.*, MAP codeword for min-sum and MAP coded symbol for min*-sum). Since the mapping from $\mathbf{b}$ to $\mathbf{c}$ is one to one, this can be converted to a decision on $\{b_k\}$ which is trivial if the code is in systematic form.

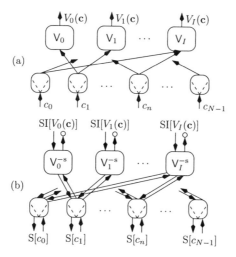

Figure 2.62. (a) A loopy graphical model for the parity check of a (N, K) block code and (b) a suboptimal iterative decoder for uniform code input bits. Note that $I = N - K - 1$ is used for compactness in the figure.

Low density parity check codes (LDPCs) were originally suggested by Gallager [Ga62, Ga63] along with this suboptimal iterative decoding technique. In an LDPC, each parity check involves only a small number of coded symbols. Thus, soft inversion of each subcode is relatively simple using the technique described in Example 2.17. Specifically, an LDPC is characterized by the number of 1's in each column and row of the parity check matrix. For example, a $w_c = 3$, $w_r = 6$ LDPC is one with 3 ones in each column of $\mathbf{P}$ and 6 ones in each row. Thus, each parity check is involves 6 codeword bits and each c_k is involved in 3 parity checks. The location of these 1s may be selected at random to design an LDPC, but a good code structure will avoid short cycles in the graph of Fig-2.62. Gallager's LDPCs were rediscovered[17] after the introduction of PCCCs [Ma99]. The performance of LDPCs is similar to that of turbo codes [MaNe96]. Recently, *irregular* LDPC constructions having a variable number of ones per row/column (*e.g.*, $w_c \in \{3, 5\}$ and $w_r \in \{6, 8\}$) have been demonstrated to perform better than the original Gallager constructions and PCCCs of comparable block length and rate [RiShUr00]. Notice that it is natural to implement the parity checks in Fig-2.62(b) in parallel. Also, decoding of an LDPC is typically com-

---

[17]Interestingly, Gallager did not simulate the iterative decoding algorithm in 1963 because the computational resources were unavailable. Had Gallager's iterative decoding algorithm been more widely appreciated, much of the material in this book may have been developed much earlier!

putationally less intensive than the decoding of the comparable PCCC. Thus, LDPCs offer attractive features for hardware implementation.

———————————————————————————————————— *End Example*

The use of validity checks can be viewed as introducing hidden or auxiliary variables (*i.e.*, $\{V_i\}$) which are viewed as system outputs in our graphical convention. One can also introduce auxiliary input variables as well. For example, suppose that the validity check in (2.54) can be written as $V(\mathbf{a}, \mathbf{x}) = \prod_j V_j(u_j)$ where $\mathbf{u}$ is a set of auxiliary variables that also determines $\mathbf{a}$ and, thus, $\mathbf{x}$. In this case the soft inversion can be performed based on this model with $a_m(\mathbf{u})$, $x_n(\mathbf{u})$, and $V(\mathbf{u})$ viewed as outputs and u_j viewed as iid, uniformly distributed (*i.e.*, zero metric) input variables. This is illustrated in the final example of this section.

Example 2.19. ————————————————————————————————
As mentioned previously, the standard trellis associated with an FSM may be viewed a set of valid configuration tests. This is inherently sequential, however, since the global validity check is factored into a sequence of validity transition checks (*i.e.*, see Fig-2.53). Specifically, if the auxiliary variables s_k are introduced, then the k-th validity check is $V_k(a_k, s_k, s_{k+1}, x_k)$, which checks to ensure that the local transition configuration is allowable under the FSM structure.

An equivalent tree-structured set of validity checks is shown in Fig-2.63. Note that, in this case the auxilary inputs are the transitions t_k with $a_k(t_k)$ and $x_k(t_k)$ considered outputs. Similarly, the states $s_k(t_k)$ and $s_{k+1}(t_k)$ may be viewed as outputs of teh subsystems at the first level. The check variables $V_k(s_k)$ are outputs of the corresponding systems in Fig-2.63, but are not shown for compactness. The validity check systems also have an additional output which is the input to another check. Each check confirms that the state value obtained from the subgraph below is consistent. For example, $V_7(s_7)$ checks to make sure that the two values of s_7 obtained from below (*i.e.*, $s_7(t_7)$ and $s_7(t_6)$) are the same). For the example in Fig-2.63, it is clear that the set of checks $\{V_i\}_{i=1}^7$ admits all valid paths in the FSM and no others. Since this graph has no cycles, message passing will provide the optimal processing.

The details of message passing using the generalized forward-backward schedule are shown in Figure 2.64 and 2.65 in min-sum processing. The inward messages are shown in Fig-2.64. Specifically, initially $\mathrm{MI}[t_k]$ is set to uniform and the algorithm begins by activating the first level of subsystems to compute $\mathrm{M}_k[t_k]$ from $\mathrm{MI}[x(t_k)]$ and $\mathrm{MI}[a(t_k)]$. The messages passed inward to the next level of the tree are $\mathrm{MSM}_k[s_k, s_{k+1}]$ which is simply $\mathrm{M}_k[t_k]$ if there are no parallel transitions. This inward

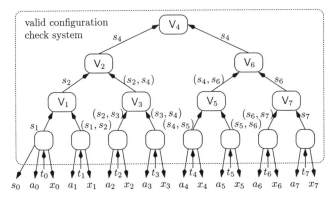

Figure 2.63. A binary tree-structured configuration check for a sequence of FSM transitions.

message passing continues with the messages shown. When the two messages on s_4 reach V_4^{-s}, the outward propagation begins and proceeds downward as shown in Fig-2.65. Again, all nodes at a given level of the tree are activated before activating any of the nodes at the next level. At the bottom level, the input metric of (s_k, s_{k+1}) is $\text{MSM}_{\{k\}^c}[s_k, s_{k+1}]$ – *i.e.*, the sum of the forward and backward state metrics in the standard forward-backward algorithm. Thus, the final activation of the nodes on the bottom level produces the desired extrinsic output metrics. Note that if an invalid set of transitions were applied to the system of Fig-2.63, one of the check variables would be zero. Thus, the corresponding tree-SISO is outputting the soft information based only on marginalization and combining over valid transition sequences.

This tree structured SISO was suggested in [ThCh00] and referred to as a *forward-backward Tree-SISO (FBT-SISO)* to distinguish it from the tree-SISO of Section 2.5.5 and [BeCh00]. This FBT-SISO has twice the latency of the tree-SISO in Section 2.5.5 because the messages must propagate both inward and outward. This modest increase in latency is accompanied by a significant reduction in computational complexity. Specifically, the FBT-Tree SISO has $\mathcal{O}(K)$ computational complexity and $\mathcal{O}(\log_2 K)$ latency. This is to be compared to $\mathcal{O}(K \log_2 K)$ computational complexity and $\mathcal{O}(\log_2 K)$ latency for the Tree-SISO of Section 2.5.5 and $\mathcal{O}(K)$ computational complexity and $\mathcal{O}(K)$ latency for message passing on the standard trellis (*i.e.*, Fig-2.53).

—————————————————————————— *End Example*

We note that validity checks were introduced only in the context of performing the soft inversion. This is different than the direct input/output explicit block diagrams used before this section. Specifi-

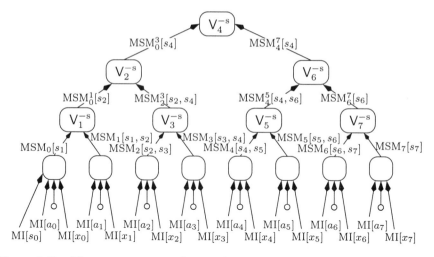

Figure 2.64. The messages passed inward using a generalized forward-backward schedule on the tree structure in Fig-2.63.

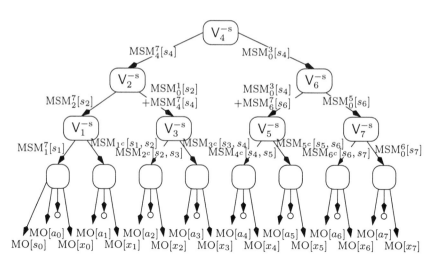

Figure 2.65. The messages passed outward using a generalized forward-backward schedule on the tree structure in Fig-2.63 with the active forward messages from Fig-2.64. The notation 2^c, for example, is shorthand for $\{2\}^c$ – *i.e.*, all indices except 2.

cally, if one were to actually build an FSM using the model in Fig-2.63 it would imply that potential transition sequences would be applied until the actual input sequence **a** was observed correctly, then the corresponding **x** could be read out. For most systems of interest this is inefficient as compared to directly building the system based on the input/output explicit block diagram. Both of these are so-called *realizations* of the system[18] since they characterize the entire system structure. However, we use the term *constructive realization* to refer to graphs that could be directly translated to a signal generation circuit (*e.g.*, an encoder). Thus, constructive realizations are those with no "leaf" variables.[19]

2.6.4 Other Graphical Models

Explicit index block diagrams are quite natural graphical models for the applications considered in the remainder of this book. They are not, however, the standard graphical tools used in most of the literature. Our reasons for using these graphs are primarily pedagogical. Specifically, they allow one to describe both implicit index and graphical approaches as one in the same. They also allowed us to present message passing on graphical models from a detection-theory point of view and without presenting a large body of material from graph theory. In this section we briefly summarize several popular graphical models and point the reader to the relevant references. The objective of each of these approaches is to solve a general combine and marginalize (CM) problem of the form in (1.37), (2.5), possibly by introducing hidden variables (*e.g.*, states) to factor the joint soft information.

2.6.4.1 Belief Networks and Pearl's Belief Propagation Algorithm

Belief networks are directed graphs without directed cycles[20] that describe the probabilistic dependencies of a set of variables. Given a set of variables, $\{u_i(\zeta)\}$, a Belief Network is defined by a factorization of the joint probability $p(\mathbf{u})$ via Bayes law. Specifically, the joint probability is written in the form $\prod_i p(u_i | \mathcal{P}(u_i))$, where $\mathcal{P}(u_i)$ is a set of *parents* for variable u_i. Conversely, u_i is referred to as a child of all $v \in \mathcal{P}(u_i)$.

The common notation for a belief network is that shown in Fig-2.66(a) for the equivalent FSM graph in Fig-2.53. In a belief network, all "outputs" of a node labeled x are the variable x – *i.e.*, in our notation, the subsystem outputs and the subsystem configuration are one in the same. Given a system diagram in the explicit index convention, the global system inputs are nodes without parents and the global system outputs are nodes with no children. For any of the systems previously modeled without validity checks, a belief network can be constructed by starting the

---

[18]We use the terms system "representation," "model" and "realization" interchangeably. The term realization is engrained in the literature.
[19]By this we mean variables that are either not the input to any system or are not the output of any system (*i.e.*, even a graph with cycles).
[20]These are called Directed Acyclic Graphs in the literature.

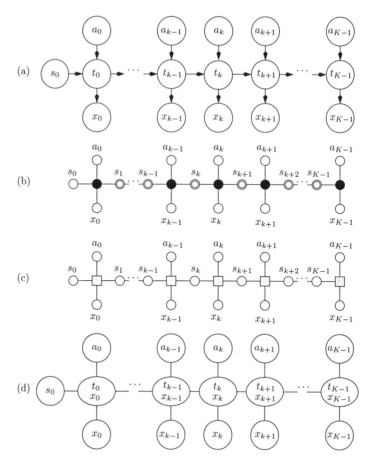

Figure 2.66. Common graphical models used to represent the system structure of the FSM illustrated in Fig-2.53: (a) belief network, (b) Tanner-Wiberg graph, (c) factor graph, and (d) junction tree .

factorization with $p(\mathbf{z}|\mathbf{a})p(\mathbf{a})$ and introducing hidden variables to characterize the subsystem configurations as necessary (*e.g.*, transitions).

The associated message passing algorithm on a belief network is referred to as Pearl's Belief Propagation algorithm (BPA).[21] It is well known that Pearl's algorithm produces the APP of all variables u_i in a cycle-free network by message passing (*i.e.*, see [Pe88] for a rigorous proof along the lines of the argument in Section 2.6.1). One distinction between the previous development and Pearl's algorithm is that, in the latter, messages are passed on the node variables themselves. This

---

[21] Pearl's algorithm usually refers to the sum-product version. Pearl referred to the min-sum version of the algorithm as "belief revision" [Pe86].

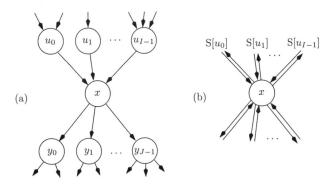

Figure 2.67.   (a) A general node in a belief network and (b) the corresponding message updating node in a belief propagation algorithm.

may be inefficient if the node variables have some overlap. For example, in Fig-2.66(a) Pearl's algorithm would pass messages on the state transitions t_k instead of the state variables themselves which, from the previous development, is clearly inefficient.[22] If, however, the belief network of Fig-2.66(a) is modified using the states s_k and s_{k+1} in place of the transition t_k, an undirected cycle will be introduced and would void the optimality of Pearl's original algorithm. Thus, in general, is possible that the messages passed in Pearl's algorithm can be simplified to account for overlap between variable conditional values.

Finally, we note that, except for this minor difference, Pearl's algorithm is the same as the message passing described above (*i.e.*, iterative detection). Specifically, each node in the belief network is replaced by the equivalent of the soft inverse node and messages are passed between these soft inverse nodes. The notation used for the soft inverse node is given in Fig-2.67.[23] Two final caveats should be pointed out. First, the soft-inverse of a variable corresponding to a global system input/output is a source and sink of soft information. Second, due to the convention of a single variable node producing several copies of the variable, the activation of a node in Pearl's BPA includes the equivalent of a soft broadcaster. This is illustrated in Fig-2.68 where the information in Fig-2.67 is expressed in the explicit index block diagram notation.

The correspondence between Pearl's algorithm and the original turbo decoding algorithm was pointed out in [McMaCh98, KsFr98].

---

[22]This is similar to the early description of the forward-backward algorithm described by Chang and Hancock [ChHa66].
[23]Although, as has become clear in our notation as well, stating the belief network is sufficient to determine the BPA except for scheduling.

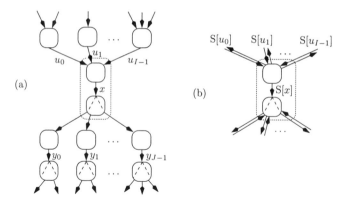

Figure 2.68. (a) The explicit index block diagram graphical model corresponding to the belief network node in Fig-2.67(a) and, (b) the associated soft inverse processing producing the messages of the BPA node in Fig-2.67(b).

2.6.4.2   Factor Graphs and Tanner-Wiberg Graphs

Factor graphs and Tanner-Wiberg graphs are closely related undirected graphs used to describe systems and/or the associated CM problem. Tanner introduced a graphical notation for describing error correction codes [Ta81] which was extended by Wiberg to include the representation of hidden variables (*i.e.*, state) [Wi96]. *Tanner-Wiberg graphs* are bi-partite graphs with variable nodes and check nodes. Variable nodes can only be connected to check nodes and vice-versa. Hidden variable nodes are sometimes denoted slightly differently. For example, the Tanner-Wiberg graph corresponding to the graph in Fig-2.53 is shown in Fig-2.66(b) with Wiberg's notation for the hidden variable nodes. In the Tanner-Wiberg graphs, the check nodes enforce valid configurations for all of the connected variables. In Fig-2.66(b), for example, the check node corresponding to transition t_k verifies that s_k, s_{k+1}, a_k and x_k are all consistent with a valid transition. Thus, establishing the Tanner-Wiberg graph Fig-2.66(b) requires factorization of the validity check for the transition sequence, into a set of checks for each individual transition.

Factor graphs are similar to Tanner-Wiberg graphs with the check nodes replaced by factor nodes [KsFr98]. This is shown in Fig-2.66(c) for the FSM example. These factor nodes represent the associated factor in the product function (or total metric) to be marginalized. In Fig-2.66(c), for example, the factor node corresponding to transition t_k represents the contribution to $p(\mathbf{z}, \mathbf{a})$ from those variables associated with t_k. Namely in the sum-product version this is $p(z_k|x_k)p(a_k)p(s_{k+1}|s_k)$ where s_k, s_{k+1}, a_k, and x_k are all consistent with the same transition.

The message passing algorithms for Tanner-Wiberg graphs and factor graphs are the same as that described in the context of explicit index block diagrams. Specifically each check (factor) node in the Tanner-Wiberg (factor) graph is replaced by the analogous soft-inverse node which passes soft-in and soft-out messages on all variables involved in that check (factor) node. The associated processing at the soft-inverse is identical to that described in the context of explicit index block diagrams. Again, variable nodes involved in several checks or factors have implicit soft broadcasters in the soft inverse processors and the soft inverses for nodes corresponding to the global system inputs/outputs are sources and sinks of soft-in/soft-out information. Wiberg identified the optimality of message passing on cycle-free graphs and interpreted the method of turbo decoding as the same on loopy graphs [Wi96]. Similar results for factor graphs have been reported with generalization to arbitrary semi-ring marginalization and combining operators [WiLoKo95, KsFr98, KsFrLo00].

Note that, in contrast to our convention and, for the most part, belief networks, factor graphs and Tanner-Wiberg graphs are not constructive in nature. In other words, they generally do not represent a constructive realization and were developed only based on a factorization of the joint soft information to be marginalized. As a result, these graphs are undirected and handle validity checks more naturally than the implicit index block diagram convention.

2.6.4.3   Junction Trees

Like factor graphs and Tanner-Wiberg graphs, junction trees [JeJe94, Je96, AjMc00] are undirected, and are not necessarily based on constructive realizations of the system. Junction trees are, however, cycle-free by definition (*i.e.*, they are trees not graphs). In a junction tree, each node corresponds to a collection of variables and all nodes that are involved in a common factor in the CM problem are connected. The notation used is illustrated via the FSM example in Fig-2.66(d), where t_k is, as usual, shorthand for (s_k, a_k, s_{k+1}). The message passing algorithm is, again, identical to that already described with each node replaced by the associated soft-inverse. Once again, the global system input/output nodes are sources and sinks with a soft broadcaster operation as appropriate. Messages are passed on the intersection of the node variables (*e.g.*, the states in Fig-2.66(d)).

Since junction trees are cycle-free, message passing is optimal. This has been shown in [AjMc00] where arbitrary semi-ring marginalizing and combining operators have been considered (*i.e.*, the generalized distributive law) and the general CM (equivalently the MPF) problem has shown to be solved by message passing. Again, this is a rigorous proof of the argument in Section 2.6.1.

As described above, junction trees are similar to factor graphs or Tanner-Wiberg graphs without cycles. However, junction trees are associated with a graphical factorization procedure as described in [AjMc00, Aj99]. This procedure allows one to start with a joint soft information measure (*e.g.*, $p(\mathbf{z}|\mathbf{x}(\mathbf{a})p(\mathbf{a}))$) and obtain a (hopefully efficient) factorization of this function. For a simple FSM, starting from a graph like that in Fig-2.58, one can obtain the junction tree in Fig-2.66(d) through a graphical procedure. Thus, the primary advantage of junction trees is that they are part of a larger framework for establishing valid factorizations or, equivalently, trees for which message passing is an efficient solution. Once the valid factorization has been established, the graphical model can be described in another convention. For example, the tree in Fig-2.63 follows directly from the junction tree approach [ThCh00]. Although it is clear that the set of validity checks introduced in Example 2.19 fully characterize the FSM and are non-redundant (*i.e.*, no single check may be removed without admitting invalid configuration), following the junction tree approach always ensures that this is the case.

Some qualifications should be stated with regard to the above procedure for graphical factorization. First, the junction tree for a given system is not unique (*e.g.*, the junction tree associated with Fig-2.63 and that in Fig-2.66(d) represent the same system). There is no known method for producing some form of "optimal" junction tree. For example, while the procedure described in [AjMc00, Aj99] can transform Fig-2.58 to Fig-2.66(d) for a non-recursive FSM, it is less clear how the procedure could be applied to obtain Fig-2.66(d) for a recursive FSM (*e.g.*, the encoder of Fig-2.24) without prior knowledge of the appropriate state variables. Second, graphical models with cycles are not addressed under the domain of junction trees and the associated graphical factorization procedure.[24] Armed with the knowledge that message passing is effective, if suboptimal, for loopy graphs, often times we are interested in efficient graphical models with cycles. In general, efficient realizations of a system is a well studied problem in computer engineering and computer science (*e.g.*, [Ha65, Gi62, CeMaSa79, Mo82, Fo00]) with no simple procedure solving all cases.

---

[24]The notion of a junction graph (*i.e.*, a junction-tree-like graph with cycles) has been introduced in [Aj99] in an effort to analyze message passing on single-cycle graphs.

2.7   On the Non-uniqueness of an Iterative Detector

For a given system, there may be several representations, each of which implies a different iterative detector. In fact, for systems with loopy graphical models, there are inherently many equivalent system models, each of which represents the system. Furthermore, because message passing is generally suboptimal for such graphs, one can modify the types of messages passed. This may result in a complexity reduction, performance improvement, or change the convergence characteristics of the iterative detector. Because of the suboptimal nature of iterative detection, the selection of a good system model and modification of the message passing rules remains somewhat of an art. Chapters 3 and 5 focus on this material. We conclude this chapter with a few examples that highlight this non-uniqueness.

Example 2.20. ――――――――――――――――――――――――――――

Consider the simple (7,4) Hamming linear block code [LiCo83] with iid, uniform inputs b_k and outputs c_n. Three message passing algorithms are considered for this code. The loopy graphical models for the parity check and generator matrices are shown in Fig-2.69. The iterative detectors implied by these models have different complexities. For example, a min-sum, parity check based decoder will have three 4-way ACS-type operations and four soft broadcasters. The corresponding generator matrix based iterative detector will have three 2-way ACS units, one 3-way ACS unit and similar soft broadcasters. The corresponding decoders for larger block codes may differ more dramatically in complexity. The number of inputs to either of these is also an important complexity consideration (*e.g.*, LDPCs have manageable complexity despite their large size because there are only a small number of codeword bits involved in each component check).

A third graphical model is shown in Fig-2.70. This is a cycle-free trellis representation of the code. This is a generalization of the graph found in Fig-2.61 and can be obtained using the procedure in [BaCoJeRa74]. The forward-backward algorithm may be run on this graph to obtain the optimal decoding.

The performance of these three decoders is shown in Fig-2.71 for min-sum decoding. Both of the iterative detectors perform near optimally with the generator version obtaining slightly better performance. This improvement is due to the fact that the generator-based decoder continues to gain up to the eighth iteration whereas convergence was observed after five iterations of the iterative parity check decoder. This may have

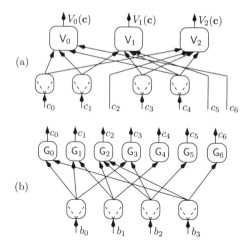

Figure 2.69.   Loopy graph representations of the (7,4) Hamming code: (a) the parity check structure and (b) the generator matrix graph.

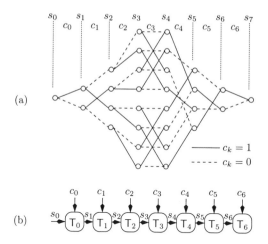

Figure 2.70.   A cycle-free graphical model for the (7,4) Hamming code.

been predicted since the parity check matrix has several cycles of length four while the minimum cycle length of the generator graph is six. In general, iterative detection algorithms converge quickly on graphs with short cycles. A method of slowing the convergence for graphs with short cycles is introduced in Chapter 3 (*i.e.*, message filtering) which can provide significant performance improvements. Note that, as in Example 2.15, the suboptimal iterative processors are actually more complex than the optimal algorithm. For larger codes with sparse structure, however, an iterative decoder can be significantly less complex (*e.g.*, LDPCs).

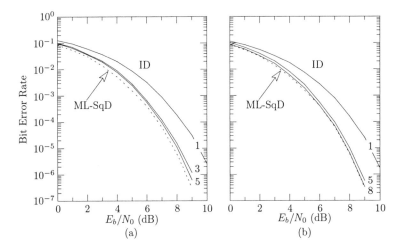

Figure 2.71. Performance of the iterative decoders implied by the (a) parity check graph and (b) generator graph from Fig-2.69.

Finally, for block codes, an iterative decoder can be based on a graph representing the structure of the dual code [HaOfPa96, HaRu76, Ri98, Fo00]. Thus, our list of decoders in this example is still incomplete.

———————————————————————————— *End Example*

Next we consider different decompositions of the block diagram of a given system.

Example 2.21. ————————————————————————————
Consider the convolutionally encoded, ISI channel in Example 2.9. The natural decomposition of this system is into the code and ISI subsystems. For this decomposition, it is natural to pass messages on the 4-ary QPSK symbols a_k that connect the two subsystems. In this example, we show that both the decomposition and the messages are not unique.

First, we consider the standard decomposition, but alter the messages. Messages are passed on subsets of transitions in the super-trellis that defines the joint trellis of the code and ISI channel. This 64-state super-trellis has states $s_k = \mathbf{b}_{k-6}^{k-1}$. Since the code trellis has states $s_k = \mathbf{b}_{k-4}^{k-1}$, each state in the code trellis corresponds to four states in the super-trellis. This is illustrated in Fig-2.72. The 128 transitions in the super-trellis can then be partitioned into 16 sets $\{\mathcal{D}_i(j)\}_{i,j=0}^{3}$, each containing 8 transitions. The individual ISI and code trellises can then be labeled by these 16 subsets. These subsets may be thought of as 16-ary hidden symbols that play the same role as the 4-ary (hidden) QPSK symbols in Example 2.9. The labeling of the code and ISI trellises by the subsets

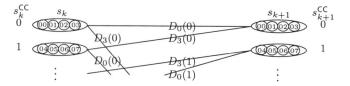

Figure 2.72. The mapping from the super-trellis to the 16-ary hidden symbols as outputs of the convolutional code.

is summarized in Fig-2.72 and Table 2.4. Specifically, the ISI trellis state is defined by $s_k^{(ISI)} = (a_{k-1}, a_{k-2})$ and indexed according to the standard 4-ary expansion $4a_{k-1} + a_{k-2}$, where a_k is mapped to the QPSK constellation via the Gray labeling convention (*i.e.*, $0(00) \leftrightarrow \theta = 0$, $1(01) \leftrightarrow \theta = \pi/2$, $3(11) \leftrightarrow \theta = \pi$, $2(10) \leftrightarrow \theta = 3\pi/2$). Under this convention, $\{D_a(j)\}_{j=0}^3$ all correspond to the same QPSK symbol value $a_k = a$.

$D_a(j)$	$(s_k^{(CC)}, s_{k+1}^{(CC)})$		$(s_k^{(ISI)}, s_{k+1}^{(ISI)})$	
$D_0(0)$	(0, 0)	(1, 8)	(0, 0)	(3, 0)	(12, 3)	(15, 3)
$D_0(1)$	(3, 1)	(2, 9)	(1, 0)	(2, 0)	(13, 3)	(14, 3)
$D_0(2)$	(4, 2)	(5, 10)	(4, 1)	(7, 1)	(8, 2)	(11, 2)
$D_0(3)$	(7, 3)	(6, 11)	(5, 1)	(6, 1)	(9, 2)	(10, 2)
$D_1(0)$	(8, 4)	(9, 12)	(0, 4)	(3, 4)	(12, 7)	(15, 7)
$D_1(1)$	(11, 5)	(10, 13)	(1, 4)	(2, 4)	(13, 7)	(14, 7)
$D_1(2)$	(12, 6)	(13, 14)	(4, 5)	(7, 5)	(8, 6)	(11, 6)
$D_1(3)$	(15, 7)	(14, 15)	(5, 5)	(6, 5)	(9, 6)	(10, 6)
$D_2(0)$	(9, 4)	(8, 12)	(0, 8)	(3, 8)	(12, 11)	(15, 11)
$D_2(1)$	(10, 5)	(11, 13)	(1, 8)	(2, 8)	(13, 11)	(14, 11)
$D_2(2)$	(13, 6)	(12, 14)	(4, 9)	(7, 9)	(8, 10)	(11, 10)
$D_2(3)$	(14, 7)	(15, 15)	(5, 9)	(6, 9)	(9, 10)	(10, 10)
$D_3(0)$	(1, 0)	(0, 8)	(0, 12)	(3, 12)	(12, 15)	(15, 15)
$D_3(1)$	(2, 1)	(3, 9)	(1, 12)	(2, 12)	(13, 15)	(14, 15)
$D_3(2)$	(5, 2)	(4, 10)	(4, 13)	(7, 13)	(8, 14)	(11, 14)
$D_3(3)$	(6, 3)	(7, 11)	(5, 13)	(6, 13)	(9, 14)	(10, 14)

Table 2.4. Definition of the 16-ary hidden symbol, D, from Example 2.21

Iterative detection for the same system partitioning, can be done in the standard way using the 16-ary hidden symbols in place of the 4-ary QPSK symbols. The simulated performance of this iterative detector is shown in Fig-2.73(a) with the curves from Example 2.9 shown for reference (all curves use min-sum processing). Using the 16-ary messages improves the performance on the first iteration, but the iteration gain is

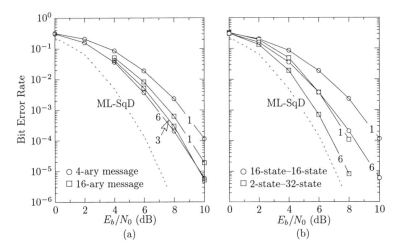

Figure 2.73. The performance of modified iterative detectors for the convolutional code, ISI channel example from Example 2.21. (a) 16-ary message passing on the standard (16 state, 16 state) trellises and (b) a 2-state, 32-state decomposition.

reduced. The computational complexity of these two detectors is comparable since the number of states in the forward-backward processors remains fixed, but the 16-ary version requires more storage.

One can also modify the actual decomposition. For example, the state of the super-trellis may be split in any way desired. Specifically, we can define the state of the outer system as $\mathbf{b}_{k-L_o}^{k-1}$ and the inner system as $\mathbf{b}_{k-6}^{k-L_o-1}$. The outputs of the outer system are then the transitions $\mathbf{b}_{k-L_o}^{k}$. This yields a 2^{L_o}-state, 2^{6-L_o}-state decomposition. A standard iterative detector can then be run using the forward-backward SISO for each subsystem that exchanges soft information on the 2^{L_o+1}-ary hidden symbols. The performance of this iterative detector is shown in Fig-2.73(b) for $L_o = 1$. Thus, the hidden symbols are 4-ary and the outer and inner trellises have 2 and 32 states, respectively. The performance of the 2-32 decomposition is significantly better than the standard 16-16 decomposition and approaches that of the optimal detector. The complexity is roughly characterized by the sum of the number of states in the inner and outer trellis, so that the two iterative detectors have roughly the same complexity.

———— End Example

Next we consider a single system described by several equivalent block diagrams.

Example 2.22. ───────────────────────────────

There is an inherent asymmetry in the system model and decoder associated with the PCCC in Section 2.4.3.1. This is because the systematic information is used for encoder 1, but not for encoder 2. This apparent asymmetry may be removed by considering the systematic bit as the output of the broadcaster and the two encoders to produce parity bits only. This view leads to the same decoder considered in Section 2.4.3.1 (see Problem 2.18). It is tempting, however, to try to use the channel information on the systematic bit to decode both constituent codes. In this example, we give system models that imply different methods for utilizing the channel information on the systematic bit and compare the performance of the associated iterative decoders.

Specifically, consider the rate 1/4 binary code in Fig-2.74. This is

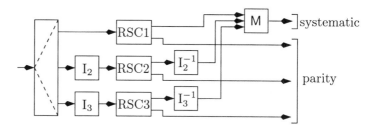

Figure 2.74.   The encoder for the rate 1/4 PCCC example.

a PCCC with three constituent codes, taken to be the same as that used in the PCCC of Section 2.4.3.1, with all of the parity information sent. With the addition of the deinterleavers after each encoder, the systematic bit can be viewed as the output of any of the encoders. This is summarized by the system M that maps the three versions of the systematic information onto a single systematic bit. Obviously, this mapping is trivial because all of the inputs are equal. The iterative decoding algorithm, however, changes with the definition of this mapper.

We consider three different cases: (i) a "pass to 1", (ii) a "funnel", and (iii) a switch. Let the systematic and parity bits of the i-th encoder be denoted by $c_k^{(i)}(0)$ and $c_k^{(i)}(1)$, respectively. Also, denote the output of M by $x_k(0)$. Then the mapper is defined by the function $x_k(0) = m_k(c_k^{(1)}(0), c_k^{(2)}(0), c_k^{(3)}(0))$. For case (i), $x_k(0) = c_k^{(1)}(0)$. For the funnel, the mapper checks that each are equal and passes their common value – i.e., $x_k(0) = c_k^{(1)}(0) = c_k^{(2)}(0) = c_k^{(3)}(0)$ and is undefined if they are unequal. Finally, the switch just cycles through the different constituent codes – i.e., $x_0(0) = c_0^{(1)}(0)$, $x_1(0) = c_1^{(2)}(0)$, $x_2(0) = c_2^{(3)}(0)$, $x_3(0) = c_3^{(1)}(0)$, etc. The soft inverse of (i) and (iii) are trivial since they just invert the distribution of that mapper. As a result, in the corresponding iterative detector, M^{-s} needs only to be activated once in cases (i) and (iii) since further activation does nothing. The funnel is akin to a broadcaster run in reverse, and the soft inverse of the funnel is equivalent to a soft broadcaster (see Problem 2.27).

Simulation results for the three systems are shown in Fig-2.75. Unless

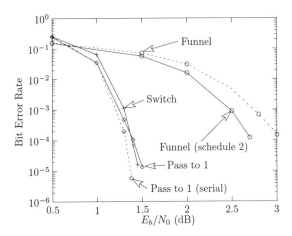

Figure 2.75. The performance of the min-sum iterative decoders associated with the encoder of Fig-2.74.

indicated otherwise, the activation schedule used is: $M^{-s} \to$ (SISO1, SISO2, SISO3) in parallel $\to$ soft broadcaster (SOBC) $\to$ (SISO1, SISO2, SISO3), with the interleavers/deinterleavers activated as required. This defines one iteration. For the "pass to 1" case, we also consider a serial activation of the SISOs: $M^{-s} \to$ SISO1 $\to$ SOBC $\to$ SISO2 $\to$ SOBC $\to$ SISO3 $\to$ SOBC. This serial schedule is equivalent to viewing the systematic bit as the direct output of the broadcaster with all RSCs producing only parity. Only the funnel-based decoder shows a significant performance difference, showing a loss of nearly 2 dB. Some improvement is gained by modifying the activation schedule. Specifically, the curve labeled with schedule2 was generated by activating the SOBC and soft inverse funnel in parallel after each set of SISO activations.

—— *End Example*

2.7.1 Additional Design Guidelines

In addition to the conventions described in Section 2.3, we describe several rules of thumb based on the above examples and experience gained from other applications:

1. Avoid system models with short cycles whenever possible.

2. Avoid marginalizing information when it is not necessary.

3. Exploiting the channel information "strongly" in one subsystem is preferred to exploiting the information weakly in several subsystems.

4. Avoid over characterization of the system which tends to add cycles without describing any further structure.

Several examples illustrating the first point have been given. As an example of the second rule, consider a SISO for a simple FSM that produces the soft-out on the inputs a_k using the standard forward-backward algorithm, but produces the soft-out on the FSM output in a modified fashion. Specifically, the soft-out for x_k is obtained by combining the associated marginal soft information on a_k. For example, $P[x_k(t_k)] = \prod_{m=0}^{L-1} P[a_{k-m}(t_k)]$ is one way to construct intrinsic information on x_k from intrinsic information on the consistent $\mathbf{a}_{k-L}^k$. While this may work, it is generally preferable to obtain the soft-out information for x_k directly from t_k.

The last two rules were illustrated in Example 2.22. Specifically, the "pass to 1" model is preferred over the "switch" (illustrating rule 3) and the funnel model yields poor results (illustrating the last rule). More precisely, the structure added by the funnel has already been accounted for by the soft broadcaster, so its addition adds only unnecessary loops into the associated decoder. Similarly, the 2-32 decomposition of Example 2.21 illustrates rule 3.

It appears that it is most important to avoid short and/or unnecessary cycles. For example, according to rule 2, the 16-ary message passing in Example 2.21 should be preferred over the 4-ary message passing. However, because there is no interleaver, this graph has short cycles. As a result, the iterative decoder tends to converge quickly to a local minimum. Using the 16-ary messages increases the convergence further which is undesirable. This is not predicted by the above set of rules because the system model has short cycles. In other words, the rules above are heuristics and they should be applied disjointly (*e.g.*, less marginalization is better when there are no short cycles).

2.8   Summary and Open Problems

Based on the notion of a marginal soft inverse of a system, which updates marginal likelihood information on both the system input and output, iterative or "turbo" detection was developed in a systematic manner. Several important examples of iterative detection were demonstrated using only these few conventions. In general, it is observed that iterative detection techniques provide significant gains over traditional segregated detection methods. Also, iterative detection may be used as a complexity reduction method (with near optimal performance) by decomposing a system into subsystems. Thus, iterative detection is useful even in systems that are not constructed as concatenated networks and/or do not contain subsystem isolating interleavers. While a significant list of applications were considered, it is not exhaustive by any means (*e.g.*, [ChDeOr99, LiRi99, MoAu00]). Our goal has been to illustrate the method so that the reader can apply it where it will be fruitful.

The soft inverse of an FSM was given special attention due to the importance of the FSM subsystem in many encoder and channel models. The exact soft inverse can be computed using the (fixed-interval) forward-backward algorithm. While replacing the exact marginal soft inverse by a fixed-lag, minimum-lag, sliding-window, or minimum half-window SISO changes the iterative algorithm (*i.e.*, one may get different final decisions), these variations are important because they allow trade-offs in practical implementations with minor performance degradations. Two tree-structured exact soft inverses for an FSM were also given as example alternatives to the standard forward-backward SISO.

Using explicit index block diagrams allowed the optimality conditions for concatenated detectors to be more clearly identified. Specifically, if a graphical model has no (undirected) cycles, then the concatenated detector, defined by the node soft inverses, provides MAP detection by inward likelihood message passing. Allowing the soft inverse nodes to pass messages in both directions with appropriate scheduling, leads to a solution of the soft inverse problem. Specifically, after each node has received a message from all other nodes, the soft-out information is mature and does not change with further activation of the soft inverse nodes.

This view made it clear that iterative detection is equivalent to message passing on graphs with cycles. With cycles present, the soft-outputs are not guaranteed to stabilize. In practice, however, message passing on graphs with long cycles has been empirically observed to be quite stable.

Some open research issues related to this chapter are in the areas of system modeling, performance analysis, and modified message passing rules. Determining systematic approaches to obtain system models has been applied mainly to error correction codes [BaCoJeRa74, Fo00, AjMc00]. The tree-SISOs described in this chapter are an example of an alternative system model for an arbitrary FSM that has a desirable property (*i.e.*, an associated soft inverse with low-latency). Application of alternative system models to many practical detection problems appears to be a rich area for future research. Similarly, exploring encoder constructions that result in simplified iterative decoders is another interesting area.

Convergence of message passing on graphs with cycles is a very active area of research. Recent results address the dynamics of message passing in PCCC decoders using non-linear system analysis methods [Ag99] or SNR transformations [HaEl00]. Also, a practical limit on performance for turbo-like codes has been obtained which is akin to the computational cut-off rate for sequential decoding [Ag99, RiUr00]. Similarly, recent advances have been made in obtaining good upper bounds on the performance ML decoding of turbo-like codes [Di99]. An interesting area for future research is to obtain good lower bounds, potentially using the method of uniform side information introduced in Chapter 1.

Modified message passing rules may be useful when the system is modeled by a graph with short cycles. For example, if no interleavers separate subsystems, then the sequence of soft information exchanged has a high degree of temporal correlation. This correlation is scrambled when an interleaver is present so that, on a small time scale, the soft information appears uncorrelated. In such cases, it may be useful to model the statistics of the soft information sequence and compute messages accordingly. For example, instead of the marginal soft inverse based on a memoryless channel, one may consider a SISO based on a colored noise channel. In general, modified rules may be obtained by viewing the marginal likelihoods as an observation and deriving likelihood combining rules (*e.g.*, see Problem 2.28). It may be possible to draw upon results from the distributed detection and data fusion fields for these rules.

2.9   Problems

2.1. Suppose that $\mathcal{S}(\zeta)$ is a set of statistics for deciding on $H(\zeta)$ from $\mathbf{z}(\zeta)$ that is not sufficient. Show that any other set of decision statistics $\mathcal{J}(\zeta)$ obtained by a deterministic transformation on $\mathcal{S}(\zeta)$ is also insufficient.

2.2. Consider making a MAP decision for a hypothesis $H(\zeta)$ based on the observation $\mathbf{z}_1(\zeta)$. Suppose that you already have access to a processing unit which takes as its inputs the observation $\mathbf{z}_1$ and the a-priori probabilities $\{p(H_m)\}$ to produce the optimal decision. This processing unit is shown in Fig-2.76. Suppose that you now have access to a second, independent observation $\mathbf{z}_2(\zeta)$ and you would like to use this same processing unit. Show that the optimal decision based on the observation $\{\mathbf{z}_1, \mathbf{z}_2\}$ may be obtained by using this processor with $p'(H_m) = \alpha p(\mathbf{z}_2|H_m)p(H_m)$ where α is a constant selected so that $p'(H_m)$ sums to unity over all possibilities. What happens if you replace $p(\mathbf{z}_2|H_m)$ by $p(H_m|\mathbf{z}_2)$?

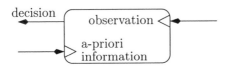

Figure 2.76. The processing unit available in Problem 2.2

2.3. Consider the parallel concatenation analogous to the serial concatenation of Fig-2.4. Specifically, consider the binary input sequence a_m driving a serial to parallel converter (1 to 2), followed by the 4PAM mapper to produce x_n. Also, a_m drives a simple FSM with memory L to produce y_m – *i.e.*, y_m and a_m are on the same time scale and there are half as many x_n. Determine the MAP received for a_m if y_m and x_n are both transmitted over independent AWGN channels. **Hint:** the entire system is an FSM.

2.4. Repeat the calculations of Example 2.5 for the min*-sum version using the numerical values in Example 1.12 and Fig-1.15 – give the MO$[a_3]$ and MO$[x_7]$ (*i.e.*, extrinsic output metrics).

2.5. Using the results of Problem 1.22, modify your program for the forward-backward algorithm to compute extrinsic output metrics on both the FSM inputs and outputs. You may want to pass flags to allow you to disable one or more of the four SISO ports.

2.6. Based on a BPSK modulation scheme and an ISI-AWGN channel, create histograms of the output metrics generated by a min-sum forward-backward algorithm. Specifically, assume that the input metrics MI$[a_k]$ are zero (uniform) and use the channel $\mathbf{f}^T = [\,1\ 2\ 1\,]$. For various SNR values, generate a histogram of $|\text{MO}[a_k = +1] - \text{MO}[a_k = -1]|$. Repeat this for the min*-sum case.

2.7. Suppose that one used the output of a SISO to bias another run of that same SISO. Specifically, for an FSM with uniform inputs, after running the forward-backward algorithm to produce $\text{MO}^{(1)}[a_k]$, discuss the effects of setting $\text{MI}^{(2)}[a_k] = \text{MO}^{(1)}[a_k]$ to produce $\text{MO}^{(2)}[a_k]$ with another run of the forward-backward SISO – *i.e.*, use MI$[x_k]$ for each of these "self-iterations." For example, are $\text{MO}^{(1)}[a_k]$ and $\text{MO}^{(2)}[a_k]$ threshold consistent? What about the implied sequence decisions? Consider the same method for an APP-based SISO – how do your conclusions vary? Also

$k =$	0	1	2	3	4
$I(k) =$	3	1	4	0	2

$k =$	0	1	2	3	4	5	6
$z_k(0) =$	0.414	0.564	-1.66	-1.26	-0.457	0.455	-0.539
$z_k(1) =$	-2.643	-0.631	1.567	-0.820	-0.997	1.889	-0.346

Table 2.5. The interleaver definition and observation sequence for Problem 2.11.

consider iterating intrinsic information in place of extrinsic information.

2.8. Show that, for an AWGN channel, in order to use the metrics $MI[x_k] = |z_k - x_k|^2$ and $MI[a_k] = -N_0 \ln(p(a_k))$, a log-domain version of the APP-SISO should use the marginalization operator

$$\min^*_{N_0}(x, y) = \min(x, y) - N_0 \ln \left(1 + e^{\frac{-|x-y|}{N_0}} \right) \qquad (2.60)$$

Show that after processing, the soft-output produced by a soft inverse is $N_0 M^* SM[\cdot]$.

2.9. Consider a concatenated system with two output sequences $x_k(0)$ and $x_k(1)$ that are transmitted over independent AWGN channels with noise spectral level $N_0(0)$ and $N_0(1)$, respectively. If the a-priori probabilities on the global system inputs is uniform, show that an MSM-based iterative decoder requires knowledge of $N_0(0)/N_0(1)$ and an APP-based iterative decoder requires knowledge of $N_0(0)$ and $N_0(1)$.

2.10. Show that the encoders in Fig-2.79 are realizations of the same FSM as that in Fig-2.25.

2.11. Consider the PCCC from Section 2.4.3.1 with $K = 7$ and the sequence of channel observations and interleaver as show in Table 2.5. This data was generated using $\sqrt{E_s} = 1$ and $E_b/N_0 = 1$ dB. Create tables for $F_{k-1}[s_k]$, $B_k[s_k]$ for each of the four states and $k = 0,\ldots 7$. Also, create a table for $MO[b_k]$ and $MO[a_k]$ and $M[b_k]$ (the soft-broadcaster output to be thresholded) for $k = 0, 1, \ldots 4$. Hand compute (or use a spread-sheet) the values of these quantities for two iterations of the (min-sum) PCCC decoder described in Section 2.4.3.1 and fill-in these tables (*i.e.*, four $F_{k-1}[s_k]$ and $B_k[s_k]$ tables, and two $MO[a_k]$, $MO[b_k]$, and $M[b_k]$ tables).

2.12. Suppose that the two output bits of the rate 1/2 PCCC considered in Section 2.4.3.1 were mapped onto a 4-ary constellation. Specifically, consider the 4-PAM labeling of Fig-1.4 and a Gray-labeled QPSK constellation. Show that for the 4PAM case, a SOMAP should be included in the decoding iteration. Also show that, in contrast, for the QPSK case the corresponding SOMAP need be activated just once and then the decoding proceeds as described in the example considered in Section 2.4.3.1. Why is there a difference?

2.13. Consider the FSM defined by the RSC constituent codes in Section 2.4.3.1 with initial state set to $s_0 = 0$ with probability one. Show that for the sliding window SISO, if $k - D = 1$, $\text{APP}_{k-D}^{k+D}[u_k]$ should be computed using non-uniform initialization of the forward recursion based on $p(s_0)$. Specify this initialization.

2.14. Repeat the SW-SISO simulations of Example 2.11 for $D = 4$ using the initialization of forward and backward state metrics suggested in Example 2.12.

2.15. Consider the multiuser channel with $L = 1$, but without a specific structure for $\mathbf{R}(1)$.

    (a) Find an expression for $M_{k,m}[\cdot]$ analogous to that in (2.41) by applying (2.39) to all three terms inside the real-part operator in (2.37c).

    (b) Show that this allows processing on a trellis with $|\mathcal{A}|^M$ states, but with state transitions corresponding to individual user data transitions. Sketch the trellis corresponding to the example in Fig-2.35 for the case of general $\mathbf{R}(1)$. Is there any advantage to using this trellis and $M_{k,m}[\cdot]$ over the trellis in Fig-2.35(a) and $M_k[\mathbf{a}_k, \mathbf{a}_{k-1}]$?

    (c) Assuming the upper-triangular structure for $\mathbf{R}(1)$, simplify this expression for $M_{k,m}[\cdot]$. Does this enable processing on a $|\mathcal{A}|^{M-1}$-state trellis?

2.16. Show that the equivalent discrete-time multiuser channel model at the output of matched filter bank is

$$\mathbf{r}_k(\zeta) = \mathbf{y}_k(\mathbf{a}(\zeta)) + \mathbf{n}_k(\zeta) \qquad (2.61a)$$

$$\mathbf{y}_k = \sum_{m=0}^{L} \mathbf{R}(m)\mathbf{a}_{k-m} \qquad (2.61b)$$

with $\mathbb{E}\{\mathbf{n}_{k+n}(\zeta)\mathbf{n}_k^{\mathrm{H}}(\zeta)\} = N_0\mathbf{R}(n)$.

2.17. Consider the FSM defined by the RSC constituent codes in Section 2.4.3.1 with the tail bits added to drive the encoder to the zero state. Verify that, for the purposes of computing MO[b_k] for $k = 0, \ldots (K-3)$, one can account for the tail bits by either setting MI[b_{K-1}] and MI[b_{K-2}] to zero or infinity accordingly or by setting the backward state metrics $B_K[s_K]$ to zero for $s_K = 0$ and infinity otherwise. Are these two approaches the same if extrinsic soft-out information is desired for the tail bits?

2.18. Consider the equivalent system block diagram for the PCCC system of Fig-2.24 as shown in Fig-2.77. Give an activation schedule for the associated iterative decoder that produces exactly the same decoding algorithm considered in Section 2.4.3.1.

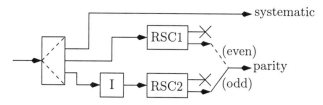

Figure 2.77.   An equivalent model of the PCCC encoder of Fig-2.24.

2.19. [LiRi99] Consider the bit interleaved coded modulation system shown in Fig-2.78. Is the non-iterative concatenated detector optimal in this case? Draw the associated explicit index diagram.

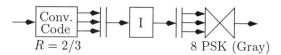

Figure 2.78.   The system consider in Problem 2.19.

2.20. Based on (1.70), the forward and backward parameters in the APP algorithm are

$$F_{k_1}^k[s_k] \equiv APP_{k_1}^{k-1}[s_k] \equiv p(z_{k_1}^{k-1}, s_k) \qquad (2.62a)$$

$$B_k^{k_2}[s_k] \equiv APP_k^{k_2}[s_k] \equiv p(z_k^{k_2}|s_k) \qquad (2.62b)$$

One method suggested in [BeMoDiPo96] for a fixed-lag algorithm was the forward-backward sum-product algorithm in which the backward recursion parameters were initialized using the forward recursion parameters. Specifically, it was suggested that $B_{k+D+1}^{k+D}[s_{k+D+1}] = F_0^{k+D}[s_{k+D+1}]$ for each conditional value of s_{k+D+1}. What is the extrinsic soft information produced by this algorithm?

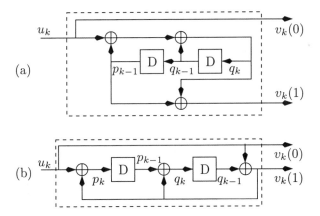

Figure 2.79.   Alternate structures for the encoder of Fig-2.25.

2.21. Show that, for a simple FSM, a $D = L$ fixed-lag SISO can be implemented using only a (single) forward recursion with a modified completion operation.

2.22. [Vi98] Draw figures analogous to Fig-2.41 and Fig-2.44 for the minimum-lag forward-backward SISO with $H = D$.

2.23. [BeCh00] Explain how tree-SISOs of the form in Fig-2.48 can be tiled without any overlap to provide a minimum half window SISO.

2.24. Show that activation of the soft inverse of a transition subsystem is as shown in Fig-2.54. Specifically, for min-sum processing, show that this does one forward ACS step, one backward ACS step, and the completion steps for the system inputs and outputs.

2.25. For the iterative detector of Example 2.15 and determine the computational complexity of the activation of each X_k^{-s} node in terms of 4-way ACS operations. The forward-backward algorithm requires roughly $4 \times 2 \times K = 8K$ 2-way ACS operations for the forward and backward recursions and K 4-way ACSs for the completion operations (on a_k only). Discuss how many iterations of the iterative detector can be made with complexity comparable to the forward-backward algorithm. Is there any other advantage to the structure of the iterative detector?

2.26. The goal of this problem is to relate the tree-SISO in Section 2.5.5 to the forward-backward tree-SISO (FBT-SISO) of Example 2.19.

   (a) Show that the inward message passing in Fig-2.64 is the same as that for the "middle tree" in Fig-2.49 – *i.e.*, the tree in Fig-2.49 producing F[s_4] and B[s_4].

(b) Show that after the inward message pass in Fig-2.64, one could obtain $\text{MSM}_0^7[s_4]$.

(c) Show that the computation of $F[s_5]$ in Fig-2.49 may be viewed as a forward message passing over a tree similar to that in Fig-2.63, but not including any validly check for s_6, s_7, and s_8.

(d) Generalize the above to show that the tree-SISO in Fig-2.63 can be obtained by six different validity check trees. Explain why an outward message pass is not required. **Hint:** are the validity checks redundant?

(e) Show that, if only hard decisions are desired that a "tree-Viterbi algorithm" can be defined by the inward message passing in Fig-2.64. Specifically, describe the "survivor" information that would need to be stored during this inward recursion and the associated traceback. For similar ideas, see [FeMe89].

2.27. Show that the marginal soft inverse of the "funnel" defined in Example 2.22 is the same as a soft broadcaster. Specifically, consider a broadcaster with input $x_k(0)$ and outputs $c_k^{(1)}(0) = c_k^{(2)}(0) = c_k^{(3)}(0) = x_k(0)$ and show that the soft inverse of this system is the same as that of the funnel.

2.28. Consider a binary hypothesis testing problem with N observations: $z(\zeta, 0) \ldots z(\zeta, N-1)$. Recall that the MAP receiver will form $p(z_0^{N-1}|H_m)p(H_m)$ and maximize over $m = 0, 1$. In this problem we consider the case where instead of $z(\zeta, n)$, the observations are pre-processed and in place of $\{z(\zeta, n)\}$ the receiver has access to the random variables $L_m(\zeta, n) = g_m(z(\zeta, n))$ for $m = 0, 1$ and $n = 0, \ldots N-1$, where

$$g_m(z_n) = p_{z(\zeta,n)}(z_n|H_m) \qquad (2.63)$$

In other words, only the *marginal channel likelihoods* are available.

(a) In general, is the set of observations $\{L_m(\zeta, n)\}$ for $m = 0, 1$ and $n = 0, 2 \ldots N-1$ a set of sufficient statistics for the decision problem? Explain.

(b) Consider the special case of $N = 2$ and

$$H_0: \quad \mathbf{z}(\zeta) = \sqrt{E/2} \begin{bmatrix} +1 \\ +1 \end{bmatrix} + \mathbf{w}(\zeta) \qquad (2.64a)$$

$$H_1: \quad \mathbf{z}(\zeta) = \sqrt{E/5} \begin{bmatrix} -2 \\ -1 \end{bmatrix} + \mathbf{w}(\zeta) \qquad (2.64b)$$

where $\mathbf{w}(\zeta)$ is a mean-zero Gaussian vector with covariance

$$\mathbf{K_w} = \frac{N_0}{2} \begin{bmatrix} 1 & \rho \\ \rho & 1 \end{bmatrix} \qquad (2.65)$$

Show that the optimal MAP receiver based on observing $\mathbf{z}(\zeta)$ can be realized in terms of $L_0(\zeta, 0)$, $L_1(\zeta, 0)$, $L_0(\zeta, 1)$, and $L_1(\zeta, 1)$? Diagram the *likelihood combining* required to implement the MAP receiver for observation $\mathbf{z}(\zeta)$ in terms of the four likelihoods.

(c) Compare the performance of the optimal likelihood combining and the receiver that simply multiplies likelihoods –*i.e.*,

$$L_0(\zeta, 0) L_0(\zeta, 1) \underset{H_1}{\overset{H_0}{\gtrless}} L_1(\zeta, 0) L_1(\zeta, 1) \qquad (2.66)$$

Characterize the performance difference as a function of ρ and E/N_0

(d) Revisit your answer to part (a). If you believe that the marginal likelihoods are not a sufficient statistic, can you provide a counterexample? Otherwise can you prove their sufficiency? Given a condition on the signals $\mathbf{s}_0$ and $\mathbf{s}_1$ (associated with the hypotheses) which ensures that the optimal receiver based on the marginal likelihoods in the same as that based on the joint likelihood.

Chapter 3

ITERATIVE DETECTION FOR COMPLEXITY REDUCTION

As outlined in the introduction of this book and demonstrated in Chapter 2, iterative detection itself may be viewed as a complexity reduction technique. In fact, the vast majority of complexity reduction obtained in the examples of this book and chapter results from modeling a given system efficiently. Specifically, a decomposition into a concatenated network of subsystems or, equivalently, an efficient (possibly loopy) graphical model is sought for which the associated iterative detector is relatively simple. Comparing the iterative detectors in Section 2.4 against the optimal receivers that directly exploit the global system structure, it is apparent that a huge decrease in complexity has been achieved. Furthermore, all evidence indicates that this complexity reduction comes at the cost of little performance degradation.

In this chapter we exploit this view of iterative detection to obtain significant complexity reduction (even for systems in isolation). First, a set of tools based primarily on decision feedback concepts is introduced that can yield complexity reduction beyond that inherent to the modeling gain described above. When these techniques are applied or iterative detection is performed on graphs with short cycles, modification of the iterative detection rules can improve performance. We demonstrate these approaches on (coded) ISI channels.

3.1 Complexity Reduction Tools

Even though each marginal soft inverse block considers only local structure, it does so using locally optimal methods. For example, for an FSM subsystem, the marginal soft inverse has complexity determined by the number of FSM states. For a simple FSM with input a_k, therefore, the complexity grows as $|\mathcal{A}|^L$, where L is the memory. Although this is typically a small fraction of the complexity of the globally optimal processing, it may still be prohibitive as the following simple example illustrates.

Example 3.1. ─────────────────────────────────
Consider a TCM-Interleaver-ISI serial concatenation with the associated iterative detector as shown in Fig-2.20. If the TCM-encoded sequence is QPSK modulated, and the ISI channel has 10 taps ($L = 9$), an inner SISO based on the forward-backward algorithm will run on a trellis with $4^9 =262,144$ states with 1,048,576 different transitions at each time.
─── *End Example*

In fact, while an iterative detector will typically have several types of soft inverses (*e.g.*, interleaver/deinterleavers, SOMAPs, SOBCs, etc.), it is common for the computational and storage complexities of the FSM soft inverse to dominate that of the overall detector. Thus, much of our focus in this chapter is on the soft inverse of the FSM, specifically on reducing the complexity of the forward-backward algorithm. Below is a brief list of methods that can be used to reduce complexity.

3.1.1   Operation Simplification

Several modifications to the baseline forward-backward SISO (*i.e.*, the fixed-interval) algorithm were described in Chapter 2 that are primarily intended to decrease the computation and/or storage requirements of the SISO. For example, fixed-lag approaches yield a significant reduction in memory requirements. The minimum-lag modification yields further reduction in complexity, as does L-early completion in the case of a simple FSM. Similarly, the primary allure of SOVAs is a potential complexity reduction relative to the forward-backward algorithm. Many references also present the min-sum algorithm[1] as an approximation to the min*-sum version using

$$\sum_{\mathbf{a}:u_k} p(\mathbf{z}, \mathbf{a}) \cong \max_{\mathbf{a}:u_k} p(\mathbf{z}, \mathbf{a}) \tag{3.1}$$

3.1.2   Decision Feedback Techniques

Decision feedback is a well-known approach for complexity reduction. Several complexity reduction methods for iterative detection have been

─────────────────────────────

[1]According to the view presented in Chapters 1-2, both may be justified from a MAP detection point of view. Both are suboptimal when used on loopy graphs, but approximate each other at high SNR.

suggested based on decision feedback. In fact, one may view iterative detection itself as a method of (soft) decision feedback. Techniques suggested differ in several aspects. First, either soft or hard decision information can be fedback. Initially, we focus on hard decision feedback (HDF) which may be viewed as replacing soft information on a quantity by a hard decision (*e.g.*, zero metric for one conditional value and infinite for the others). Second, the scope over which the feedback decisions are enforced is another variation. For example, one may make a hard decision on an input bit after the first iteration and enforce that hard decision in all subsequent processing. Alternatively, one may enforce a hard decision only locally to simplify processing and then release this hard decision condition. For example, a hard decision on an input bit may be made after the first iteration and used to simplify the activation of a single iteration of a SISO, with this decision released for subsequent processing. This may eventually lead to soft information on this bit that changes the implied hard decision. Third, based on the implicit index block diagram view, it may be useful to distinguish between internal versus external decision feedback. The latter corresponds to enforcing decision feedback at the soft-in ports of a soft inverse and the former corresponds to some decision feedback internal to the (approximate) soft inverse processor. These concepts are illustrated by the following examples.

Internal Hard Decision Feedback for State Reduction   Hard decision feedback is a well-known technique for complexity reduction in forward- (backward-) only hard-decision algorithms. Specifically, the reduced-state sequence estimation (RSSE) algorithm [EyQu88], and the similar delayed decision feedback sequence estimation (DDFSE) algorithm [DuHe89] both use decision feedback on a per-survivor basis to approximate the Viterbi algorithm using a smaller trellis. A RSSE algorithm with reverse-time structure was suggested in [McKe97]. In [MuGeHu96] HDF has been applied to the L^2VS (fixed-lag) algorithm. This is conceptually straightforward since the L^2VS has a forward-only structure (see Problem 3.12). Similarly, the same approach has been applied to the forward-backward SISO of Chapter 2 [CoFeRa00]. This is accomplished by running a reduced state forward recursion (*e.g.*, as in DDFSE) and storing all transition metrics. These transition metrics are then used for the backward recursion and completion operation (see Problem 3.12).

Note that these reduced-state techniques apply hard decision feedback internal to a SISO and enforce these hard decisions only for the current activation of the SISO.

Internal Probability Truncation   Other complexity reduction schemes based on soft information have been suggested. Exploiting the fact
that many conditional values are highly unlikely, the complexity can
be reduced by truncating this soft information. For example, in the
probability domain with "sum-to-unity" normalization (*e.g.*, see (2.15))
a small probability can be truncated to zero. This was suggested in
[FrAn98] for the forward and backward state APPs.[2] After truncation,
paths emanating from that state need not be considered in the subsequent recursion. This is illustrated in the following example.

Example 3.2. ————————————————————————————
Consider a forward sum-product recursion on a four-state trellis with
sum-to-unity normalization where all state probabilities below 0.1 are
truncated to zero. For the scenario shown in Fig-3.1, the forward state
APPs of state 1 and 3 at time k are truncated to zero. Consequently
the transitions departing from these states can be dropped from the following calculations.

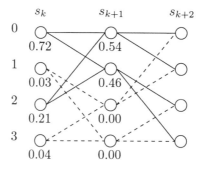

Figure 3.1.   Illustration of the small probability truncation scheme in an sum-product
SISO algorithm. The dotted lines represent the transitions killed due to the probability truncation.

———————————————————————————— *End Example*

Similar to the reduced-state approaches, this technique applies hard
decision feedback (on the states) internal to the SISO and enforces these
decisions only during the given activation. In contrast to the reduced-
state approaches described above, however, this probability truncation
yields a variable-complexity algorithm with a reduction in the *average*
computational complexity. This presents unique challenges for hardware
implementation. For example, the chip area for such an algorithm is

———————————————————————
[2]This is similar to the concept underlying the so-called T-algorithm for reduced complexity
sequence detection [Si89].

typically as large as the standard fixed-complexity algorithm but the power consumption may be reduced depending on the aggressiveness of the design (*e.g.*, computational units not being used may be powered down). Another potentially challenging aspect of variable-complexity algorithms is that the amount of effort expended is typically a function of the SNR and other parameters which results in a variable processing time (throughput) as well.

External Hard Decision Feedback In [FrKs98], a so-called *early detection* scheme is based on external HDF which, once performed, is enforced over all subsequent processing. Any quantity which has high reliability (*i.e.*, dominance of the soft information by one conditional value) is decided early and held fixed from that point on in the processing. These early decisions result in trellis splicing in subsequent iterations and therefore reduced complexity. This is also a variable complexity method.

Example 3.3. ——————————————————————————————
Suppose that after the n-th iteration in the decoding procedure for a turbo code, we obtain $P[a_k = 0] = 0.98$ and $P[a_k = 1] = 0.02$ as the intrinsic information used to make decisions. Based on this, it appears reliable to decide $\hat{a}_k = 0$. A decoder using the early decision technique will fix the final decision for a_k to 0 and run all subsequent iterations with this value fixed. In the next iteration, therefore, the transition from s_k to s_{k+1} is deterministic due to the early decision $\hat{a}_k = 0$. The k-th transition can then be dropped out of the forward and backward recursions and the two trellis segments between s_k and s_{k+2} can be "spliced" into a single segment. This results in computational and storage savings as more transitions get spliced out.
——————————————————————————————— *End Example*

External Soft Estimate Feedback In contrast to hard decision feedback, which is based on selecting one of the conditional values for a digital quantity, soft estimate feedback can be used to replace a quantity by its average value. Computation of this average value can be based on available soft information. For example, consider a linear ISI channel with input a_k and output x_k. Furthermore, assume that some of the ISI coefficients are small. In this case, it may be effective to average out the effects of these taps using the current soft information available for a_k. Specifically, we can apply the following approximation

$$\sum_{|f_i|<\epsilon} f_i a_{k-i} \cong \sum_{|f_i|<\epsilon} \sum_a f_i a P[a_{k-i} = a] = \sum_{|f_i|<\epsilon} f_i \mathbb{E}\{a_{k-i}(\zeta)\} \quad (3.2)$$

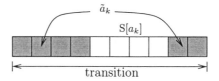

Figure 3.2.   Soft decision feedback on a simple FSM transition.

where the expectation is over the pmf implied by the properly normalized soft information (*i.e.*, sum-to-unity normalization in the probability domain). In this case $\tilde{a}_i = \mathbb{E}\{a_i(\zeta)\}$ may be viewed as the soft estimate being fedback. Note that, in general, $\tilde{a}_i \notin \mathcal{A}$. This concept is illustrated in Fig-3.2. While the linearity of the channel simplifies the computation, the soft estimate feedback concept may be applied to more general problems (see Problem 3.4). For this linear ISI example, if the average value of $a_{k-i}(\zeta)$ is zero under the current beliefs, soft estimate feedback corresponds to ignoring those taps (*i.e.*, this would be the case if $\mathcal{A}$ is a symmetric constellation and the current beliefs are uniform).

Connection Cutting   As the above special case of soft estimate feedback suggests, it may be reasonable to simply ignore a variable associated with a soft inverse processor for the purposes of performing the local combining and marginalization. We refer to this as *connection cutting*. This may be a reasonable approach for complexity reduction, especially when the local dependency on that variable is weak. Furthermore, this may have the effect of alleviating the problem of short cycles in some message passing algorithms. This is illustrated in the following example.

Example 3.4. ———————————————————————————
Consider the (suboptimal) SISO in Fig-2.59. For a given k, consider cutting some connections corresponding to message passes from X_k^{-s} back to the soft broadcaster. In fact, considerable complexity reduction can be realized by cutting all but one of these message passing connections. In this case, we refer to the connection not cut as the *pivot* location. In Fig-3.3, the connection-cut SISO is shown based on the pivot of tap 1 (*i.e.*, the "2" tap in $[1, 2, 1]$). How many and which connections are cut are design options. It is reasonable to expect, however that the connections corresponding to "heavier" taps (*i.e.*, larger energy in the taps) should not be cut.
————————————————————————————————— *End Example*

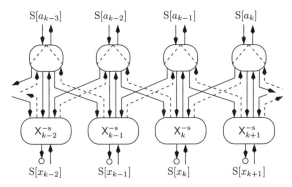

Figure 3.3. The SISO algorithm from Fig-2.58 with all but the "center" connections to the soft broadcasters cut. Dashed lines indicate a cut connection – *i.e.*, no messages are passed along these edges and uniform soft information is used instead.

3.2 Modified Iterative Detection Rules

When working with graphical models having short cycles and/or using decision feedback methods, it may be useful to slightly modify the iterative detection rules to alleviate rapid convergence to a local minimum or even divergence. Below, we describe several reasonable modifications to the standard conventions.

3.2.1 Altering the Convergence Rate

Many iterative algorithms (*i.e.*, iterative detection being a special case) exhibit a trade-off between convergence rate and accuracy of the eventual solution. Specifically, faster convergence typically comes at the expense of a less accurate solution (*e.g.*, convergence to a local minimum is more likely). This concept is illustrated in Fig-3.4.

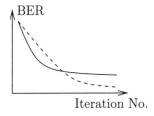

Figure 3.4. Illustration of relationship between convergence rate and system performance typical of iterative algorithms.

We have observed that, when using aggressive complexity reduction techniques and/or graphical models with short cycles, iterative detectors can convergence rapidly with relatively poor performance. We have

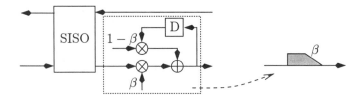

Figure 3.5. A single-pole low pass filter for soft information (shown filtering the soft-out on the subsystem output). Note the unit-delay is applied across activations. The symbol shown to the right is adopted to indicate such a filter in the following.

found that applying techniques to slow the convergence of the iterative detector can improve the performance in such cases.

The method that we have utilized most effectively is *soft information filtering* which simply filters out large variations in the soft information of a particular quantity from one activation to the next. For example, a single-pole soft information filter was proposed in [ChChNe98] to slow down the convergence. A similar scheme has been suggested in [MuWeJo99] to alleviate steady-state oscillations (*e.g.*, limit cycles). This concept is illustrated in Fig-3.5. The actual soft output information after n-th activation that is passed to other processors in[3]

$$\text{SO}^{(n)}[a] \leftarrow \beta \times \text{SO}^{(n)}[a] + (1 - \beta) \times \text{SO}^{(n-1)}[a] \qquad (3.3)$$

where $\text{SO}^{(n)}[a]$ on the right hand side of (3.3) is the standard extrinsic information produced by the processor. The parameter β can be used to adjust the bandwidth of the filter. Specifically, when $\beta = 1$, there is no filtering. The smaller β is selected, the smaller the filter bandwidth which is expected to slow convergence. Clearly, this concept generalizes to more complicated filter designs, although we have found the performance to be relatively insensitive to the specific filter choice. This concept is demonstrated in the next example.

Example 3.5. ───
In the (suboptimal) SISO of Example 2.15, a filter can be used for each soft-out port of the soft broadcaster nodes in Fig-2.59. As shown in Fig-3.6, filtering with $\beta = 0.3$ provides more than a 2 dB gain in E_b/N_0 at high SNR. This gain is achieved only after the unfiltered iterative detector has converged (*i.e.*, after the 6-th iteration). By trial and error, it was found that $\beta = 0.3$ provided the best performance.

---

[3]The superscript (n) used in notation $\text{S}^{(n)}[\cdot]$ denotes the number of iteration with which the soft quantity is associated.

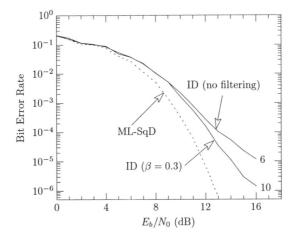

Figure 3.6. Impact of soft information filtering on the performance of SISO. The numbers attached to the curves are the iteration numbers used.

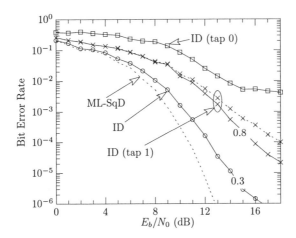

Figure 3.7. Impact of connection cutting on the performance of SISO. The value of β is attached to each curve if the soft information filtering is employed. 20 iterations has been used for ID (tap 0). All other iterative schemes use 10 iterations.

This same approach can be applied in conjunction with connection cutting. For example, the effect of the above filtering was considered with two connection cutting schemes of the form described in Example 3.4. Specifically, all but one connection from each X_k^{-s} back to the soft broadcasters was cut. We consider two pivot locations: tap 0 (*i.e.*, the first "1" tap in $(1, 2, 1)$), and tap 1 (*i.e.*, the "2" tap). The results are shown in Fig-3.7 with comparison to the fully-connected version with filtering. It was observed that filtering did not significantly improve the performance of the connection-cut version using tap 0 as the pivot. When tap 1 is used as the pivot, filtering with $\beta = 0.8$ provides an improvement of approximately 2 dB at high SNR. As expected, using tap 1 as the pivot yields better performance, but this is still nearly 4 dB worse than the fully-connected version. Compared to the fully-connected version in Example 2.15, however, the connection cutting scheme reduces the marginalization operations at each X^{-s} by a factor 3 and no computation is required at the soft broadcasters.

——————————————————————————— *End Example*

Other methods for slowing the convergence are conceivable. For example, we use the term *belief degradation* to refer to methods that modify the standard soft-out information so that it is less reliable. For example, let $PO[b]$ be the standard extrinsic information associated with a binary variable and assume that it has been normalized to sum to unity. This information can be degraded as

$$PO[b] \leftarrow \begin{cases} 0.5 & \text{if } |PO[b] - 0.5| < \Delta \\ PO[b] & \text{otherwise} \end{cases} \quad (3.4)$$

where $\Delta < 0.5$ is a belief threshold. The larger Δ is chosen, the slower the convergence is expected to be. Other choices for the degrading function are possible (*i.e.*, see Problem 3.6).

Complexity reduction can be achieved by performing fewer iterations. Thus, just as one can slow the convergence of an iterative detector, an attempt can be made to increase this rate of convergence. Thus, some performance degradation may be traded to obtain fewer iterations (*i.e.*, complexity reduction). For example, instead of removing all of the soft-in information on a quantity to create extrinsic information, one could remove only part of this information. Similarly, one could apply the opposite of the belief degrading concept above (*i.e.*, belief enhancing) to increase the convergence rate.

3.2.2   Modified Initialization Schemes

In an effort to increase the reliability of the initial soft-in information, one can attempt to use some additional information to form this bias. Below we describe two methods along these lines.

Cross Initial Combining   In Example 2.15 the a-priori information $p(z_k|x_k(t_k)) = p(z_k|t_k)$ has been used to initialize $PI[x_k]$. Since the objective of the iterative algorithm is to converge to the likelihood of a_k based on the whole observation sequence $\mathbf{z}$, it may desirable to compute $PI[x_k]$ based on a larger region of the observation data to achieve stronger initial combining effects. In this specific example (3-tap ISI), $\text{PI}[x_k]$ can be set using

$$\text{PI}[x_k] = p(\mathbf{z}_{k-1}^{k+1}|x_k(t_k)) = p(z_k|t_k) \times \sum_{a_{k+1}} p(z_{k+1}|t_{k+1})p(a_{k+1})$$

$$\times \sum_{a_{k-L-1}} p(z_{k-1}|t_{k-1})p(a_{k-L-1}) \quad (3.5)$$

As illustrated in Fig-3.8, three observations are used in this initialization scheme. This approach was proposed in [ChChNe98] for two dimensional

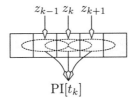

Figure 3.8. Illustration of a 3-term cross initial combining scheme.

applications. Thus, we refer to this approach as cross initial combining (*i.e.*, in 2D, the region is shaped like a cross). More than three observations can be included if desired. However, using more than three observations in this example requires significantly more complexity and we have observed that most of the benefits of cross combining are obtained with small combining regions.

Example 3.6.
The cross initial combining scheme in (3.5) was applied to the iterative processors in Example 3.4 with the results presented in Fig-3.9. For the connection-cut versions, the cross initial combining helps to improve the performance significantly. In fact, the version using tap 0 as the pivot outperforms the fully-connected SISO at high SNRs with a lower complexity. Thus, the cross initial combining may be useful to compensate for the performance degradation due to complexity reduction techniques.
———————————————————————— *End Example*

Self-Iteration If an approximation to the marginal soft inverse is used as a SISO, it may be reasonable to reactivate the SISO applying the soft-out information as soft-in information for the next activation. We will use this technique for reduced-state SISOs in this chapter and (suboptimal) adaptive SISOs in Chapter 4. In this context, we refer to this technique as *self-iteration*. It is interesting to note, however, that for a given concatenated system, the standard iterative detector is an

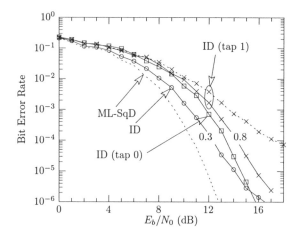

Figure 3.9. Impact of cross initial combining on the performance of SISO. The value of β is attached to each curve if soft information filtering is employed. The fully-connected version (labeled ID) uses 10 iterations and all other curves are based on 20 iterations.

approximation of the marginal soft inverse of the global system. Thus, applying self-iteration to this suboptimal SISO for the global system may be meaningful. This concept is illustrated in Fig-3.10 for a serially concatenated system. In this context, $SO[a_k]$ can be viewed as the output of a suboptimal SISO for the global system. Thus, self-iteration applied to this system would entail using the $SO^{(n-1)}[a_k]$ for $SI^{(n)}[a_k]$.

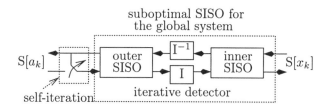

Figure 3.10. Self-iteration applied to a suboptimal (global) SISO for a serially concatenated system.

3.3   A Reduced-State SISO with Self-Iteration

In this section we present a reduced-state SISO (RS-SISO) based on separate decision feedback for the forward and backward recursions of the forward-backward algorithm [ChCh00]. This results in an approximate soft inverse for an FSM. We apply self-iteration to this RS-SISO to refine the soft information produced. The performance is demonstrated for ISI channels both in isolation and as part of a concatenated system.

3.3.1   Reduced-State SISO Algorithm

Hard decision feedback of the type used in DDFSE can be applied to both the forward and backward recursions of the standard forward-backward SISO. However, the completion step must be modified to take into account the HDF. We refer to the resulting (suboptimal soft inverse) SISO algorithm as the reduced-state (RS) SISO algorithm. For the simplicity of presentation, we consider the simple FSM with state $s_k = \mathbf{a}_{k-L}^{k-1}$. With little modification, however, this complexity reduction scheme is also applicable to other types of FSMs.

To reduce the number of states, the RS-SISO truncates the state and rebuilds the corresponding trellis. The *truncated trellis state* at time k is defined as $v_k = \mathbf{a}_{k-L_1}^{k-1}$, where $L_1 \leq L$. The corresponding *truncated trellis transition* at time k is $\tau_k = (v_k, v_{k+1})$. Note that all of the information in τ_k (the trellis transition) can be obtained from t_k (the FSM transition). We denote this relation by $\tau_k \sqsubset t_k$.[4] If $L_1 = L$, then $v_k = s_k$, and the following derivation will simply result in the standard forward-backward SISO algorithm. We present the derivation using min-sum computations for concreteness. The forward and backward state metrics $\mathrm{F}[v_k]$ and $\mathrm{B}[v_k]$ are defined for the RS-SISO algorithm recursively as

$$\mathrm{F}[v_{k+1}] = \min_{v_k : v_{k+1}} \left(\mathrm{F}[v_k] + \mathrm{M}^{(\mathrm{f})}[\tau_k] \right) \tag{3.6a}$$

$$\mathrm{B}[v_k] = \min_{v_{k+1} : v_k} \left(\mathrm{B}[v_{k+1}] + \mathrm{M}^{(\mathrm{b})}[\tau_k] \right) \tag{3.6b}$$

where $\mathrm{M}^{(\mathrm{f})}[\tau_k]$ and $\mathrm{M}^{(\mathrm{b})}[\tau_k]$ are the truncated-transition metrics in the forward and backward recursions, respectively. Specifically, the forward recursion is initialized by

$$\mathrm{F}[v_0] = \min_{s_0 : v_0} \mathrm{MI}[s_0] \tag{3.7}$$

where $\mathrm{MI}[s_0]$ is the a-priori soft information on s_0. As usual, the backward recursion is begun with uniform state metrics. If, as described in Chapters 1 and 2, one would like to account for tail bit information by initializing the backward state metrics, one may use $\mathrm{B}[v_K] = \min_{s_K : v_K} \mathrm{MI}[s_K]$.

Due to state reduction, the trellis transition metrics required in (3.6a) and (3.6b) are not directly available. However, this soft information is

---

[4]This is essentially a subset relation. However, since our notation implies that, for example, t_k is a variable and not a set, we use this modified notation. In the following, we will use $\sqsubset$, $\sqsupset$, $\sqcup$, and $\sqcap$, in place of the standard set operations/relations $\subset$, $\supset$, $\cup$, and $\cap$ to describe the analogous relations between variables.

closely related to the transition metrics $M_k[t_k]$ that are given. Since $\tau_k \sqsubset t_k$, one can obtain the required metrics by marginalization

$$M^{(f)}[\tau_k] = M^{(b)}[\tau_k] = \min_{t_k:\tau_k} M_k[t_k]. \qquad (3.8)$$

However, it can be shown that the complexity of using (3.8) with (3.6a–3.6b) will be close to that of the standard forward-backward SISO algorithm. In order to achieve the goal of complexity reduction hard decision feedback is employed in the calculation of $M[\tau_k]$. Define the survivor state associated with v_k in the forward (or backward) recursion as the state contributing most in the marginalization operation in (3.6a) (or (3.6b)). For the min-sum version, this survivor state is exactly the same as in DDFSE [DuHe89]. On the other hand, since the marginalization in the APP version is the summation operation, this definition of survivor state is not the only reasonable choice. Therefore, at this point in the development, it is clear that the RS-SISO discussed is not a semi-ring algorithm. Associated with each trellis state v_k a survivor path can be obtained in the forward and backward recursions (see Fig-3.11)

$$\check{\mathbf{a}}_k^{(f)}(v_k) = [\check{a}_{k-L}^k(v_k), \check{a}_{k-L+1}^k(v_k), \dots \check{a}_{k-L_1-1}^k(v_k)] \qquad (3.9a)$$

$$\check{\mathbf{a}}_k^{(b)}(v_k) = [\check{a}_k^k(v_k), \check{a}_{k+1}^k(v_k), \dots \check{a}_{k+L_2-1}^k(v_k)] \qquad (3.9b)$$

where $L_2 = L - L_1$. Note that $\check{a}_l^k(v_k)$ denotes the symbol at time l on

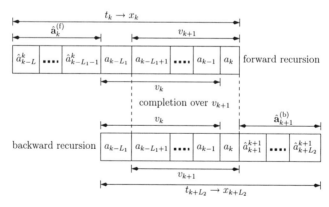

Figure 3.11. The reduced states in the forward and backward recursions at time k in the RS-SISO algorithm.

the survivor path terminating into trellis state v_k. When $k > l$, $\check{a}_l^k(v_k)$ is on the survivor path in the forward recursion. Otherwise it is on the survivor path in the backward recursion. Based on the definition of survivor paths, the desired soft information about τ_k can be calculated simply using

$$M_k^{(f)}[\tau_k] = MI[x_k = x(\check{a}_k^{(f)}(v_k), \tau_k)] + MI[a_k] \qquad (3.10a)$$

$$M_{k+L_2}^{(b)}[\tau_k] = MI[x_{k+L_2} = x(\tau_k, \check{a}_{k+1}^{(b)}(v_{k+1}))] + MI[a_k] \qquad (3.10b)$$

Note that a subscript has been added to $M^{(f)}[\tau_k]$ ($M^{(b)}[\tau_k]$) to denote the observation index used in its computation. This is similar to the standard convention for $M_k[t_k]$. Similarly, the notations for forward and backward state metrics are modified from this point forward to $F_k[v_{k+1}]$ and $B_{k+L_2}[v_k]$, respectively (*i.e.*, denoting the largest index used in the forward recursion and the smallest index included in the backward computation). The key notion used in (3.10a) and (3.10b) is the rebuilding of the full FSM transition t_k and t_{k+L_2} using the survivor paths and truncated trellis transitions (see Fig-3.11). When $L_2 = 0$, $M_k^{(f)}[\tau_k] = B_{k+L_2}^{(b)}[\tau_k] = M_k[t_k]$, yielding the standard full-state forward-backward SISO algorithm.

The complexity in (3.6a–3.6b) is determined by $|\mathcal{V}| = |\mathcal{A}|^{L_1}$. Compared to the standard SISO algorithm, both the computational complexity and memory requirements of the RS-SISO algorithm are approximately reduced by $|\mathcal{A}|^{L_2}$ times. After executing both the forward and backward recursions, the extrinsic soft-out metric for a_k can be obtained by

$$MO[a_k] = \min_{v_{k+1}:a_k} (F_k[v_{k+1}] + B_{k+L_2+1}[v_{k+1}]) - MI[a_k] \qquad (3.11)$$

The completion step in (3.11) is also illustrated in Fig-3.11. Due to the state truncation, $SO[x_k]$ cannot be obtained directly by marginalizing over $F_k[v_{k+1}] + B_{k+L_2+1}[v_{k+1}]$. Nevertheless, there are reasonable approaches to obtain some form of soft-out information for x_k. For example, the soft-information on the corresponding inputs can be combined

$$MO[x_k] = \min_{t_k:x_k} \sum_{l=k-L}^{k} M[a_l(t_k)] - MI[x_k] \qquad (3.12)$$

where $M[a_l(t_k)] = MO[a_l(t_k)] + MI[a_l(t_k)]$ represents intrinsic soft information obtained by (3.11) without subtracting the soft-in metric. Numerical results in [ChCh98b] suggest that this approach is reasonable although extensive experimentation in this RS-SISO context has not been conducted. It is notable that (3.12) can be calculated recursively due to the temporal relationship between t_k and t_{k+1} (see Problem 3.15). If a final decision is sought from the soft information produced, one should threshold $M[a_k]$.

As the notation used in (3.11) suggests the soft-out metric for a_k is computed based only using soft-in metrics with indices in $\{0, 1, \ldots k\} \cup \{(k + L_2 + 1), \ldots K - 1\}$. The omission of $\{\mathrm{MI}[x_{k+1}], \cdots, \mathrm{MI}[x_{k+L_2}]\}$ in the completion operation for $\mathrm{MO}[a_k]$ can also be seen in Fig-3.11. This omission is a consequence of the completion strategy selected. Other completion approaches which use all of the observations are feasible for an RS-SISO, but this may again result in a SISO algorithm with complexity dominated by the "full (FSM) state" completion. Regardless of the completion technique used, an RS-SISO algorithm is suboptimal due to the decision feedback used in the forward and backward recursions. The simple completion scheme defined in (3.11) results in further performance degradation and complexity reduction. The smaller L_1, the simpler the RS-SISO algorithm while the more severe the performance degradation. Adjusting this tradeoff between complexity and performance is sometimes difficult using only the parameter L_1 since the RS-SISO complexity grows exponentially with L_1. In an attempt to improve performance with a less substantial increase in complexity, we suggest *self-iteration* for the RS-SISO algorithm. This self-iteration is realized by running the RS-SISO algorithm with $\mathrm{MO}[a_k]$ recirculated through the soft-in port of the RS-SISO several times before passing on the final soft-out metric. Thus, the external interface of the RS-SISO is the same as that of a standard SISO with several self-iterations of the RS-SISO performed in place of a single iteration of a full-state SISO. Self-iteration of the RS-SISO algorithm allows the metrics $\{\mathrm{MI}[x_{k+1}], \cdots, \mathrm{MI}[x_{k+L_2}]\}$ to affect the soft-out information $\mathrm{MO}^{(n)}[a_k]$ after $n > 1$ self-iterations. Furthermore, these self-iterations result in a linear increase in complexity which allows one to trade-off the size of the trellis used and the number of self-iterations to achieve the desired trade-off between complexity and performance. Specifically, with N self-iteration activations, the RS-SISO algorithm has a complexity determined by $N|\mathcal{A}|^{L_1}$ and the parameters N and L_1 can be adjusted separately.

Viewing the RS-SISO with self-iteration as a replacement for the full-state forward-backward SISO, several observations can be made. First, the RS-SISO can be substituted for the SISO in an iterative detector that dominates the overall receiver complexity without affecting the rest of the receiver design. Second, an RS-SISO may be an effective method for complexity reduction even for systems considered in isolation (*e.g.*, as an alternative to DDFSE for a reduced complexity approximation to the Viterbi algorithm).

3.3.2   Example Applications of the RS-SISO

Two example transmission systems – an ISI-AWGN system and a TCM-ISI-AWGN system – are used to investigate the features of the RS-SISO. In the former case, the RS-SISO is used in the place of hard decision algorithms. In the latter, it replaces the standard SISO in an iterative detector. For the applications considered, min-sum processing and min*-sum processing yield nearly identical results so all processing is done using min-sum algorithms.

ISI-AWGN System   This example application is summarized in Fig-3.12. Two 12-tap ($L = 11$) ISI-AWGN channels are used (Fig-3.12(a)). Channel A has equal entries and Channel B is chosen to be

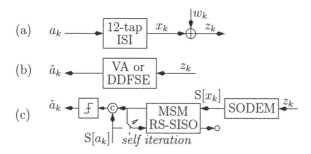

Figure 3.12.   (a) The tested 12-tap ISI-AWGN channel model, (b) hard decision detectors and (c) a RS-SISO based detector.

$(c, 2c, \cdots, 12c)$, where c is a positive constant. Both channels are normalized to have unit power, *i.e.*, $\|\mathbf{f}\| = 1$. BPSK modulation of an iid-uniform source (*i.e.*, $a_k = \pm\sqrt{E_b}$) is used at the transmitter. The output of the ISI channel is corrupted by AWGN with $\mathbb{E}\{w_k^2(\zeta)\} = N_0/2$. For comparison, two hard decision algorithms are also tested under identical conditions (Fig-3.12(b)): the VA [Fo73] and DDFSE [DuHe89]. As described earlier, DDFSE is equivalent to the forward ACS recursion of the RS-SISO described with a traceback used to provide the final decisions. Thus, similar to the RS-SISO, a truncated state of length L_1 is defined in DDFSE. The detector using the RS-SISO with self-iteration is illustrated in Fig-3.12(c).

In all numerical experiments conducted, the convergence of the RS-SISO with self-iteration has been observed. A typical example is shown in Fig-3.13 which indicates that convergence occurs after 4–5 iterations in this case. Although any number of self-iterations may be used to trade complexity and performance, we select the number of self iterations to be the minimum number for which convergence is achieved and denote this by N_c. The performance of the three algorithms on channel A is

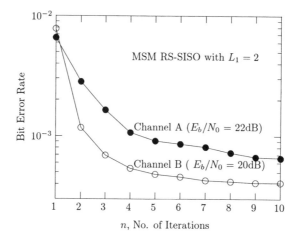

Figure 3.13. The convergence of a min-sum RS-SISO with self-iteration ($L_1 = 2$) for an isolated ISI channel.

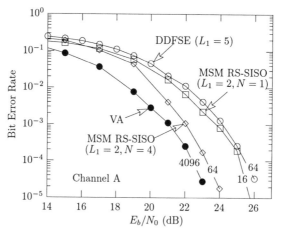

Figure 3.14. The performance comparison of various detection algorithms for Channel A. The number attached to each curve is the complexity index of the corresponding algorithm.

compared in Fig-3.14. Attached to each curve is a complexity index C defined as the product of the number of transitions $|\mathcal{A}|^{L_1+1}$, the self-iteration number N and the recursion number r, *i.e.*, $C = rN|\mathcal{A}|^{L_1+1}$ ($|\mathcal{A}| = 2$ in this example). For the forward-only algorithms (*e.g.*, VA, DDFSE) $r = 1$, while $r = 2$ for the forward-backward SISO. The performance of the VA and its complexity index $C_{VA} = 4096$ are presented as a baseline. It can be shown that while thresholding the soft-output of the full-state MSM-SISO yields ML-SqD as the VA, thresholding the RS min-sum SISO does not yield the same result as DDFSE using the same L_1 (see Problem 3.14). For this example, DDFSE performs roughly 3 dB

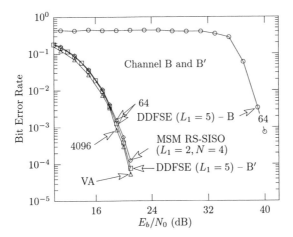

Figure 3.15. The robustness of the iterative detector using the MSM RS-SISO algorithm. Note the MSM RS-SISO performs the same for Channel B and B'.

worse than the VA at a BER of 10^{-4} when $L_1 = 5$, *i.e.*, $C_{\text{DDFSE}} = 64$. Even without self-iteration ($N = 1$), the RS-SISO with $L_1 = 2$ ($C = 16$) performs 0.3 dB better than a DDFSE algorithm with 4 times less complexity. This gain can be attributed to the bidirectional architecture of RS-SISO – *i.e.*, the use of decision feedback in both directions provides robustness to decision feedback error propagation. Moreover only 4 self-iterations ($N_c = 4$) improve the performance by an additional 1.9 dB. Thus, with the same complexity index, the RS-SISO outperforms DDFSE by 2.2 dB. Even compared to the VA, this RS-SISO performs only 0.8 dB worse but is 64 times less complex. As is typical with decision feedback algorithms, the degradation of the RS-SISO relative to the VA is smaller at high SNR.

The use of separate decision feedback in the forward and backward recursions makes the RS-SISO insensitive to non-minimum phase channels. The DDFSE, however, is very sensitive to non-minimum phase channels since it is based on the assumption that HDF is performed on weak taps of the channel. Channel B is a non-minimum phase channel. Channel B' is the time-reversed version of Channel B, *i.e.*, $\{12c, 11c, \cdots, c\}$ and is a minimum phase channel. The results in Fig-3.15 show that DDFSE with $L_1 = 5$ virtually fails for Channel B but works well for Channel B'. However, the detector based on RS-SISO with the same complexity ($L_1 = 2$ and $N = 4$) performs the same for both Channel B and B'. This is also true for the optimal Viterbi algorithm.

An Interleaved TCM-ISI-AWGN System In this experiment, an iid-uniform binary source is encoded by an 8-state, rate $R = 2/3$

Ungerboeck 8-PSK TCM code with the following generator matrix.

$$G(D) = \begin{bmatrix} D & D^2 & 1 \\ 0 & 1 & D \end{bmatrix}$$

The 8-PSK signals from the TCM encoder are fed into a 32 × 32 block interleaver. The interleaved 8-PSK signals pass through a 5-tap $(L = 4)$ ISI channel with equal entries (normalized to unit power), and the output is corrupted by complex circular AWGN w_k with $\mathbb{E}\{|w_k(\zeta)|^2\} = N_0$. This system is illustrated in Fig-3.16(a). Similar to the examples in

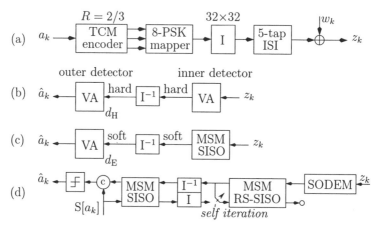

Figure 3.16.   (a) The tested TCM-ISI-AWGN channel model, (b) the VA-VA detector, (c) the SISO-VA detector and (d) the iterative RS-SISO/SISO detector. Note that the symbol attached to each outer detector represents the type of metric used. d_H denotes "Hamming distance" while d_E denotes "Euclidean distance".

Section 2.4.2, we consider the receivers shown in Fig-3.16. The detector in Fig-3.16(b) employs the VA at both the inner and outer stages with hard-in decoding (HID) of the TCM. In Fig-3.16(c) a better solution is shown, which replaces the inner VA by a forward-backward SISO which enables soft-in decoding (SID) of the TCM. Moreover, one can apply iteration between the inner and outer processors in Fig-3.16(c) as described in Section 2.4.2. The inner FSM of this concatenated system, however, has $8^4 = 4,096$ states which dominates the complexity of the overall receiver. Therefore, we consider using the RS-SISO with self-iteration in place of the inner SISO in Fig-3.16(c). For this RS-SISO based iterative detector shown in Fig-3.16(d), the complexity index is $rN_oN_i|\mathcal{A}|^{L_1+1}$, where N_i is the number of inner iterations (*i.e.*, the self-iterations of the RS-SISO at inner stage), N_o is the number of outer iterations (*i.e.*, the iterations between the two stages), and $|\mathcal{A}| = 8$ for the current example.

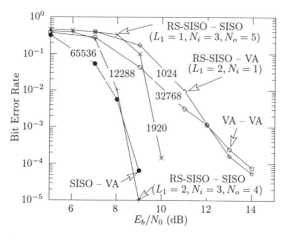

Figure 3.17. The performance comparison of various detection algorithms for the interleaved TCM-ISI channel. The number associated with each curve is the complexity index of the corresponding detector.

The performance of various detection schemes for this concatenated system is shown in Fig-3.17. First, replacing the VA by the MSM-SISO at the inner stage, a 5 dB gain in E_b/N_0 at a BER of 10^{-4} is achieved. Instead of the full-state MSM-SISO, min-sum RS-SISO with $L_1 = 2$ and without self-iteration (*i.e.*, $N_i = 1$ and $N_o = 1$) yields only a 0.3 dB gain at this BER. This comparable performance, however, is achieved with a complexity that is 32 times smaller than the full-state system of Fig-3.16(b). Two different RS-SISO based receiver configurations with iteration are considered. One such detector employs $L_1 = 1$ (8 states), and 3 inner iterations ($N_i = 3$). Hard decisions are made after five outer iterations ($N_o = 5$). Compared to the SISO-VA scheme in Fig-3.16(c), the performance degradation is only 1.1 dB while the complexity reduced by a factor of 34. In order to obtain better performance, the second iterative detector uses $L_1 = 2$ and $N_i = 3$. After four outer iterations ($N_o = 4$), a 0.3 dB gain is obtained over the SISO-VA approach while the complexity saving is roughly a factor of 5. For comparison, note that applying any reduced complexity hard-decision processor (*e.g.*, RSSE [EyQu88], DDFSE [DuHe89]) will perform worse than the VA-VA scheme in Fig-3.17.

3.4 A SISO Algorithm for Sparse ISI Channels

3.4.1 Sparse ISI Channel

The RS-SISO is a method of complexity reduction applicable to the soft inverse of a generic FSM channel. In this section we consider a suboptimal marginal soft inverse (SISO) for a specific type of ISI channel

– the sparse ISI (S-ISI) channel. We refer to this as the sparse SISO (S-SISO) algorithm [ChCh00b]. An S-ISI channel has an impulse response with large delay spread, but with energy concentrated in several small regions within this overall delay spread. This channel model may be applicable to various data communication systems. Two typical S-ISI channels are shown in Fig-3.18. In a high frequency (HF) radio channel

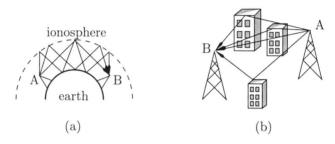

(a)   (b)

Figure 3.18.   Sparse ISI channels between site A and site B. (a) a high frequency radio channel, (b) a point-to-point high data rate wireless channel.

(3-30 MHz) [Go66, WaJuBe70], the signal is transmitted from one site to another by ionospheric reflection. This phenomenon usually results in an S-ISI channel since more than one ionospheric reflection path exists in practice and the spread in arrival time could be on the order of 3-5 ms due to the long distance between the two sites. A fixed point high data rate wireless connection can be modeled as an S-ISI channel because of the multipath transmission and the short symbol time. The S-ISI channel model can also be applied to certain mobile radio channels [Pr91]. In addition, the S-ISI structure is similar to certain error correction encoder structures. Specifically, the S-ISI channel may be viewed as a parallel concatenation of several ISI channels with small delay spread. A similar convolutional encoder with sparse generator polynomials was considered in [BeBeMa98]. The LDPCs considered in Chapter 2 have a similar graphical structure.

An example of S-ISI channel (baseband equivalent model) is shown in Fig-3.19. Usually a real-world S-ISI channel has many taps that are nearly zero which may be neglected in the system analysis. As a general ISI channel, the noisy observation of an S-ISI channel is

$$z_k = x_k + w_k = \sum_{i=0}^{L} f_i a_{k-i} + w_k \qquad (3.13)$$

where w_k is AWGN. The model in (3.13) may be viewed as the output of a whitened matched filter [Fo72b] or, with slight modification, may be viewed as a fractionally-spaced (over-sampled) model [ChPo96]. For an S-ISI channel, most of the terms in (3.13) are either zero or are small

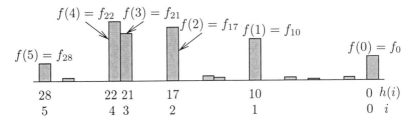

Figure 3.19. A sparse ISI channel with $L = 28$, $L_s = 5$. Those taps that are almost zero (not indexed) are not modeled in the detector design. The delay is shown increasing to the left.

enough that they may be modeled as such. Thus, for an S-ISI channel, we denote the i-th non-zero tap (NZT) by a 3-tuple $(i, h(i), f(i))$. Given the NZT index i, $h(i)$ represents the original index of this NZT in the S-ISI channel, and $f(i)$ is its value, *i.e.*, $f(i) = f_{h(i)}$. The total number of NZTs is $L_s + 1$. Note $L_s \ll L$ since the channel is sparse. We assume that f_0 and f_L are non-zero – *i.e.*, $(0, 0, f_0)$ and (L_s, L, f_L) are present in the S-ISI model. With this notation, (3.13) can be specialized for the S-ISI channel to

$$x_k = \sum_{i=0}^{L_s} f(i)a_{k-h(i)} \qquad (3.14)$$

In order to describe classes of S-ISI channels, some definitions are required. First, define a tap distance function $d(i, j) \triangleq |d_1(i, j)|$ where $d_1(i, j) \triangleq h(i) - h(j)$. Then, we say that NZT i_1 and i_2 are *adjacent* if $d(i_1, i_2) = 1$. The i-th NZT is *isolated* if it is not adjacent to any other NZT, *i.e.*, $d(i, i-1) > 1$ and $d(i+1, i) > 1$. A set of NZTs $\{i : i_1 \leq i \leq i_2\}$ is called *grouped* if $d(i, i-1) = 1$ for $i_1 + 1 \leq i \leq i_2$. An S-ISI channel is called *regular* if the NZT set $\{i : nl_g \leq i \leq (n+1)l_g - 1\}$ is grouped for $n = 0, 1, \cdots, M_g - 1$ (*i.e.*, $M_g l_g = L_s + 1$) and $d(i, i-1) = d_g > 1$ for $i = l_g, 2l_g, \cdots, (M_g - 1)l_g$; otherwise, it is called *irregular*. An S-ISI channel is called *discrete* if all its NZTs are isolated. An S-ISI channel is called *simple* if it is regular and discrete (*i.e.*, $l_g = 1$). The class of simple S-ISI channels is also called the zero-pad channel class in [McKeHo98]. These definitions are explained in the following example.

Example 3.7. ————————————————————
The S-ISI channel in Fig-3.19 is irregular with $L = 28$ and $L_s = 5$. The tap distance between NZT 0 and NZT 1 is $d(0, 1) = |0 - 10| = 10$. Also $d_1(0, 1) = 0 - 10 = -10$. Therefore $d_1(\cdot, \cdot)$ is not a distance. The 0th, 1st, 2nd and 5th NZTs are isolated. The 3rd and 4th NZT are grouped since $d(3, 4) = 1$. Thus, this S-ISI channel is not discrete.

Suppose that another S-ISI channel has 10 NZTs ($L_s = 9$) and its $h(i)$ set is $\{0, 1, 10, 11, 20, 21, 30, 31, 40, 41\}$ ($L = 41$). Then it is regular and has $l_g = 2$, $d_g = 9$ and $M_g = 5$. However, this regular S-ISI channel is not simple since it is not discrete. As another example, the S-ISI channel with NZT at locations $h(i) \in \{0, 7, 14, 21, 28\}$ for $i = 0, \ldots 4$ is simple.

— *End Example*

The S-SISO algorithm is based on message passing similar to that described in Example 2.15. Specifically, for a discrete S-ISI channel, the loops associated with the graphical model will be relatively long and the associated message passing algorithm should work well while avoiding the state explosion associated with the optimal trellis processing (*i.e.*, see Fig-2.53). In order to include the case of grouped taps, however, we also introduce models that are hybrids of the graphical models shown in Fig-2.53 and Fig-2.58 and therefore lead to algorithms that are hybrids of the forward-backward algorithm, and the distributed algorithm described in Example 2.15. Furthermore, to reduce the complexity, we also would like to use connection cutting. In the following, we introduce the notation required to concisely describe these different schemes.

We define the *neighborhood* of a_k associated with the NZT set $\{i, (i + 1), \ldots j\}$ (*i.e.*, having $L_g = j - i + 1$ elements) as a set of $J = L_s - L_g + 1$ input symbols

$$\mathrm{N}_i^j(k) \triangleq \{a_{k+n} | n = d_1(i,l), l \in \{0, \ldots i-1\} \cup \{j+1, \ldots L_s\}\} \quad (3.15)$$

In words, the neighborhood $\mathrm{N}_i^j(k)$ is the set of input symbols, excluding $a_k, \cdots, a_{k+d_1(i,j)}$, which are included in (3.14) when the term $f(i)a_k$ appears. For each neighborhood set, a *fusion* set is defined as

$$\mathrm{F}_i^j(k) \triangleq \bigcup_{l=i}^{j} \{a_{k+d_1(i,l)}\} \quad (3.16)$$

It can be shown that their union, called the *support* set

$$\tau_k \triangleq \mathrm{N}_i^j(k) \cup \mathrm{F}_i^j(k), \quad (3.17)$$

determines the output $x_{k+h(i)}$ uniquely. The support set τ_k plays the role of the "transition" for (3.14) and corresponds to the components of the transition t_k for the $(L + 1)$-tap ISI channel associated with the NZTs.

The i-th NZT is called the *pivot NZT* associated with $\mathrm{N}_i^j(k)$ (and $\mathrm{F}_i^j(k)$) and $\{i, i + 1, \ldots j\}$ is called the associated *pivot NZT set*. For example, when the NZT set has only one element i ($L_g = 1$), the neighborhood of a_k associated with this NZT is denoted as $\mathrm{N}_i(k) = \mathrm{N}_i^i(k)$. In this case, the size of the neighborhood is $J = L_s$ and the fusion set is $F_i(k) = \{a_k\}$. This corresponds to the case analogous to that shown in Fig-3.3, where all but one of the connections departing each X_k^{-s} have been cut. More generally, soft information is passed back to the SOBCs from X_k^{-s} for the members of the fusion set.

Example 3.8. ──

The concept of the neighborhood is illustrated in Fig-3.20 for the S-ISI channel in Fig-3.19. The neighborhood set $N_2(k)$, shown in Fig-3.20(a),

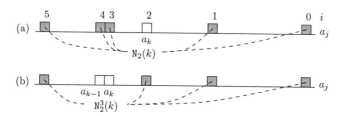

Figure 3.20.   Neighborhood of a_k for an S-ISI channel. (a) $N_2(k)$, (b) $N_3^4(k)$.

is a collection of 5 input symbols (shaded squares). For this case, NZT 2 is the pivot NZT. When $f(2)a_k$ appears in (3.14), a_{k+17}, a_{k+7}, a_{k-4}, a_{k-5} and a_{k-11} are also involved in (3.14) – these symbols constitute the neighborhood $N_2(k)$. The fusion set for this example consists of a_k only (the hollow square). The support set τ_k is the union of $N_2(k)$ and $F_2(k)$.

Similarly, as shown in Fig-3.20(b), when $f(3)a_k$ (also $f(4)a_{k-1}$) is included in (3.14), a_{k+21}, a_{k+11}, a_{k+4} and a_{k-7} constitute $N_3^4(k)$. The corresponding fusion set $F_3^4(k)$ consists of a_k and a_{k-1}. Again, the support set is the union of $N_3^4(k)$ and $F_3^4(k)$.

── *End Example*

3.4.2   Existing Algorithms for S-ISI Channels

Decision feedback equalization (DFE) [BePa79] is a generic solution for ISI mitigation. For the S-ISI channels, both the linear feedforward and feedback filter in a conventional DFE can be fairly long. In [FeGeFi98], a fast algorithm is developed to compute optimal DFE settings for the S-ISI channels. The complexity of DFE is relatively low as compared to trellis-based algorithms. In [BeLuMa93, BeSa94, BeMa96], a so-called multi-trellis VA (MVA) was developed for S-ISI channel that reduces the complexity significantly as compared to that of the Viterbi algorithm. The MVA is based on an irregular trellis construction. This trellis construction and the resulting algorithm is *ad hoc* because it ignores some data dependencies and uses early decisions. This is exacerbated when the S-ISI channel has any two adjacent strong taps. Furthermore, when the structure of the S-ISI channel is complicated (*e.g.*, ≥ 5 non-zero taps) and a reasonable traceback depth is used (usually $5L$ - $7L$ [HeJa71]), the construction of MVA can become fairly complicated. In [McKeHo98], the so-called parallel trellis VA (PTVA) is developed which is applicable only to simple S-ISI channels. It can be shown that the PTVA yields the MAP-SqD solution. Both the PTVA and MVA have a complexity determined by the number of non-zero taps instead of the length of memory. Also, based on the development in Chapters 1 and 2, modification of the PTVA to a forward-backward version which produces soft-out information is straightforward. In [CuMa99] an sum-product SISO algorithm is developed for S-ISI channels which has an L^2VS structure with lag $D = (L-1)$. This algorithm uses soft decision feedback to obtain a soft output that approximates the desired APP. No iteration was used in [CuMa99].

3.4.3   The Sparse SISO Algorithms for S-ISI Channels

We first present an S-SISO algorithm based on the graphical model of Fig-2.58 which also includes connection cutting as an option. We refer to this as the distributed S-SISO (DS-SISO) algorithm. Second, we consider a hybrid of the graphical models in Fig-2.53 and Fig-2.58 applied to the S-ISI channel. This allows use to model grouped taps together as an FSM subsystem and the isolated taps in the parallel fashion of Fig-2.58. This results in what we refer to as the grouped sparse SISO (GS-SISO).

3.4.3.1   Distributed S-SISO Algorithms Associated with a Single NZT

Consider an algorithm based on a single pivot NZT. Any single NZT in an S-ISI channel, *e.g.*, the i-th NZT, can be assigned as the pivot NZT. Associated with this given pivot NZT, a SISO algorithm can be developed. To distinguish soft information on a given quantity associated with different pivots, a subscript i is used in $S_i[\cdot]$ for the corresponding algorithm. As illustrated in Fig-2.58 for a 3-tap ISI channel, a generic ISI channel can be modeled graphically as a collection of subsystems $\{X_k\}$. This graphical model is also applicable to arbitrary S-ISI channels. As described in Section 2.6, therefore, the corresponding SISO can be obtained by message passing as shown in Fig-2.59. However, this SISO may be prohibitively complex when the number of NZTs is large. When the j-th NZT is relatively small, the correlation between the related $x_{k+h(j)}$ and a_k (*e.g.*, $x_{k+h(j)} = f_{h(j)}a_k + \cdots$) is weak. Thus, the soft information obtained by $X^{-s}_{k+h(i)}$ for a_k is relatively unreliable. By selecting a single pivot NZT properly, a simplified SISO algorithm is obtained by connection cutting as described in Example 3.4 and shown in Fig-3.3.

In terms of the general notation described above, the marginal soft-out information on the elements of the fusion set $F_i(k) = \{a_k\}$ is generated by marginalizing the soft information collected by $X^{-s}_{k+h(i)}$ from the neighborhood $N_i(k)$. Clearly, this is an approximation to the marginal soft inverse because of the loops and the connection cutting. Since the underlying graphical model ignores the temporal index originally embedded in the ISI channel, this S-SISO has a distributed architecture and is referred to as the distributed S-SISO (DS-SISO) algorithm.

Based on the above discussion, given the pivot NZT i, the operation of the DS-SISO can be specified directly and is illustrated in Fig-3.21. First, consider the n-th activation of $X^{-s}_{k+h(i)}$. At this point, $X^{-s}_{k+h(i)}$ has

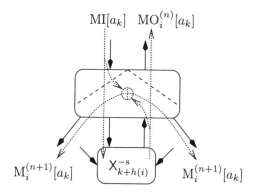

Figure 3.21.   Illustration of message passing in the SOBC associated with a_k in Fig-3.3 at the n-th iteration.

$MI[x_{k+h(i)}]$ from the channel, and $M^{(n)}_i[a_j]$ from the SOBCs for all a_j corresponding to τ_k. Activation of $X^{-s}_{k+h(i)}$ produces soft output only for the fusion set $F_i(k) = \{a_k\}$ via

$$MO^{(n)}_i[a_k] = \min_{\tau_k:F_i(k)} \left(MI[x_{k+h(i)}(\tau_k)] + \sum_{N_i(k)} M^{(n)}_i[a_j(N_i(k))] \right) \quad (3.18)$$

which is sent back to the SOBC associated with a_k. Notice that this is in extrinsic form because a_k is not part of the neighborhood. The following activation of the SOBC node associated with a_k is trivial due to the connection cutting. This SOBC has $MI[a_k]$ from the source and $MO^{(n)}_i[a_k]$ from $X^{-s}_{k+h(i)}$. The SOBC sends $MO^{(n)}_i[a_k]$ back to the source (*i.e.*, the desired soft-out information on a_k), sends $\text{MI}[a_k]$ to $\mathsf{X}^{-\text{s}}_{k+h(i)}$ corresponding to the pivot NZT and sends $\text{M}_i^{(n+1)}[a_k] = \text{MI}[a_k] + \text{MO}_i^{(n)}[a_k]$ to all $\mathsf{X}^{-\text{s}}_{k+h(n)}$ for $n \neq i$ corresponding to the neighborhood.

The iteration process may be continued until some stopping criterion is met or simply stopped after N iterations. If necessary, soft-out information for $x_{k+h(i)}$ can be obtained via the final activation of $\mathsf{X}^{-\text{s}}_{k+h(i)}$ – i.e.,

$$\text{MO}_i^{(N)}[x_{k+h(i)}] = \min_{\tau_k : x_{k+h(i)}} \sum_{\tau_k} \text{M}_i^{(N)}[a_j(\tau_k)] - \text{MI}[x_{k+h(i)}] \qquad (3.19)$$

Notice that (3.19) is similar to (3.18) with soft-in on a_k included and the soft-in contribution on $x_{k+h(i)}$ removed to convert to extrinsic form. Similarly, the soft information passed back from $\mathsf{X}^{-\text{s}}_{k+h(i)}$ to the SOBC for a_k is the extrinsic soft-out information for a_k due to connection cutting.

3.4.3.2   S-SISO Algorithms for a Grouped NZT Set

When a grouped NZT set is assigned as the pivot NZT set, a corresponding graphical model for the S-ISI is illustrated in Fig-3.22. In Fig-

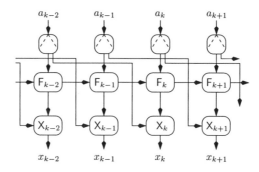

Figure 3.22.   Another graphical model for generic ISI channels (illustrated by a 3-tap ISI channel and the pivot NZT set consists of the first two taps).

3.22, Tap 0 and 1 of this ISI channel are chosen as the pivot set, which determines the fusion set $\text{F}_0^1(k) = \{a_{k-1}, a_k\}$. In Fig-3.22 this set corresponds to the subsystem F_k. As illustrated, $\text{F}_0^1(k)$ is uniquely determined by the current input a_k and the previous fusion set $\text{F}_0^1(k-1)$. However, $\text{F}_0^1(k)$ alone is not enough to determine x_k, which also depends on the current neighborhood $\text{N}_0^1(k)$. This dependence is illustrated in Fig-3.22 as the connection from the broadcaster of a_{k-1} to X_k. If not for this connection, the rest components in Fig-3.22 would constitute a cycle-free graph on which the forward-backward SISO can be run. The SISO associated with Fig-3.22 is shown in Fig-3.23. Again, connection cutting

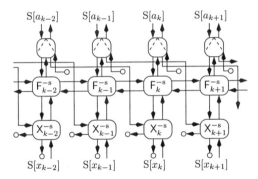

Figure 3.23. The SISO algorithm obtained by message passing on the model in Fig-3.22. Notice that connections from X_k^{-s}'s to soft broadcasters have been cut and are not shown.

is used for complexity reduction. Specifically, the connections from X_k^{-s} back to the soft broadcasters are cut. We refer to this as the grouped sparse SISO (GS-SISO) algorithm. One major difference between Fig-2.58 and Fig-3.22 is that the second model has not ignored the temporal axis completely. Consequently, the corresponding GS-SISO has a recursive architecture. Analogous to Fig-2.53(a), a cycle-free graphical model for the same ISI channel can be established when the fusion set includes all taps in the ISI channel. The corresponding SISO is just the standard forward-backward FI-SISO, which is the exact marginal soft inverse, but is prohibitively complex for most S-ISI channels. The DS-SISO and GS-SISO are developed by modeling the S-ISI channel as a graph with cycles and applying a connection-cut message passing algorithm. The only difference is that the GS-SISO is based on a model which treats certain grouped NZTs as an FSM subsystem.

Given a grouped NZT set $\{i, i+1, \ldots j\}$ where $L_1 = j - i$, a local state is defined as $s_k = \{a_{k-L_1}, \cdots, a_{k-1}\}$ (*i.e.*, this is the state of the FSM subsystem in Fig-3.22. The corresponding transition in the FSM subsystem trellis is the fusing set $F_i^j(k) = (s_{k-1}, s_k)$. Then a standard forward-backward SISO can be applied along this subgraph. This can be used to compute extrinsic information on the corresponding S-ISI inputs and the subsystem FSM transitions $F_i^j(k)$. Analogous to (3.18), the combining and marginalization performed at the n-th activation of the node $X_{k+h(i)}^{-s}$ is

$$M_i^{(n)}[F_i^j(k)] = \min_{\tau_k : F_i^j(k)} \left(MI[x_{k+h(i)}(\tau_k)] + \sum_{N_i^j(k)} M_i^{(n)}[a_l(N_i^j(k))] \right) \quad (3.20)$$

which is passed back to F_k^{-s} as the soft-in information on "transition" $F_i^j(k)$ for the subsystem FSM. Note this is in extrinsic form because the neighborhood and the fusion set are disjoint and the $\mathrm{M}_i^{(n)}[a_l]$ corresponding to the latter are excluded from (3.20). A similar message passing procedure occurs and a similar soft information $\mathrm{M}_i^{(n+1)}[a_k]$ yields in the SOBC of a_k as illustrated in Fig-3.21. Specifically, the SOBC here has $\mathrm{MI}[a_k]$ from the source and $\mathrm{MO}_i^{(n)}[a_k]$ from the forward-backward algorithm that runs on the subsystem FSM and is activated once after each activation of the SOBCs and $\mathsf{X}_k^{-\mathrm{s}}$'s. Again, on the final activation of $\mathsf{X}_{k+h(i)}^{-\mathrm{s}}$, soft-out information on $x_{k+h(i)}$ can be produced. Similarly, the final soft-out information on a_k is produced by the final activation of the corresponding SOBC in Fig-3.23.

3.4.3.3   Decision Feedback S-SISO and Multiple S-SISO Algorithms

In the DS-SISO or GS-SISO, the combining is performed over the whole neighborhood. This procedure (*i.e.*, (3.18) or (3.20)) determines the complexity of these SISO algorithms. A reduction in complexity can be obtained by applying hard decision feedback to a proper subset of the neighborhood. We refer to the resulting algorithm as the decision feedback S-SISO (DFS-SISO) algorithm. The complexity of the processing at $\mathsf{X}_k^{-\mathrm{s}}$ is reduced exponentially with the number of connections that use HDF. The hard decisions are enforced only during the combining and marginalization associated with $\{\mathsf{X}_k^{-\mathrm{s}}\}$ after which they are released.

For a given S-ISI channel, various DS-SISOs or GS-SISOs can be defined by assigning different pivot NZTs or NZT sets. For example, if an S-ISI channel has two significant NZTs 2 and 5, two distinct DS-SISOs can be developed based on these two pivot choices. Since each of these D-SISOs operate on different soft information, these two algorithms can be executed independently and simultaneously and the resulting soft outputs can be combined to yield a higher quality. In general, if a DS-SISO is run for each NZT in set $\mathcal{I}_d$ and a GS-SISO for each pivot NZT set in set $\mathcal{I}_g$, the final soft output for this multiple S-SISO algorithm can be obtained as

$$\mathrm{MO}[a_k] = \sum_{i \in \mathcal{I}_d} \mathrm{MO}_i^{(N_i)}[a_k] + \sum_{j \in \mathcal{I}_g} \mathrm{MO}_j^{(N_j)}[a_k] \tag{3.21}$$

where N_i is the number of iterations used in the corresponding S-SISO algorithm. This algorithm will be referred to as the multiple S-SISO (MS-SISO) algorithm since it consists of multiple S-SISO submodules. Intuitively, the MS-SISO algorithm should yield a more reliable soft output than any single submodule since it exploits more information on the structure of the S-ISI.

3.4.4   Features of the S-SISOs

Again, we define an index of the complexity index $C \triangleq TNr$, in which T is the number of transitions (*i.e.*, the cardinality of t_k or τ_k as appropriate), N is the number of iterations and r the number of recursions. Table 3.1 tabulates T, N, r and C for several algorithms that are applicable to an S-ISI channel and list some other relevant features. With the broadest applicability and highest complexity, both the VA

	VA	FBA	PTVA	MVA	DS-SISO	GS-SISO
Applicable S-ISI	arbitrary	arbitrary	simple	short	arbitrary (single)	arbitrary (grouped)
Optimality In/Out	opt s/h	opt s/s(h)	opt s/h	sub-opt s/h	sub-opt s/s(h)	sub-opt s/s(h)
T	M^{L+1}	M^{L+1}	M^{L_s+1}	$\sim M^{L_s+1}$	M^{L_s+1}	M^{L_s+1}
N	1	1	1	1	N	N
r	1	2	1	1	1	2
C	M^{L+1}	$2M^{L+1}$	M^{L_s+1}	$\sim M^{L_s+1}$	NM^{L_s+1}	$2NM^{L_s+1}$

Table 3.1.   Comparison of algorithms for S-ISI channels. Note "s" stands for soft, "h" for hard, and "(h)" means that the hard decision is available if necessary. FBA is short for the forward-backward algorithm.

and the standard forward-backward SISO algorithm yield the optimal sequence decision, but their complexity is prohibitive in practice for S-ISI channels. Both the PTVA and the MVA can reduce this complexity significantly with acceptable performance, but their applicability is limited to a small subset of the S-ISI channels expected in practice. Also, they do not directly yield soft outputs, although modification using the concepts of Chapter 2 is conceivable. The S-SISOs described herein are applicable to arbitrary S-ISI channels, have a good deal of flexibility for trading complexity for performance and may be used in place of the standard forward-backward SISO in a given iterative detection network.

3.4.5   Design Rules for the S-SISO Algorithms

Due to the variations possible in the graphical models, there can be many versions of the S-SISO algorithm. Several rules can be established to streamline the algorithm specification. In the following numerical experiments, the transmitter uses a BPSK modulation of an iid-uniform binary source (*i.e.*, $a_k = \pm\sqrt{E_b}$), and the output of the S-ISI channel is corrupted by an AWGN w_k with $E\{w_k^2(\zeta)\} = N_0/2$. All S-ISI channels are normalized (*i.e.*, $\|\mathbf{f}\| = 1$). For compactness, the results will be labeled by the following convention: (i) the algorithm used, (ii) the channel simulated, (iii) the pivot NZT[5], and (iv) the number of iterations used. Usually the algorithm used is represented by the initial letter of its name. Further explanation of other labels will be given as needed. Some S-ISI channels to be used are listed in Table 3.2. Others will be specified in the following. Only the min-sum version of a given S-SISO algorithm is used in these experiments.

	i	0	1	2	3	4	5
A	$h(i)$	0	13	14	20	27	/
	$f(i)$	0.32	-0.25	-0.12	0.3	0.17	/
B	$h(i)$	0	4	10	11	17	21
	$f(i)$	0.72	-0.64	-0.85	-0.52	1.3	0.67
C	$h(i)$	0	6	12	18	24	/
	$f(i)$	0.29	0.5	0.59	0.5	0.29	/
E	$h(i)$	0	4	5	/	/	/
	$f(i)$	0.22	0.41	0.29	/	/	/
G	$h(i)$	0	10	26	27	89	103
	$f(i)$	0.36	-0.24	0.38	1.0	-0.23	1.19

Table 3.2.   S-ISI channels used in numerical experiments.

3.4.5.1   The Pivot NZT and the Type of S-SISO

To apply the S-SISO for an S-ISI channel, the pivot NZT (set) must be assigned first, which in turn determines the type of S-SISO to be used. The choice of the pivot NZT (set) is strongly dependent on the structure of the S-ISI channel. Define a modified version of Channel A, Channel A_i $(0 \le i \le 4)$, by replacing the value of i-th NZT Channel A by 0.8. The NZT with value 0.8, *i.e.*, the i-th NZT for Channel A_i, dominates the channel in energy – *i.e.*, by this we mean that $f^2(i) \gg f^2(j)$ for all $j \ne i$. The BER results for various S-SISO simulations are plotted in Fig-3.24. For comparison, the BER curve for a channel without ISI is also included as a lower bound (*i.e.*, this is the ISI-free lower bound from Example 1.21). In Fig-3.24(a) the DS-SISO is executed for Channel A_3 with all possible choices of the pivot NZT. The results show that the detector

---
[5]For a GS-SISO, only the first NZT in the pivot NZT set will be listed. The corresponding NZT set is the largest grouped NZT set in which the listed NZT is involved unless otherwise specified.

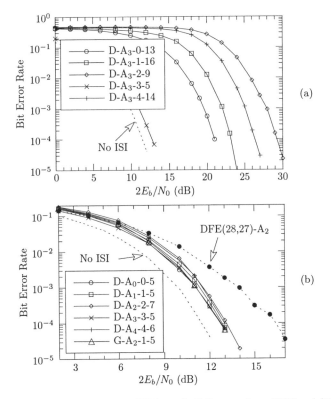

Figure 3.24. Performance of an S-SISO with different pivot NZTs. (a)Use different pivot NZTs for Channel A_3, (b)use the dominant NZT (set) as the pivot NZT (set). Note the pivot NZT set for G-A_2-1-5 is $\{1, 2\}$.

using the dominant 3rd NZT significantly outperforms other detectors. Moreover, the ranking (in terms of performance) of the curves in Fig-3.24(a) is consistent with the order of the energy in the pivot NZT used. Thus, when an S-ISI channel has a dominant NZT, it can be assigned as the pivot NZT for the DS-SISO. The results shown in Fig-3.24(b) also support this conclusion. Regardless of the location of the dominant NZT in the channel impulse response, the DS-SISO associated with it performs well in all experimental results.

Also shown in Fig-3.24(b) are the results for A_2, which is dominated by the NZT set $\{1,2\}$. For this channel, the GS-SISO associated with this set outperforms the DS-SISO associated only with the 2nd NZT. Thus, when a grouped NZT set dominates an S-ISI channel, improved performance is achieved by using this as the pivot NZT set for the GD-SISO. In general, when a detector can only afford a single S-SISO, the NZT (set) with the largest weight can be chosen as the pivot NZT (set) for the DS-SISO (GS-SISO).

3.4.5.2   Convergence of S-SISOs

The complexity of the S-SISOs developed grows linearly with the number of iterations used. Again, we use N_c to denote the observed minimum value of N that obtains virtually all of the iteration gain. The convergence of various S-SISOs is illustrated in Fig-3.25 for Channel B. It follows from Fig-3.25(a) that the closer to the center of the S-ISI chan-

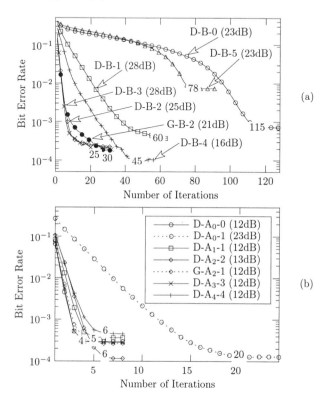

Figure 3.25. Convergence of various S-SISO algorithms. (a) D-SISO with different pivot NZTs for Channel B, (b) with the dominant NZT (set) for Channel A_i used for the pivot NZT. The number with dB-unit is the value of E_b/N_0 at which the corresponding algorithm is simulated. The number attached to each curve is the corresponding N_c.

nel the pivot NZT is located, the faster the convergence. Specifically, the DS-SISO associated with the 2nd or 3rd NZT has the same convergence rate as the GS-SISO associated with the NZT set {2,3}. A similar trend can be observed in Fig-3.24(a) where N_c is used for each curve. However, when the pivot NZT (set) dominates an S-ISI channel, N_c becomes almost independent on its location. This is demonstrated in Fig-3.25(b). Specifically, although the NZT 1 is much closer to the channel center than the dominant NZT 0, D-A_0-1 has an N_c 3 times as large as that of

D-A_0-0. Note that in all cases the BER curves become very stable after the convergence occurs. In summary, convergence occurs more rapidly when the pivot NZT (set) is located closer to the channel center, or it contains more weight.

3.4.5.3 Impact of Decision Feedback

For an S-ISI channel we can apply the hard information feedback on those NZTs with relatively small weights, *e.g.*, the NZT 1 and 3 in Channel B. The hard information used is the tentative hard decision obtained by thresholding $S_i^{(n)}[a_k]$ after each iteration. Using NZT 4 as the pivot NZT, several DFS-SISOs were simulated for Channel B with the results shown in Fig-3.26(a). These simulations suggest that the convergence of the DS-SISO is significantly slowed when hard decision feedback is used. Specifically, the more weight the feedback entry set contains, the

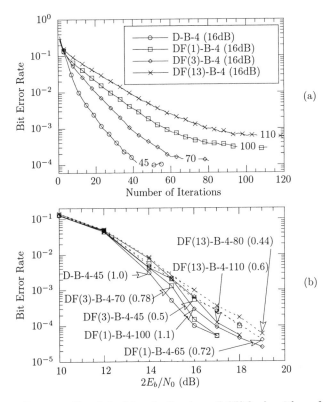

Figure 3.26. Impact of hard decision feedback on S-SISO algorithms for Channel B. (a) Convergence property of the DFS-SISO. (b) BER performance of the DF-SISO. The number following "DF-" is the feedback entry set. The number following a label is the corresponding complexity index normalized to that of D-B-4-45.

slower the DFS-SISO converges and the more performance loss the DFS-SISO experiences. Compared to the standard S-SISO these DFS-SISO algorithms only observe a fractional dB performance degradation (see Fig-3.26(b)). This actually implies that iteration greatly compensates for the effect of decision feedback. The effective complexity reduction is neutralized partially by the increase in N_c. Sometimes a DFS-SISO using N_c (*e.g.*, DF(1)-B-4-100) can be even more complex than the corresponding standard S-SISO. For each DFS-SISO, an $N < N_c$ is properly assigned and the resulting performance is shown in Fig-3.26(b). Specifically for DF(13)-B-4-80 and DF(3)-B-4-45, the complexity has been reduced by a factor of 1.5 with a performance loss of approximately 1 dB. If a large feedback entry is used, it is possible that error propagation will cause very poor performance of the DFS-SISO.

3.4.5.4   Performance Improvement by MS-SISO

When an S-ISI has a few NZTs of similar weight, an individual DS-SISO or GS-SISO may not provide satisfactory performance and the MS-SISO should be considered. Channel C is a simple but severe S-ISI channel. The performance of several MS-SISO algorithms on Channel C is shown in Fig-3.27. First, it is found that the more submodules an

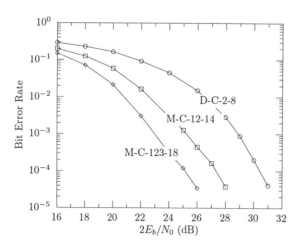

Figure 3.27. Performance improvement associated with the MS-SISO for Channel C. The notation "M-C-12-14" means that this MS-SISO consists of two DS-SISO submodules associated with NZT 1 and 2.

MS-SISO consists of, the slower the convergence. This is expected since the MS-SISO will not converge until all its submodules have converged. Therefore, N_c for an MS-SISO will increase when more submodules are included. Compared to D-C-2-8, M-C-12-14 gains 3.2 dB in SNR with the complexity increased by a factor of 3.5. Thus, it also appears that the more submodules, the better the performance.

3.4.5.5   Summary of Design Rules

In summary, when an S-ISI channel is specified, an S-SISO based detector can be designed using the following rules

1. If the S-SISO channel has a dominant NZT (set), choose it as the pivot NZT (set). Go to step 4.

2. If two NZTs (sets) have similar tap weight, choose the one closer to the channel center as the pivot NZT (set). Go to step 4.

3. If the S-ISI channel does not have a dominant NZT (set), an MS-SISO is suggested. Choose the heaviest NZTs and NZT sets as the pivot NZTs (sets).

4. If the S-ISI channel has some "light" NZTs, choose these as the feedback entry set and use the DFS-SISO for each pivot NZT (set). Otherwise, use the corresponding S-SISOs for each pivot NZT (set).

5. Determine N_c numerically. Use this value or choose a smaller value for N for further complexity reduction if desired.

Note that these are heuristics based on our numerical experiments and we expect them to lead to a reasonably good design. It is possible, however, that in some cases a better design could be found by another method.

3.4.6   Using the Sparse SISO Algorithms

The S-SISO can be applied to arbitrary S-ISI channels. For comparison, the DFE, MVA and PTVA are also considered when applicable. Specifically, the DFE(K_f,K_b) has a linear feed-forward filter of length K_f and a feedback filter of length K_b. The MMSE criterion [SmBe97] is used to determine the filter coefficients. The inputs to the feedback filter are the previously detected symbols.

First, the S-SISO can be applied to any channel to which the MVA is applicable. Channel E, referred to as the channel 2 in [BeSa94], is an S-ISI channel with only 3 NZTs. In Fig-3.28, we reproduce the performance of the MVA designed for this channel from [BeSa94]. The MS-SISO algorithm M-E-12-16 outperforms this MVA by approximately 0.5 dB at high SNRs, and betters the performance of the DFE(11,4) receiver by more than 5.5 dB. Also, this specific S-SISO algorithm performs within approximately 1 dB of the ISI-free lower bound (see Fig-3.28).

When the S-ISI structure becomes more complicated, developing the corresponding MVA becomes more difficult. Channel F is a regular S-ISI

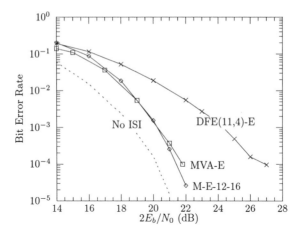

Figure 3.28.   Simulation results for S-ISI Channel E (8-PSK signal).

channel, defined by $f_0 = f_2 = f_{40} = f_{42} = 0.0993$, $f_1 = f_{20} = f_{22} = 0.352$ and $f_{21} = 1.0$. To this channel The GS-SISO algorithm is directly applicable to this channel while the MVA is not. For Channel F, the grouped NZT set $\{20, 21, 22\}$ contains most of the channel energy. Following the design rules developed in Section 3.4.5, a GS-SISO associated with this NZT set is a good choice. The performance of this GS-SISO is presented in Fig-3.29. The performance curves in Fig-3.29(a) show that $N_c = 7$ is a reasonable choice for the number of iterations. Furthermore, most of the performance gain is achieved by the first 4 iterations. Thus, using $N = 4$, the complexity can be reduced by almost 50% at a cost of approximately 1 dB in SNR. Moreover, Fig-3.29(b) shows that G-F-3-7 performs only 1.2 dB away from the ML-SyD lower bound described in Section 3.4.7. By comparison, the DFE(43,42) performs more than 2.1 dB worse than this GS-SISO.

Channel G is a fairly long and complicated S-ISI channel with 6 NZTs spread over a memory length of 104. For this channel, application of the MVA is complicated and is not considered. It can be shown that Channel G is dominated by NZT 3 and 5, so, according to the design rules, a reasonable choice can be either a DS-SISO associated with 3 or 5, or an MS-SISO consisting of both of them. The results for these three algorithms are shown in Fig-3.30. Similarly, Fig-3.30(a) shows that D-G-5 obtains a large gain through 10 iterations. In this case, using fewer than 10 iterations may not provide a good complexity/performance trade. For example, compared to $N_c = 10$, using $N = 5$ reduces the complexity by half but incurs a 9 dB performance loss. Using a DFS-SISO may be a better choice to reduce complexity. The performance of

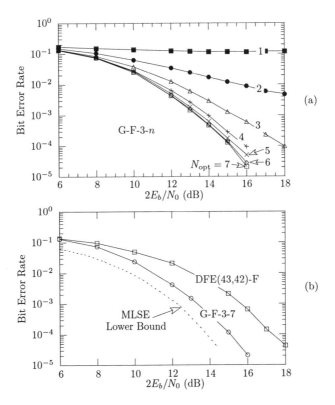

Figure 3.29. Simulation results for S-ISI Channel F. (a) Convergence property of G-F-3, (b) the BER performance of various algorithms.

various algorithms is shown in Fig-3.30(b). For this channel, the filters in the DFE are fairly long. By 12 iterations the D-G-3 works as well as the DFE(104,103). By assigning the heavier NZT 5 as the pivot NZT, D-G-5-10 obtains a 1 dB gain with a slightly lower complexity. As expected, an MS-SISO consisting of these two DS-SISOs has a better performance as shown in Fig-3.30(b). Specifically, M-G-35-10 performs 2 dB better than D-G-5-10 and more than 3 dB better than either D-G-3-12 or DFE(104,103). On the other hand, M-G-35-10 performs only 1.5 dB worse than the ISI-free performance bound. It should be noted that at low SNR, DFE(104,103) outperforms both DS-SISOs.

3.4.7   On Performance Bounds for S-ISI Channels

In theory, the performance bounds developed in Section 1.4.3-1.4.4 can be applied to the S-ISI channel. In practice, however, the memory length L is very long making evaluation of the bounds difficult at best. In particular, the upper bound is not practical for arbitrary S-ISI channels.

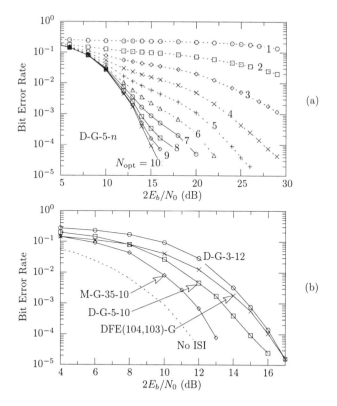

Figure 3.30. Simulation results for S-ISI Channel G. (a) Convergence property of D-G-5, (b) BER performance of various algorithms.

The lower bound can be used for S-ISI channels if one can determine a set of small distances that can be used for a uniform side information scheme. One obvious choice results in the ISI-free bound of Example 1.21. For a tighter upper bound, we can attempt to find error sequences that yield normalized distance less than unity. One approach is to use an S-SISO to solve the shortest path problem associated with the minimum distance (*i.e.*, see (1.131)). Specifically, a good detection algorithm can be used to find small distances to produce a lower bound. This method is used for the two-dimensional (2D) ISI channel in Section 5.2.1.

Simple S-ISI Channel   A simple S-ISI channel has the same distance spectrum as the "compact" ISI channel with coefficients $\{f_{h(0)}, f_{h(1)}, \cdots f_{h(L_s)}\}$. This can be seen by appealing to Fig-3.31 which shows that the two channel produce exactly the same output sequence. Thus, these

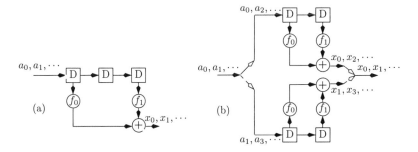

Figure 3.31.   Two equivalent models for a simple S-ISI channel.

two channels have the same achievable performance.[6] As a result, upper and lower bounds for the simple S-ISI channel can be obtained using standard techniques applied to the equivalent compact ISI channel.

Regular S-ISI Channel   For a regular S-ISI channel, sequences resulting in small distances may be obtained by appealing to a close relationship with 2D ISI channels. Specifically, as illustrated Fig-3.32, a regular S-ISI may be viewed as the equivalent of raster scanning a 2D ISI channel with NZTs that are concentrated in 2D. If the number of zeros between NZTs is large for the S-ISI channel, then this analogy becomes quite a good approximation. Thus, lower bounds for the regular S-ISI channel may be obtained using the methods described in Chapter 5.

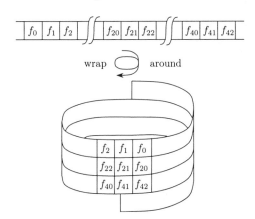

Figure 3.32.   Equivalence between a regular S-ISI channel and a two-dimensional ISI channel on a cylinder surface.

---

[6]This equivalence relationship can be used to directly inspire the PTVA and show that the PTVA is optimal.

3.5   Summary and Open Problems

Complexity reduction can be obtained using iterative detection in a number of ways. Possibly the most powerful method is the basic idea underlying iterative detection – *i.e.*, the efficient modeling of a system by a graph with loops. In this chapter we explored one application, the sparse ISI channel, where this modeling trick provides significant complexity reduction while providing near optimal performance. In addition to inherent gain, we also described a number of techniques that can be used to further reduce complexity. Most of these techniques may be viewed as decision feedback in one form or another.

When applying these tools for complexity reduction in an aggressive manner, it may be helpful to alter the iterative detection principles introduced in Chapter 2. In particular, filtering of soft information has been found to improve the performance of some iterative detectors operating on graphs with short cycles and parallel activation schedules.

Two methods were considered in detail. First, we demonstrated that a reduced state SISO can be used as a replacement for the forward-backward SISO for a generic FSM. This provides a method not only for complexity reduction in iterative detectors, but an attractive reduced complexity substitute for the Viterbi algorithm in some cases. Second, the S-ISI channel was considered in detail. Various modeling options, connection cutting, and hard decision feedback were all used to develop effective low complexity SISOs for this channel.

Although we have not attempted to give a single solution for complexity reduction, the concepts explored in this chapter are fairly generic. Some reasonable directions for future research include the adaptation of the RSSE (set partitioning) approach to the forward-backward SISO algorithm, and using RS-SISOs in place of several SISOs in an iterative detection network. Furthermore, a careful comparison of the various RS-SISO approaches suggested in the literature (especially with the self-iteration concept) is lacking. This is especially the case when complexity is measured not in terms of computation and storage counts, but in the ease of implementation. Finally, it may be possible to obtain reduced complexity SISOs for FSMs using modified linear and decision feedback equalizers, sequentially decoders, and other known reduced complexity hard-out data detection algorithms.

The general modeling methods used in the S-SISO remain an interesting and open area of research. For example, the discussion surrounding Fig-3.32 suggests that good 2D detection algorithms may be applicable to the S-ISI problem (for regular S-ISI channels at least). It may be possible to model more general S-ISI channels (and analogous error correcting codes) using higher dimensional models and/or collections of linked-graphs.

BER	Bit Error Rate
BPSK	Binary Phase Shift Keying
DDFSE	Delayed Decision Feedback Sequence Estimate
DFE	Decision Feedback Equalization
DFS-SISO	Decision Feedback S-SISO
DS-SISO	Distributed S-SISO
GS-SISO	Grouped S-SISO
HDF	Hard Decision Feedback
MS-SISO	Multiple S-SISO
MVA	Multi-trellis VA
NZT	Non-Zero Tap
PTVA	Parallel trellis VA
QPSK	Quadratic Phase Shift Keying
RS-SISO	Reduced State SISO
RSSE	Reduced State Sequence Estimate
S-ISI	Sparse InterSymbol Interference
SNR	Signal-to-Noise Ratio
S-SISO	Sparse SISO
VA	Viterbi Algorithm

Table 3.3. Table of abbreviations specific for Chapter 3

3.6 Problems

3.1. Let $M[a] \equiv -\ln P[a]$ be the metric domain equivalent of a given soft information measure in the probability domain. Show that sum-to-unity normalization can be accomplished in the metric domain using

$$\min{}^* \left(M[a = 0], M[a = 1], \ldots M[a = |\mathcal{A}| - 1] \right) = 0 \qquad (3.22)$$

Specifically, show that the above normalization implies that the sum of $\exp(-M[a])$ is 1. Can you find a simple method to implement the normalization implied by (3.22)? Can you adapt the "internal probability truncation" technique to the metric domain?

3.2. Show how trellis splicing can be applied to the PCCC decoder in Section 2.4.3.1. Specifically, suppose after the 4-th iteration, with sum-to-unity normalization and sum-product processing, the extrinsic soft-out from the SOBC is $P[b_k]$. Furthermore, consider the specific case of $P[b_{10} = 1] = 0.05$, $P[b_{11} = 1] = 0.56$, $P[b_{12} =$

1] $= 0.77$, $P[b_{13} = 1] = 0.95$, $P[b_{14} = 1] = 0.2$, and $P[b_{15} = 1] = 0.81$. Show the (spliced) trellis diagram for SISO1 of Fig-2.28 using early detection based on a probability threshold of 0.1. Repeat for a probability threshold of 0.25.

3.3. Given a pmf $P[a_j]$ for an iid sequence $\{a_k(\zeta)\}$, find the soft estimate $\tilde{a}$ that minimizes the mean square error (MSE)

$$\mathbb{E}\left\{\left|\sum_{|f_i|<\epsilon} f_i a_{k-i}(\zeta) - \sum_{|f_i|<\epsilon} f_i \tilde{a}\right|^2\right\}$$

3.4. Consider a nonlinear mapper with inputs $a_k, a_{k-1} \ldots a_{k-L}$ and output x_k. Suppose that soft information on $\{a_i(\zeta)\}$ is available in the probability domain with sum-to-unity normalization (*i.e.*, $P[a_k]$). Consider using soft estimate feedback to simplify the combining and marginalization associated with the soft inverse of this mapper. Specifically, consider the case where combining and marginalization is done only over a_{k-i} for $i = 0, 1, \ldots L_1$ with $P[a_{k-i}]$ for $i = L_1 + 1, \ldots L$ used to average out the effects of the other inputs. How does this differ from the linear case considered in(3.2)? Describe how this method could be used in a sum-product RS-SISO.

3.5. Consider the TCM-ISI system in Example 2.9 (*i.e.*, without interleaving). Draw the explicit index (graphical) diagram for this system using a standard graphical model for each FSM (*i.e.*, as in Fig-2.53(a)). What is the minimum cycle length on this model? Try using soft information filtering in the iterative detector of Example 2.9.

3.6. For soft information on a binary variable $b(\zeta)$, the belief degrading function maps the given soft information $p_0 = P[b = 0]$ and $p_1 = P[b = 1]$ into q_0 and q_1, both in the sum-to-unity normalization convention – *i.e.*, $\mathbf{q} = g(\mathbf{p})$. Show that any $g(\cdot)$ that is convex on the interval $[0.5, 1]$ and symmetric around $\mathbf{p}^T = [0.5\ 0.5]$ is a reasonable choice. Generalize this to the case of a nonbinary variable.

3.7. To slow convergence in detector of Example 3.5, try using belief degradation in place of soft information filtering.

3.8. For an $(L + 1)$-tap ISI channel tabulate the computation and storage requirements for the algorithm of Example 2.15 and the connection cut version with all but one of the connections back to the SOBCs cut. Roughly, what is the percentage of computational saving for $|\mathcal{A}| = 4$ and $L = 4$? Repeat Problem 2.25 for the connection cut version.

3.9. Show that the cross combining in Example 3.6 can be viewed as running an SW-SISO with $D = 1$ and using the resulting soft-in information on $x_k(t_k)$.

3.10. For the RS-SISO of Section 3.3, determine the values of n such that the trellis state v_n can be used for completion on a_k. Discuss how combining of soft-out information on a_k, obtained by different values of n could be combined to produce MO$[a_k]$ which uses all available observations.

3.11. For the RS-SISO of Section 3.3, the completion is done via (3.11). Describe why modification of this completion scheme to exclude the MI$[a_k]$ term (*i.e.*, as (2.13) replaces (2.12)) is not straightforward for the RS-SISO.

3.12. Describe other options for a RS-SISO in detail and discuss the relative advantages and disadvantages of these approaches:

(a) Show that the reduced state forward recursion in (3.6a) can be applied to both the unconstrained and constrained forward recursions in the L^{2}VS structure on an $|\mathcal{A}|^{L_1}$-state trellis.

(b) Consider a DDFSE (forward algorithm) that runs across the observation records and stores a sequence of hard decisions. Explain how these hard decisions can be fedback during a backward recursion on an $|\mathcal{A}|^{L_1}$-state trellis.

(c) Consider an algorithm that runs the forward recursion in (3.6a) across the entire observation interval and stores the forward truncated trellis transition metrics. A backward recursion is then run using these truncated transition metrics with completion performed on a_k using the truncated trellis transition τ_k.

3.13. Show that, for the RS-SISO presented in Section 3.3, the soft-out information of a_k is the same for a given channel and its time-reversed version. What does this imply about the robustness of the RS-SISO to non-minimum phase channels? For which variations in Problem 3.12 is this also the case? Is this the case for the full-state VA and forward-backward algorithm?

3.14. Explain why thresholding the RS-SISO of Section 3.3 yields different decisions than the corresponding DDFSE algorithm with the same number of states. Which of the variations in Problem 3.12 are threshold consistent with DDFSE?

3.15. Develop a recursive approach to simplify the computation of $\text{MO}[x_k]$ in (3.12).

3.16. Using the rules in Section 3.4.5, design an S-SISO for the S-ISI channel defined by a set of 3-tuples: $\{(0,0,-0.31), (1,3,0.52), (2,9,-0.9), (3,10,0.25), (4,21,0.15), (5,27,1.67), (6,35,-0.35)\}$.

3.17. Use Fig-3.31 or Fig-3.32 separately to prove that the MAP-SqD problem associated with the simple S-ISI channel can be split into M_g uncorrelated sub-problems. Show that finding the MAP-SqD for each sub-problem via a Viterbi algorithm yields the PTVA. Describe how a similar forward-backward algorithm for the simple S-ISI can be constructed.

Chapter 4

ADAPTIVE ITERATIVE DETECTION

In most practical situations, perfect channel state information (CSI) is not available at the receiver. Consequently, an iterative receiver should be able to deal with unknown, and possibly time varying parameters. Applications that can potentially benefit from these adaptive receivers include Trellis Coded Modulation (TCM) in interleaved frequency-selective fading channels, and Parallel and Serial Concatenated Convolutional Codes (PCCCs and SCCCs) with carrier phase tracking (or in the presence of flat fading).

A concise description of systems that include unknown parameters can be based on the graphical models discussed in Chapter 2. The fundamental difference between this and the perfect CSI case is that the unknown parameters are almost always *continuous*[1] in nature (*e.g.*, carrier phase, or fading amplitude). Based on this graph representation, adaptive iterative receivers can be constructed in a way similar to the perfect CSI case. The exact partitioning of the graph into subgraphs, as well as the exact algorithm used for soft-decision generation for each subgraph, lead to different receiver structures.

Example 4.1. ───────────────────────────────────
Consider the SCCC system discussed in Section 2.4.3.2 in the presence of unknown time-varying carrier phase. The graph describing the original system need only be augmented with the subgraph describing the dynamics of the unknown parameters, as shown in Fig-4.1. In this example,

───────────────────────
[1]It has been suggested however, that digitized models for these parameters be adopted, which allows the parameter estimation process to be viewed as another detection problem [Ca74, LeChPo99, KoWe99].

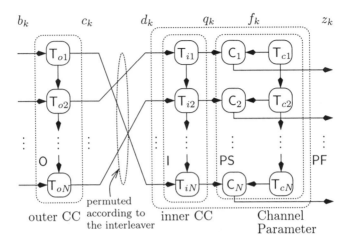

Figure 4.1. Graph representation of an SCCC in the presence parametric uncertainty. Subgraphs O and I correspond to the Outer and Inner code, respectively. Subgraphs PS and PF are related to the adaptive soft inverse blocks A-SODEM and A-SISO, respectively.

the unknown parameters are interacting with the rest of the system only through the transmitted symbols. Possible approaches for constructing adaptive receivers include:

- use of an external estimator – which is activated only once before the iterative detection processing, providing an initial estimate of the unknown parameter – followed by the standard perfect-CSI iterative detection network, using the estimate in place of the true value of the parameter, as shown in Fig-4.2(c).

- iterative detection with an external estimator which iteratively updates its estimate by utilizing decisions (soft or hard) on the output symbols. This scheme is suggested from the partitioning of the overall graph of Fig-4.1 into subgraphs O, I, and PS (*i.e.*, Parameter and Symbol), and the corresponding receiver block diagram is shown in Fig-4.2(d). The soft inverse block corresponding to PS will be referred to as the Adaptive Soft Demodulator (A-SODEM).

- iterative detection using an adaptive soft inverse algorithm corresponding to the partitioning of the overall graph into subgraphs O and PF (*i.e.*, Parameter and FSM). In this receiver, depicted in Fig-4.2(e), the inner block, which will be referred to as Adaptive Soft-Input Soft-Output (A-SISO) module, jointly estimates the parameters and generates soft information on the inner coded symbols.

End Example

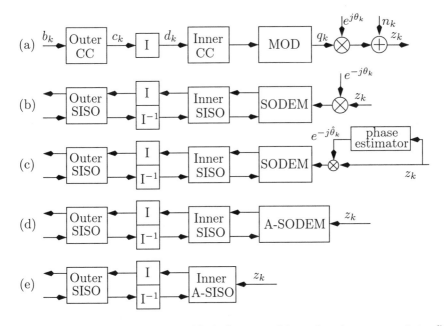

Figure 4.2.   (a) SCCC transmitter block diagram with carrier-phase uncertainty. (b) Iterative detection network for perfect CSI. (c) External estimator feeding a perfect-CSI iterative detector. (d) Iterative receiver resulting from partitioning the graph of Fig-4.1 into subgraphs PS, I and O. (e) Iterative receiver resulting from partitioning the graph of Fig-4.1 into subgraphs PF and O.

The derivation of optimal and sub-optimal (practical) A-SISO algorithms for the FSM model described in Section 1.3 is the main topic of this Chapter. In particular, the observation equations (1.44),(1.51) for the two modeling options for the unknown parameter Θ, namely the deterministic $(\Theta = \mathbf{f})$ and the stochastic Gauss-Markov $(\Theta(\zeta) = \{\mathbf{f}_k(\zeta)\})$, are the starting points for our derivations. We are particularly interested in A-SISOs, since A-SODEMs can be considered a special case. Indeed, the A-SODEM is just an A-SISO corresponding to a single-state FSM (*i.e.*, memoryless sequence).

For a generic quantity u derived from the input-output pair $(\mathbf{a}, \mathbf{x})$ of an FSM, soft-information of the form APP, MSM, GAP, and M*SM can be defined in a way similar to (1.32). In addition to the sequence related nuisance parameters, the unknown channel parameter Θ needs to be marginalized using either expectation or minimization (maximization) for the probabilistic and deterministic parameter models, respectively. The order of marginalization is important when the sequence and parameter marginalization operators do not commute (*e.g.*, $\min_{\mathbf{a}:u}$ and $\mathbb{E}_{\Theta(\zeta)}$), since it leads to different soft information. We will only derive exact expressions for the soft information defined with parameter marginalization performed first. We note that different, but meaningful soft metrics can also be defined by interchanging the marginalization order. These options will not be pursued in this work, mainly because they don't appear to lead to rigorously expressed optimal structures. To distinguish between different parameter marginalization options, the subscript "p" will denote marginalization (*i.e.*, averaging) over a probabilistic parameter model, and the subscript "d" will denote marginalization (*i.e.*, maximization or minimization) over a deterministic parameter model. The APP and MSM soft information of interest are defined as follows

$$\mathrm{APP_p}[u] \triangleq \sum_{\mathbf{a}:u} \mathbb{E}_{\{\mathbf{f}_k(\zeta)\}} \left\{ p(\mathbf{z}, \mathbf{a} | \{\mathbf{f}_k\}) \right\} = \sum_{\mathbf{a}:u} p(\mathbf{z}, \mathbf{a}) \tag{4.1a}$$

$$\mathrm{MSM_p}[u] \triangleq \min_{\mathbf{a}:u} [-\ln \mathbb{E}_{\{\mathbf{f}_k(\zeta)\}} \left\{ p(\mathbf{z}, \mathbf{a} | \{\mathbf{f}_k\}) \right\}] = \min_{\mathbf{a}:u} [-\ln p(\mathbf{z}, \mathbf{a})] \tag{4.1b}$$

$$\mathrm{APP_d}[u] \triangleq \sum_{\mathbf{a}:u} \max_{\mathbf{f}} p(\mathbf{z}, \mathbf{a}; \mathbf{f}) \tag{4.1c}$$

$$\mathrm{MSM_d}[u] \triangleq \min_{\mathbf{a}:u} \min_{\mathbf{f}} -\ln p(\mathbf{z}, \mathbf{a}; \mathbf{f}) \tag{4.1d}$$

4.1   Exact Soft Inverses – Optimal Algorithms

Equation (4.1) clearly suggests several options in manipulating $p(\mathbf{z}, \mathbf{a} | \{\mathbf{f}_k\})$ and $p(\mathbf{z}, \mathbf{a}; \mathbf{f})$ to obtain the proposed soft metrics.

- Maintaining the conditioning over the entire input sequence, marginalization (*i.e.*, expectation or maximization) can be performed on the unknown parameter depending on the underlying model. Marginalization of the resulting metrics over the nuisance parameters $\mathbf{a} : u$ is performed as a final step, leading to the final soft metrics for u. This approach is described in Section 4.1.1.

- On the other hand, the sequence and parameter marginalization can be performed in a single step, leading to the exact expression derived in Section 4.1.2.

It would be advantageous at this point to generalize the notion of the FSM state s_k and transition t_k to larger sequence portions, by defining *trellis* states and *trellis* transitions, respectively. For instance, a super-state and a super-transition can be defined as $s_k^s = (t_{k-d}, \ldots, t_{k-1}, s_k)$ and $t_k^s = (t_{k-d}, \ldots, t_k)$ for arbitrary d. This foreshadows the result that the optimal algorithms do not "fold" [Ch98] onto a trellis as in the case of known channel and that the size of the trellis eventually used is a design parameter.

4.1.1 Separate Sequence and Parameter Marginalization

We begin by deriving optimal algorithms for the evaluation of the soft outputs defined in equations (4.1a) and (4.1b) and more precisely the quantity $p(\mathbf{z}, \mathbf{a})$. Evaluating this quantity for each sequence $\mathbf{a}$ can be done using the Estimator-Correlator (EC) structure described in (1.49), (1.53). This technique, although efficient, results in suboptimal algorithms where complexity and smoothing depth are exponentially coupled, as in [ZhFiGe97].

An alternative optimal procedure for the metric calculation, is based on the forward-backward EC structure expressed in (1.87). Recall that for a forward-backward EC, and at a particular time instant k, the $|\mathcal{A}|^{k+1}$ metrics corresponding to the nodes of the forward tree are combined with the $|\mathcal{A}|^{K-1-k}$ metrics corresponding to the nodes of the backward tree (future) and weighted by the binding factor in (1.88).

The final soft output for a generic quantity u_m is the marginalization (*i.e.*, summation or maximization) over all factors with the same u_m. Note that the choice of k, the particular point in time when the past and future metrics are combined, is *completely arbitrary* (*i.e.*, it is not related to m). In a practical algorithm, however, the reference point k is chosen to be in the neighborhood of m, in order to maximize the number of relevant sequences combined to produce the soft information on u_m. Thus, while it may seem redundant to store and update both a forward and a backward tree (*i.e.*, same result can be accomplished with a single forward tree), the fact that the two trees can be pruned independently, decouples complexity and observation length, leading to practical algorithms, as will be discussed in Section 4.2.

For the deterministic parameter model, *i.e.*, (4.1c) and (4.1d), we need to evaluate $\max_{\mathbf{f}} p(\mathbf{z}, \mathbf{a}; \mathbf{f})$ for each sequence $\mathbf{a}$, which can be done using a similar forward-backward EC (see Problem 4.4 and [An99, AnCh99]), followed by marginalization over $\mathbf{a} : u_k$.

We stress that under the *folding* condition, the forward and backward trees fold and the forward-backward recursions can be performed on the corresponding trellises. This case is illustrated in the following example.

Example 4.2. ─────────────────────────────────

We now derive expressions for the forward-backward EC, for the special case presented in Example 1.14. The innovation term used in the backward recursion becomes

$$p(z_{k+1}|\mathbf{z}_{k+2}^{K-1}, s_{k+1}, \mathbf{a}_{k+1}^{K-1}) = \mathcal{N}^{cc}\left(z_{k+1}; \frac{a_{k+1}\alpha z_{k+2}}{a_{k+2}}; |a_{k+1}|^2 \sigma_u^2\right) \quad (4.2)$$

where $s_{k+1} = a_k$ does not appear in the innovations term. The backward innovations term depends on the symbols a_{k+1}, a_{k+2}, which means that the backward recursion can be performed on a $|\mathcal{A}|$-state trellis. The binding term simplifies to

$$\int \frac{\delta(f_k - z_k/a_k)\mathcal{N}^{cc}(f_k; \alpha z_{k+1}/a_{k+1}; \sigma_u^2)}{\mathcal{N}^{cc}(f_k; 0, \frac{\sigma_u^2}{1-|\alpha|^2})} df_k =$$

$$\frac{\mathcal{N}^{cc}(z_k/a_k; \alpha z_{k+1}/a_{k+1}; \sigma_u^2)}{\mathcal{N}^{cc}(z_k/a_k; 0, \frac{\sigma_u^2}{1-|\alpha|^2})} \quad (4.3)$$

The last equation provides an intuitive explanation for the binding term. Specifically, the numerator involves the quantity $|z_k/a_k - \alpha z_{k+1}/a_{k+1}|^2$, which is the squared difference between the forward filtered estimate and the backward predicted estimate. When these two estimates do not agree, a penalty is paid by means of decreasing the sequence probability (or increasing the sequence metric). We note that this example represents a very special case, and for the rest of the examples in this chapter, folding does not occur.

———————————————————————— *End Example*

4.1.2   Joint Sequence and Parameter Marginalization

The special form of $\text{APP}_p[u_k]$ allows us to obtain alternative expressions for the optimal soft outputs by explicitly marginalizing over both the parameter and the sequence in (4.1a), to obtain

$$\text{APP}_p[u_k] = \sum_{\mathbf{a}:u_k} \mathbb{E}_{\{f_k(\zeta)\}} \{p(\mathbf{z}, \mathbf{a}|\{f_k\})\} = p(\mathbf{z}, u_k) \quad (4.4)$$

We now derive exact expressions for the soft-output $\text{APP}_p[t_k]$ for the GM channel. A straightforward expression can be derived by utilizing the fact that the process $\{(t_k, \mathbf{f}_k)\}$ is a mixed-state Markov chain.

$$p(\mathbf{z}, t_k) = \int p(\mathbf{z}, t_k, \mathbf{f}_k) d\mathbf{f}_k =$$

$$\int p(\mathbf{z}_0^{k-1}, s_k, \mathbf{f}_k) p(z_k|t_k, \mathbf{f}_k) p(a_k) p(\mathbf{z}_{k+1}^{K-1}|s_{k+1}, \mathbf{f}_k) d\mathbf{f}_k \quad (4.5a)$$

where $p(\mathbf{z}_0^{k-1}, s_k, \mathbf{f}_k)$ and $p(\mathbf{z}_{k+1}^{K-1}|s_{k+1}, \mathbf{f}_k)$ can be updated by a forward and a backward recursion respectively

$$p(\mathbf{z}_0^k, s_{k+1}, \mathbf{f}_{k+1}) = \sum_{t_k:s_{k+1}} \int p(\mathbf{z}_0^{k-1}, s_k, \mathbf{f}_k)$$

$$p(z_k|t_k, \mathbf{f}_k) p(a_k) p(\mathbf{f}_{k+1}|\mathbf{f}_k) d\mathbf{f}_k \quad (4.5b)$$

$$p(\mathbf{z}_{k+1}^{K-1}|s_{k+1},\mathbf{f}_k) = \sum_{t_{k+1}:s_{k+1}} \int p(z_{k+1}|t_{k+1},\mathbf{f}_{k+1})p(a_{k+1})$$

$$p(\mathbf{f}_{k+1}|\mathbf{f}_k)p(\mathbf{z}_{k+2}^{K-1}|s_{k+2},\mathbf{f}_{k+1})d\mathbf{f}_{k+1} \quad (4.5c)$$

Unfortunately, the storage requirement for the above equations is infinite due to the fact that $\mathbf{f}_k$ takes values in a continuous space, making it of primarily conceptual value.[2] Although it is conceivable to quantize the channel values, we will follow another approach. A derivation similar to (1.87) leads to

$$p(\mathbf{z}_0^{K-1},t_k) = p(\mathbf{z}_0^{k-1},s_k)p(\mathbf{z}_{k+1}^{K-1}|s_{k+1})$$

$$\underbrace{\int \frac{p(\mathbf{f}_k|s_k,\mathbf{z}_0^{k-1})p(z_k|t_k,\mathbf{f}_k)p(a_k)p(\mathbf{f}_k|s_{k+1},\mathbf{z}_{k+1}^{K-1})}{p(\mathbf{f}_k)}d\mathbf{f}_k}_{b'_p(\cdot)} \quad (4.6a)$$

The forward and backward recursions for the first two quantities are as follows:

$$p(\mathbf{z}_0^k,s_{k+1}) = \sum_{t_k:s_{k+1}} p(\mathbf{z}_0^{k-1},s_k)p(z_k|t_k,\mathbf{z}_0^{k-1})p(a_k) \quad (4.6b)$$

$$p(\mathbf{z}_{k+1}^{K-1}|s_{k+1}) = \sum_{t_{k+1}:s_{k+1}} p(z_{k+1}|t_{k+1},\mathbf{z}_{k+2}^{K-1})p(a_{k+1})p(\mathbf{z}_{k+2}^{K-1}|s_{k+2}) \quad (4.6c)$$

Aside from the evident similarity of (4.6) with (1.87), there are two important differences: (i) the recursions described here do not depend (at least explicitly) on the entire path history, and (ii) the off-line evaluation of the third term of (4.6a) as well as the innovation terms in (4.6b) and (4.6c) is complicated due to the fact that they are mixed-Gaussian densities. Nevertheless, assuming that the latter difficulty can be overcome, the algorithm suggested by (4.6) is much simpler: only a forward and a backward recursion is performed over a state trellis, followed by combining (*i.e.*, multiplication) of the updated quantities with an appropriate weight (binding factor). This procedure is depicted in Fig-4.3. Once more we emphasize that the trellis states s_k^s and transitions t_k^s do not have to correspond to the FSM states and transitions.

The case of deterministic parameter modeling, *i.e.*, joint marginalization of the parameter and sequence in (4.1d), is not pursued further,

---

[2]The recursions in (4.5) are basically the well-known BCJR [BaCoJeRa74] recursions for a mixed-state Markov process.

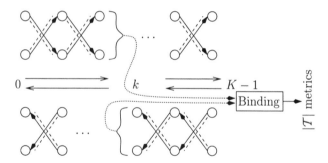

Figure 4.3.   Soft-metric evaluation in the case of joint sequence and parameter marginalization.

the reason being that the exact metric evaluation is cumbersome to explicitly express and does not offer any significant insight. Nevertheless, by utilizing the correspondence between the expectation and the maximization operator, meaningful suboptimal algorithms will be developed in the next section based on this subcase.

4.2   Approximate Soft Inverses – Adaptive SISO Algorithms

The exact evaluation of the soft metrics developed in the previous section under either modeling assumption for the unknown parameter involves likelihood updates on a forward and backward tree or trellis, assisted by per-path filters, followed by binding of the past and future metrics. In view of this fact, any suboptimal algorithm for the case of separate sequence and parameter marginalization can be interpreted as the result of applying one or more of the following simplifications to the forward-backward EC: (i) non-exhaustive tree search, (ii) non-Kalman (or non-RLS) parameter estimators, and (iii) suboptimal binding of the past and future metrics. Similarly, for the case of joint sequence and parameter marginalization, any suboptimal algorithm is the result of a simplifying assumption for the innovation terms, as well as a simpler form for the parameter estimators and binding term in (4.6). In the following, this design space is partially explored.

4.2.1   Separate Sequence and Parameter Marginalization

Regarding the tree search, all options available to prune the sequence tree in the case of hard-decisions [AnMo84] are candidates for use here as well. Breadth-first schemes seem to be the most appropriate for soft-decisions, because completion of the sequence metrics is required. Indeed, the fact that breadth-first algorithms maintain a common front in the search process facilitates the marginalizing task. Using the Viterbi algorithm (VA), and employing either the per-survivor processing (PSP) principle [RaPoTz95], or equivalently, the decision feedback assumption introduced in [SeFi95], practical A-SISO algorithms can be derived. The resulting algorithms, shown in Fig-4.4, consist of forward and backward recursions similar to the ones performed in the classical SISO. A KF (or

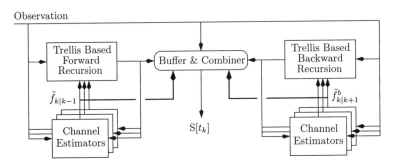

Figure 4.4. Trellis-based practical A-SISO algorithm with multiple estimators.

RLS for deterministic modeling) parameter estimate is kept for every trellis state and updated in a PSP [RaPoTz95] fashion. At this point we emphasize once more that the trellis on which this algorithm operates is not tightly related to the FSM trellis. Its size is a design parameter that determines the amount of pruning in the forward and backward trees, and eventually, the complexity of the algorithm.

Additional simplifications can be performed on the metric updates and the parameter estimates. One such simplification for the case of deterministic parameters can be achieved [An99] by approximating the information matrices used in the forward and backward RLS with $\mathbf{P}_k = (1-\rho)\mathbf{I}_{L+1}$ and $\tilde{\mathbf{P}}_{k+1}^b = (1-\rho)\mathbf{I}_{L+1}$, where $L+1$ is the parameter vector size and $\mathbf{I}_{L+1}$ is the identity matrix of size $L+1$. This simplification results in least mean square (LMS) parameter estimators and a simple and meaningful expression for the binding term $b_d(\cdot)$ as shown below

$$\tilde{\mathbf{f}}_k = \tilde{\mathbf{f}}_{k-1} + \beta \mathbf{q}_k^*(z_k - \mathbf{q}_k^{\mathrm{T}}\tilde{\mathbf{f}}_{k-1}) \tag{4.7a}$$

$$b_d(\tilde{\mathbf{f}}_k, \tilde{\mathbf{f}}_{k+1}^b) = \frac{1}{N_0}\frac{\rho}{(1-\rho^2)}||\tilde{\mathbf{f}}_k - \tilde{\mathbf{f}}_{k+1}^b||^2 \tag{4.7b}$$

In particular, (4.7a) describes the forward LMS parameter estimator with step size parameter β related to the RLS forgetting factor ρ by $\beta = (1-\rho)/(\rho + (L+1)(1-\rho))$. Regarding complexity reduction in the case of probabilistic modeling, reduced complexity KF is conceivable, although such solutions are application specific [RoSi97].

By dropping the backward recursion in the forward-backward EC, the forward-only FL algorithms proposed in the literature can be derived. The algorithm in [ZhFiGe97] calculates $\text{APP}_\text{p}[a_k]$ soft outputs in a FL configuration, using the T-algorithm [AnMo84] for path pruning and employing KF for channel estimation. To achieve the desired smoothing depth D, the forward algorithm is developed based on the trellis state $s_k^s = (t_{k-d}, \cdots, t_{k-1}, s_k)$, where d is selected such that a_{k-D} is included in s_{k+1}^s. Similarly in [AnPo98], a forward-only recursion is considered to produce $\text{APP}_\text{d}[a_k]$ and $\text{MSM}_\text{d}[a_k]$ soft outputs for the special FL case of the delay being equal to the channel length, with the VA used to prune the tree, and RLS channel estimation.

4.2.2   Joint Sequence and Parameter Marginalization

Starting from equations (4.6), suboptimal algorithms can be derived by employing a simplifying assumption for the innovation terms $p(z_k|t_k, \mathbf{z}_0^{k-1})$, and $p(z_{k+1}|t_{k+1}, \mathbf{z}_{k+2}^{K-1})$, which are in reality mixed Gaussian density functions. The Gaussian approximation leads to an attractive algorithm since only the state-conditioned/sequence-averaged forward (*i.e.*, $\tilde{\mathbf{f}}_{k|k-1}(s_k) = \mathbb{E}\{\mathbf{f}_k|s_k, \mathbf{z}_0^{k-1}\}$) and backward parameter one-step predictions together with the corresponding covariances need to be maintained and updated. Note that these estimates are only partially conditioned on the data sequence through the FSM state s_k (or more generally the trellis state s_k^s). Recursive update equations for these *partially conditioned* parameter estimates, first derived in [IlShGi94], are very similar to the KF recursions, thus we use the name partially conditioned KF (PCKF). Furthermore, in the limiting case when the trellis state represents the entire sequence, the innovation terms become precisely Gaussian and the PCKF become the sequence conditioned KF; this is the exact scenario of the separate sequence and parameter marginalization in the GM case.

In addition to the Gaussian approximation, a further simplification occurs under the assumption that the conditional means and covariances of the parameter are not functions of the states

$$\mathbb{E}\left\{\mathbf{f}_k|s_k, \mathbf{z}_0^{k-1}\right\} \cong \mathbb{E}\left\{\mathbf{f}_k|\mathbf{z}_0^{k-1}\right\} = \hat{\mathbf{f}}_{k|k-1} \tag{4.8}$$

This approximation – if valid – results in a desirable solution, since only a single forward and a single backward global estimator (averaged over the sequence) needs to be maintained and updated. Assuming that a probabilistic description $p(t_k)$ is available for the transition t_k (derived from a recent soft information on t_k), a recursion can be derived for $\hat{\mathbf{f}}_{k|k-1}$. These recursion equations, closely resemble those of the KF. The intuitive justification of this algorithm is that since a probabilistic description of t_k – and consequently $\mathbf{q}_k$ – exists, an average $\hat{\mathbf{q}}_k = \sum_{t_k} \mathbf{q}_k p(t_k)$ can be used in place of $\mathbf{q}_k$ in the KF recursions, thus resulting in what we refer to as an Average KF (AKF). The application of the AKF single-estimator idea is inhibited since (i) the independence assumption is not valid and (ii) an accurate $p(t_k)$ can only be derived from the observation $\mathbf{z}_0^k$ and is therefore tightly coupled with the estimation process. Both (i) and (ii) are alleviated by introducing a delay (advance) d in the parameter estimate to evaluate the forward (backward) transition metric at time k. Specifically, by increasing the tentative decision delay d, the accuracy of the approximation

$$\mathbb{E}\{\mathbf{f}_{k-d}|s_k, \mathbf{z}_0^{k-d-1}\} \cong \mathbb{E}\{\mathbf{f}_{k-d}|\mathbf{z}_0^{k-d-1}\} = \hat{\mathbf{f}}_{k-d|k-d-1} \qquad (4.9)$$

is improved. The resulting algorithm, that utilizes a d-lag (d-advanced) soft-decision-directed forward (backward) AKF, is depicted in Fig-4.5. The forward metrics at time k are updated using the d-delayed parame-

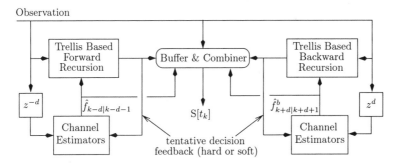

Figure 4.5. Trellis-based practical A-SISO algorithm with a single estimator.

ter estimate $\hat{\mathbf{f}}_{k-d|k-d-1}$. Starting at time k a d-step non-adaptive backward recursion is performed, at the end of which, a smoothed soft metric $p(t_{k-d}) \cong p(t_{k-d}|\mathbf{z}_0^{k-1})$ is obtained. The latter is then used in the AKF to update $\hat{\mathbf{f}}_{k-d|k-d-1}$. A similar one-step backward/d-step forward recursion is required for the update of the backward quantities.

Finally, by exploiting the correspondence between the expectation and maximization operator, a suboptimal algorithm can be derived for the case of joint marginalization and deterministic parameter model. This algorithm has a similar structure with the one described in the previous paragraph. Forward (backward) ACS operations are performed on the state trellis, aided by a single d-lag (d-advanced) hard-decision-directed forward (backward) RLS or LMS parameter estimation.

Starting from the A-SISO employing per-state PCKF, and by dropping the backward recursion, the algorithm described in [IIShGi94] is produced as a special case. Although the latter was not intended to provide soft decisions, the metric updates and parameter recursions (in the form of the PCKF) are precisely those developed therein. The A-SISO algorithm described in [BaCu98] can be regarded as a special case of the single-estimator (AKF) A-SISO presented earlier. Indeed, the latter is an FL, forward only version, operating on the trellis state $s_k^s = t_{k-1}$ with $d = 0$. Although the zero tentative decision delay eliminates the need for additional backward recursions, it seriously compromises the accuracy of the approximation in (4.9), motivating the non-zero delay d described herein. Similarly, the fixed complexity algorithm corresponding to the case of joint sequence and parameter marginalization for the deterministic parameter model can be regarded as a forward-backward extension to the conventional adaptive ML-SqD receiver [Ko71, MaPr73, Un74]. The latter is a modification to the VA, that uses a d-delayed, hard-decision directed, single external parameter estimate to update the metrics.

4.2.3   Fixed-Lag Algorithms

To apply the algorithms presented in Sections 4.2.1 and 4.2.2 in the tracking mode, it is assumed that both a forward and backward training sequences are present, providing the initial forward and backward parameter estimates. In many realistic situations, however, there is only a forward training sequence available. In addition there may be a necessity to process the received data without large delay. In these cases, an FL adaptive soft input soft output algorithm is required.

As mentioned in the previous sections, FL A-SISO algorithms have been derived in [ZhFiGe97, AnPo98, BaCu98] and can be interpreted as forward-only versions of FI A-SISOs operating on a super-trellis. In the following, we describe new FL A-SISO architectures based on the work in [HeChAn00]. In particular, the FL A-SISO is realized both in the forward-backward form [ChCh98b], as well as in the L^2VS [Le74, LiVuSa95] forward-only form. The resulting algorithms have linear complexity in D and estimate the channel without a backward channel training sequence.

Bi-directional Fixed Lag A-SISO Algorithms   An FL forward-backward A-SISO algorithm is realized by the same procedure as the FI forward-backward A-SISO, but the backward adaptive recursion is started at $k + D$ to obtain the $SO_0^{k+D}[u_k]$. In Fig-4.6(a), the backward recursion and the completion are illustrated. The challenge is how to initialize the backward channel estimates. Among the various possible solutions, the simplest method is to use the latest updated per-survivor forward channel estimate $\tilde{\mathbf{f}}(s_{k+D+1})$ as the backward channel

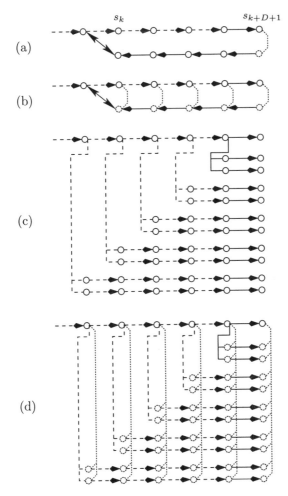

s_k s_{k+D+1}

(a)

(b)

(c)

(d)

Figure 4.6. Architectures for the FL algorithms: (a) Bi-directional recursions with forward-backward estimation, (b) Bi-directional recursions with forward-only estimation, (c) L^2VS-based A-SISO with estimation in both the constraint and unconstraint recursions, and (d) L^2VS-based A-SISO with estimation in the unconstraint recursion only. (The notation is similar to the one used in Fig-2.43).

initial coefficients $\tilde{\mathbf{f}}^b(s_{k+D+1})$. A variation on this backward initialization is also considered where the forward adaptive recursion is processed up to $k + D + d$ instead of $k + D$, d is an additional lag. When d is large enough, survivor merging will occur between times $(k + D$ and $k + D + d)$ with high probability. By storing all channel estimates for this forward recursion, one can traceback from time $k + D + d$ to find the channel estimate associated with the best state at time $k + D$. This estimate associated with the best survivor can then be used to initialize the backward channel estimates for all states.

Another variation, depicted in Fig-4.6(b) is to store the transition metrics computed during the forward adaptive-ACS processing. The backward ACS is then run using these stored metrics without channel adaptation. Since backward channel estimation is not performed in this approach, the issue of backward estimator initialization and binding is avoided.

Forward-Only Fixed Lag A-SISO Algorithms   Following the rational in Section 2.5.3, we consider two variations of the L^2VS algorithm, in the adaptive context. In the first variation, illustrated in Fig-4.6(c), both the constrained forward ACS and the unconstrained forward ACS recursions are adaptive ACS operations implemented in a PSP manner. The other variation is shown in Fig-4.6(d). When the unconstrained forward adaptive-ACS recursion is running, the transition metric (or channel estimates) are stored. As the constrained forward ACS recursions are running, the stored metrics (or channel estimates) are used in a PSP manner without adaptation. It may be shown that this second version is exactly equivalent to the forward-backward FL algorithm that utilizes the forward estimates in the backward recursion, described above. While a detailed proof of this fact is omitted, it follows similar reasoning as the known channel case. Thus, it also follows from similar reasoning that the forward-backward version of these two algorithms has less complex by a factor equal to the cardinality of u_k. However, the memory requirements for the forward-backward version are higher than the requirements for the forward-only version (see Problem 4.11). Note that the L^2VS-based A-SISO algorithm consists of forward only adaptive ACS operations. Therefore, neither the backward channel estimate initialization nor the binding term is applicable.

4.2.4   Forward Adaptive and Forward-Backward Adaptive Algorithms

The discussion of forward-backward FL A-SISO algorithms in the previous section raises some more general issues regarding A-SISO algorithm construction, that are relevant to FI structures as well. In particular when constructing a forward-backward A-SISO algorithm, two options are available. In the first one, adaptive processing is performed both in the forward, as well as in the backward direction. This is the approach suggested in all FI A-SISO algorithms discussed so far in Sections 4.2.1 and 4.2.2, and is also one of the options for building FL A-SISOs. Moreover, this is the approach that is directly implied by the optimal schemes developed in Section 4.1. We will refer to this generic processing as the forward adaptive backward adaptive (FABA) structure. In the second approach, adaptive processing is only performed in the forward direction, and all transition metrics (and/or parameter estimates) associated with the forward adaptive step are saved. It is now possible to perform the backward recursion in a non-adaptive way, by utilizing the previously saved metrics (and/or parameter estimates). This option will be referred to as the forward adaptive (FA) structure.

We would like to emphasize the distinction between FABA and FA processing only refers to the *adaptive* processing performed in an A-SISO, and does not reflect the *global* structure used to generate soft decisions, *i.e.*, forward-backward A-SISOs can be either FABA or FA. We also note that in general, FABA algorithms are more computationally intensive, while FA algorithms require more memory. Additionally, it is expected that FABA algorithms will perform better than FA, since the former can utilize binding of forward and backward metrics, which, by definition, is not applicable to the latter.

4.3   TCM in Interleaved Frequency-Selective Fading Channels

Consider the interleaved TCM system described in Section 2.4.2, where the frequency selective fading channel is unknown to the receiver. To facilitate acquisition of the channel shape, training sequences are inserted in the beginning and the end of each transmitted burst. Several options are available at the receiver front-end for pre-processing the received signal: lowpass filtering or matched filtering with the transmitted pulse shape, followed by fractionally-spaced sampling, followed by noise whitening (if necessary), as is extensively discussed in [Il92, YuPa95, Ch95, ChPo96, ChPo96b]. Regardless of the specific front-end structure, the front-end output can be modeled as an equivalent symbol-spaced vector ISI channel. Since our focus is on post-processing algorithms that are valid for any front-end processing, in order to improve the readability of the development, we choose to illustrate the concepts using a simplified symbol-spaced scalar ISI model.

An adaptive iterative receiver can be derived in a straightforward way from the non-adaptive version, by replacing the inner SISO with its adaptive equivalent, while leaving the outer SISO intact. Although there are many possible A-SISOs arising from the framework in Sections 4.1 and 4.2, we only utilize trellis-based algorithms. Several notes on the details of the implementation follow:

- Trellis-based multiple-estimator structures store and update one estimator per state with zero delay, while single-estimator schemes require d backward steps – for every forward step – to provide reliable tentative soft or hard data estimates to update their single estimator.

- Regarding the particular channel estimator used, the complexity increases in the order LMS, RLS, KF, AKF, PCKF, with the KF and the AKF having almost equal complexity.

- Optimal binding is, in general, a costly operation as shown in (1.88), while the suboptimal binding in (4.7b) results in a small increase in complexity relative to no binding.

- The trade-offs between FI and FL schemes, namely complexity vs. memory, are qualitatively the same as in the non-adaptive SISOs. The differences are amplified, however, due to the fact that channel related parameters need to be stored and updated in the A-SISOs, whereas only the forward and backward metrics are stored and updated in the perfect CSI case.

Example 4.3. —————————————————————————
Simulations were run for the transmission scheme described in Example 2.8. The convolutionally encoded sequence is interleaved using a 57×30 block interleaver. Each interleaver column is formatted into a TDMA burst together with a training sequence, equally split in 13 leading and 13 trailing symbols. Each burst is modulated and sent over a 3-tap equal power Rayleigh fading channel (each tap is assumed independent from the others) with normalized Doppler spread $\nu_d = 0.005$. Although the decorrelation time of such a channel is much larger than 57 symbols, for the purpose of simulation efficiency, a smaller interleaver depth is used in conjunction with the assumption of burst-to-burst independent channel. A rate 1/2, 16-state coded QPSK system is considered.

Comparison of FI A-SISOs Fig-4.7 presents performance curves for the iterative receiver described earlier with different options for the inner (equalizer) A-SISO. Regarding the naming of the presented algorithms, each algorithm is identified by a four-part label, each part of which denoting (i) the type of the soft decision (*i.e.*, APP or MSM), (ii) the multiplicity of the channel estimators (*i.e.*, S for single, or M for multiple), (iii) the particular channel estimator used (*i.e.*, KF, RLS, LMS, AKF), and (iv) the binding method (*i.e.*, Optimal Binding (OB), Suboptimal Binding (SB), or No Binding (NB)). Bit Error Rate (BER) curves for the first and fifth iteration are shown; no significant improvement was observed for more than five iterations. For the A-SISOs employing KF or

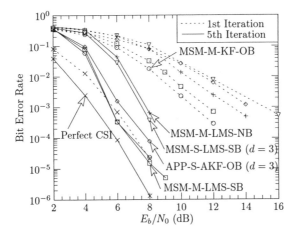

Figure 4.7. BER vs. E_b/N_0 for the TCM system and various configurations for the inner A-SISO. Performance is compared between (i) MSM-M-LMS-SB and MSM-M-LMS-NB, (ii) MSM-M-LMS-SB and MSM-M-KF-OB, and (iii) M and S.

AKF, the channel estimators were obtained by approximating the Clarke spectrum (with correlation function shown in (2.21)) with a first order model having 10dB-bandwidth equal to ν_d. Comparing the two curves corresponding to MSM-M-LMS, a loss of 2 dB (1 dB) is observed for the 5th (1st) iteration when no binding is performed. This outcome clearly indicates the significant practical – aside from the conceptual – value of the binding factor. The comparison between MSM-M-LMS-SB and MSM-M-KF-OB shows that LMS channel estimation with suboptimal binding is nearly as good as the KF with optimal – and computationally expensive – binding. In the first iteration the latter performs slightly better (by 0.7 dB at BER=10^{-3}), while in the fifth iteration no notable difference is observed. Multiple-estimator schemes are shown to be 2 to 4 dB better than single-estimator counterparts in the first iteration, while this gain is decreased to 0.5 to 2 dB after the fifth iteration as can be observed from the comparison of MSM-M-LMS-SB and MSM-M-KF-OB with MSM-S-LMS-SB or APP-S-AKF-OB. Note that the optimal value for the tentative delay was found to be $d = 3$ for both single-estimator receivers. The best A-SISO achieves performance that is just 1 dB away from that of perfect CSI. Regarding the iteration gain, as much as 6 to 7 dB can be gained using 5 iterations for both single or multiple estimator SISOs. This result is the direct antithesis with the perfect CSI case, where an iteration gain of only 1 dB does not even justify the need for ID. Simulation results that are not shown here confirm the negligible difference between APP and MSM algorithms for these operational SNRs, a fact which was noted in [AnCh97b, AnCh98] for the case of CSI as well. Finally, we note that the receiver based on PSP hard-decision inner equalization followed by hard decision VA performs approximately 9 dB worse than MSM-M-LMS-SB at a BER of 10^{-3}.

Comparison with FL A-SISOs   In this comparison, MSM algorithms with LMS estimation are utilized. Each algorithm is identified by (i) the type of observation window (*i.e.*, FI or FL), (ii) the type of FL structure (*i.e.*, FB for forward-backward or L^2VS), (iii) the type of binding (*i.e.*, SB or NB), (iv) the type of L^2VS channel adaptation (*i.e.*, CE for Constrained Estimation or UE for Unconstrained Estimation), and (v) the type of channel adaptation (*i.e.*, FABA, or FA). In Fig-4.8 the performance of the adaptive iterative detector is shown for the first and the fifth iteration. For comparison, two FL methods (FB, L^2VS) were shown with the FI algorithm (the two FI curves shown are the MSM-M-LMS-NB, and MSM-M-LMS-SB curves of Fig-4.7). As the FL-FB-FA and FL-L^2VS-UE algorithms generate the same soft output, they show the same performance in Fig-4.8. The performance of the FL-

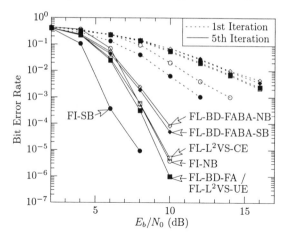

Figure 4.8.   The performance of the fixed-interval and fixed-lag adaptive SISO algorithms, D=6.

FB-FA/FL-L^2VS-UE algorithm is about 5dB in E_b/N_0 worse than that of the FI algorithm at a BER of 10^{-3} for the first iteration but is only 2dB worse for the fifth iteration. Note that the FL algorithm shows a larger iteration gain compared with the FI algorithm. For the FL-FB-FA algorithm, the backward ACS recursion was simulated with the previously stored forward channel estimates as described in the previous section. For the FL-FB-FABA algorithm we utilized the simplest backward initialization method which was to initialize the backward channel estimator with the latest estimated value of the forward channel estimator. As a result, the backward estimate has a strong dependency on the forward estimate. Therefore, we expect little inconsistency between the forward and backward channel estimates, and thus a small binding gain for the FL-FB-FABA algorithm. This is in contrast to the FI algorithm, which has a large binding gain (*i.e.*, 2dB). Although we do not present the results here, the variation of the FL-FB-FABA algorithm using channel estimates after merging was also simulated with no significant improvement observed. In conclusion, attempts to use backward channel estimation without a backward channel training sequence were not as successful as the FL-A-SISO algorithms with only forward channel estimation.

In Fig-4.8, the two different L^2VS-based adaptive FL algorithms are also compared. The FL-L^2VS-UE performs better than the FL-L^2VS-CE. The channel coefficients are estimated based on the symbol estimates in the PSP manner, and the symbol estimates come from the ACS recursion. Therefore, the constrained ACS recursion provides a constraint on the channel estimates. This constraint on the channel estimate apparently degrades the performance in the FL-L^2VS-CE. Although the FL-L^2VS-UE requires additional storage for the previous channel estimates, it yields a significant reduction in computational complexity relative to the FL-L^2VS-CE algorithm. However, the FL-L^2VS-UE is more complex by a factor of $|Q|$ which is the alphabet size of the symbol q_k (*e.g.*, $|Q|$=4 for QPSK) compared to the FL-FB-FA algorithm. As the alphabet size $|Q|$ increases, the FL-FB-FA algorithm is much less complex than the FL-L^2VS-UE algorithms while it produces *exactly* the same soft-information.

—————————————————————————————— *End Example*

Although in all previous examples the A-SISO trellis size was the same as the size of the underlying FSM trellis, it has been mentioned that the former is a design parameter and can be different from the latter. In particular, when soft information is thresholded to produce hard decisions, an A-SISO of the FA type (*i.e.*, with only forward adaptive processing) is threshold-equivalent (*i.e.*, it generates the same hard decisions) to the PSP algorithm [RaPoTz95] operating on the same trellis. On the other hand, an A-SISO of the FABA type (*i.e.*, with both forward and backward adaptive processing) produces different hard decisions. Furthermore, assuming that all these algorithms operate on the same trellis, it is expected that the latter will have a better performance, due to the utilization of two independent parameter estimators. Thus, using FABA algorithms in place of FA (or PSP) algorithms may be viewed as an alternative to increasing the trellis size for performance

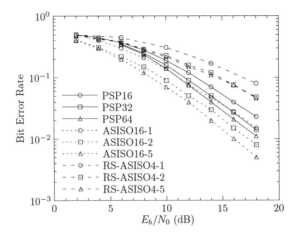

Figure 4.9. Performance comparison between different adaptive algorithms for sequence detection in an isolated ISI fading channel.

improvement. Clearly there is a trade off between number of trellis states (*i.e.*, search effort) and FA or FABA processing. This design alternative becomes even more interesting when considering additional complexity reduction techniques, *e.g.*, reduced-state A-SISOs in conjunction with self iteration, as discussed in Chapter 3 for the case of perfect CSI. In the following example, which is based on the work in [He00], we investigate these trade offs for the problem of data detection in an isolated fading ISI channel.

Example 4.4. ————————————————————————
A burst of uncoded QPSK symbols is transmitted over a 3-tap equal power Rayleigh fading channel (each tap is assumed independent from the others) with normalized Doppler spread $\nu_d = 0.005$. Each burst of length 57 symbols is padded with a training sequence, equally split in 13 leading and 13 trailing symbols. MAP-SqD for the perfect CSI case can be implemented using a VA with $4^2 = 16$ states. Three adaptive receivers are compared in the following, the last two of which are FABA type algorithms. The first is a standard PSP based receiver with $16, 32$, and 64 states. The second is an FI A-SISO (with LMS estimation and suboptimal binding) operating on a 16-state trellis and employing $1, 2$, and 5 self-iterations, similar to the one shown in Fig-3.12(c). The third is an FI reduced state A-SISO, similar to the one presented in Section 3.3.1, and shown in Fig-3.12(c), operating on a 4-state trellis and employing $1, 2$, and 5 iterations. Performance comparisons are presented in Fig-4.9. Comparing the two A-SISO based receivers, we conclude that for a given computational complexity, performance is enhanced by increasing

the state space, rather than increasing the number of self iterations. Furthermore, comparing the PSP and A-SISO receivers, it is evident that the same or better performance is obtained by running an FABA algorithm (*i.e.*, A-SISO) on a trellis, rather than running an FA algorithm (*i.e.*, PSP equivalent) in a trellis with double number of states. This result indicates that search effort can be effectively traded for FABA processing.

$\overline{}$ *End Example*

4.4   Concatenated Convolutional Codes with Carrier Phase Tracking

In Section 4.2 we derived practical A-SISO algorithms that are based on forward-backward processing and binding. We now derive similar algorithms for phase tracking in SCCC and PCCC, a nonlinear estimation problem, by appealing to the framework established for the linear estimation problem. In particular, the algorithm structure is the same as before, while the presented estimators are substituted by a first-order decision directed phase-lock loop (DD-PLL) of the form

$$\tilde{\theta}_k = \tilde{\theta}_{k-1} + \lambda \Im \left\{ z_k q_k^* e^{-j\tilde{\theta}_{k-1}} \right\} \tag{4.10}$$

where the notation is the same as in Fig-4.2.

4.4.1   SCCC with Carrier Phase Tracking

The baseline adaptive decoder of Fig-4.2(c) is derived based on the idea introduced in [LuWi98] for PCCCs. It consists of a single DD-PLL which uses decisions on the raw output symbols q_k, as well as the pilot symbols, to obtain a phase estimate and consequently derotate the observation; however, no feedback information on q_k from the inner SISO is utilized. A standard iterative decoder is then employed on the derotated observation – after discarding the pilot symbols – to produce final decisions on the source bits. The A-SODEM-based and A-SISO-based receivers of Fig-4.2(c) and (d) are constructed in a way similar to the linear estimation problem, by utilizing the DD-PLL of (4.10), and the following approximate binding term (which is implied by (4.7b)).

$$b(\tilde{\theta}_k, \tilde{\theta}_k^b) = \frac{1 - \lambda}{\lambda(2 - \lambda)} \frac{|e^{j\tilde{\theta}_k} - e^{j\tilde{\theta}_k^b}|^2}{N_0} \tag{4.11}$$

Example 4.5. $\overline{}$
The SCCC system presented in Example 2.10 is considered in this example. We generalize the transmission scheme by considering the insertion

of pilot symbols in the transmitted sequence. In particular, N_t pilot symbols are inserted in the transmitted sequence for every N_d coded symbols. The energy lost in the redundant pilot symbols is accounted for by lowering the transmitted symbol energy as

$$E_s = RR_t E_b = R_o R_i \log_2 |\mathcal{Q}| \frac{N_d}{N_d + N_t} E_b \qquad (4.12)$$

where E_b is the energy per information bit. In the development of iterative receivers for the above system, it is desirable to view the pilot symbols as part of the inner code by introducing a time-varying CC. The phase process is generated as a random walk as in [DaMeVi94]

$$\theta_k(\zeta) = \theta_{k-1}(\zeta) + \phi_k(\zeta) \qquad (4.13)$$

where $\{\phi_k(\zeta)\}$ is an iid sequence of zero mean Gaussian variables with variance σ_ϕ^2. Only APP-type soft decision algorithms are considered. The receivers consisting of the inner A-SISOs will be labeled as A-SISO-S/M-SB/NB, corresponding to single or multiple DD-PLLs and suboptimal binding of (4.11), or no binding respectively. Among the A-SODEM-based receivers two special cases are considered: (i) the single-state A-SODEM derived exactly as an A-SISO, and (ii) a single-state A-SODEM variant with forward-only recursions and no binding (labeled A-SODEM-FW). Finally, the baseline algorithm consisting of a single external DD-PLL operating on the raw 8-PSK symbols will be labeled EXT (*i.e.*, external DD-PLL). In all simulations presented here, the initial and final phase estimates are assumed ideal. Consequently, for a fair comparison between the External DD-PLL receiver and the proposed receiver structures, a forward DD-PLL starting at the beginning of the block is used to derotate the first half of the observation, while a backward DD-PLL starting at the end of the block is used for the second half of the observation. With such a scheme, the knowledge of both the initial and the final phase is utilized by the External PLL receiver. Note that interpolation between phase estimates obtained using the N_d-separated pilot symbols was found to perform poorly under all operational scenarios presented.

Fig-4.10 shows a comparison of the SCCC system with the industry standard rate 1/2, 128-state CC. The CC output is mapped on a QPSK alphabet resulting in a rate $R = 1$ (bits per channel use) code (no pilot symbols are used). ML-SqD with the aid of a VA is performed in the coherent case, while two adaptive receiver structures are considered. The first is the conventional adaptive-ML-SqD receiver of [Un74], consisting of a single DD-PLL driven by delayed tentative decisions from the VA, and the second is a PSP-based [RaPoTz95] receiver consisting of a VA with 128 DD-PLLs driven with zero-delay decisions. The

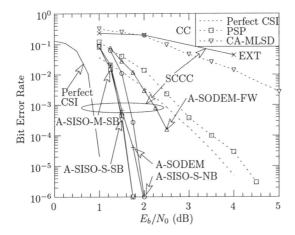

Figure 4.10. BER vs. E_b/N_0 for SCCC (interleaver size is $K = 16384$ symbols) with phase dynamics and various inner A-SISO and A-SODEM configurations (the optimal performance for S receivers was achieved for $d = 0$). The loss due to pilot insertion is $-10\log_{10}(R_t) = 0.27$dB. For comparison, the performance of CC with adaptive hard-decision detection is presented.

SCCC receivers considered are the EXT, A-SODEM, A-SODEM-FW, A-SISO-M-SB, A-SISO-S-SB, and A-SISO-S-NB. Simulations were run for $\sigma_\phi = 2$ (degrees) and λ was optimized for each E_b/N_0 value. Examining the CC performance curves, the following observations can be made. With perfect CSI, BER of 10^{-5} is achieved at $E_b/N_0 = 3.75$ dB. The PSP-based receiver operates at this BER with a loss of 0.4 dB, while the conventional adaptive ML-SqD receiver performs poorly resulting in a BER of 10^{-2} at 4 dB.

Simulation trials not shown here suggested that for the SCCC case, a reasonable pair for the training is $(N_t, N_d) = (16, 256)$ for a target BER of 10^{-5} and the mentioned phase dynamics. Regarding the performance of the proposed adaptive schemes for the SCCC, it can be noted that since the SNR loss due to the insertion of pilot symbols is $-10\log_{10} R_t = 0.27$dB, the actual loss due to the unknown phase is only 0.33dB (for the best adaptive scheme, *i.e.*, A-SISO-M/S-SB). This means that even if the state space is increased (by using super-state-based A-SISOs), the expected performance gain is very small (at most 0.33dB). As a result, in this particular applications, the adaptive algorithms built on the original state trellis suffice.

The comparison of the CC and SCCC curves clearly illustrates the importance of adaptive iterative detection. Under perfect CSI, the SCCC performs with a 2.6 dB gain over the standard CC. This gain vanishes when a PSP based ML-SqD receiver is used to decode CC and the EXT receiver is used for SCCC. By utilizing the more advanced A-SISOs or A-SODEMs proposed here, together with pilot symbols, the corresponding gain is increased to 3 dB.

——————————————————————— *End Example*

4.4.2   PCCC with Carrier Phase Tracking

We consider the PCCC system described in Section 2.4.3.1. Instead of using BPSK transmission, however, we assume that the systematic bit $c_k(0) = b_k$, together with the coded bits $c_k(1)$ or $d_k(1)$ are mapped to a QPSK constellation after alternate puncturing.

$$q_k = \left\{ \begin{array}{ll} \text{QPSK}(b_k, c_k(1)) & k \text{ even} \\ \text{QPSK}(b_k, d_k(1)) & k \text{ odd} \end{array} \right. \tag{4.14}$$

where $\text{QPSK}(\cdot, \cdot)$ maps the bits to the two-dimensional QPSK signal constellation (*e.g.*, using Gray mapping). The complex QPSK symbols q_k are transmitted over an AWGN channel which introduces phase uncertainty, modeled exactly as in the case of SCCCs. The post-correlator complex baseband observation equation is given by

$$z_k = \sqrt{E_s} q_k e^{j\theta_k} + w_k \tag{4.15}$$

Pilot symbols are inserted in the transmitted sequence in the same manner described in the previous section.

The adaptive receiver proposed in [LuWi98], consisting of a single external DD-PLL operating on the coded symbols q_k, followed by a non-adaptive turbo decoder, is shown in Fig-4.11(b). The A-SODEM-based receiver is straightforward to construct and is shown in Fig-4.11(c). Finally, for the A-SISO-based receiver, contrary to the serially concatenated examples, the PCCC has the property that the outputs of both FSMs are directly affected by the channel. Furthermore, the outputs of the constituent FSMs are coupled via the non-linear mapping (4.15), (4.14). This makes the substitution of the perfect-CSI SISO by an A-SISO insufficient for performing adaptive iterative detection in this case. Thus, adaptive iterative detection for this PCCC application requires a method for evaluating transition metrics and updating phase estimates for each A-SISO. In the following we discuss the options for doing so and demonstrate one specific approach.

Metric Evaluation   Metric evaluation in A-SISO1 can be performed by treating the output symbols corresponding to CC2 as nuisance parameters and either averaging or maximizing over them. Since APP soft metrics are typically observed to be superior compared to MSM ones

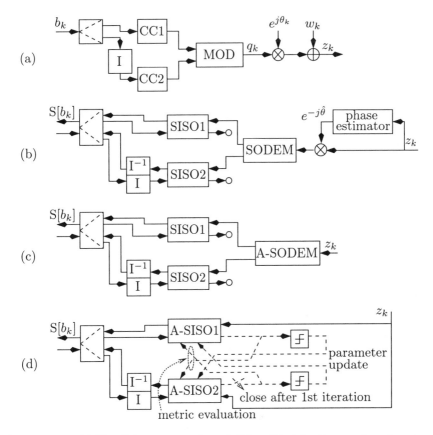

Figure 4.11. (a) Parallel concatenation of CCs. (b) Adaptive receiver based on a single external DD-PLL. (c) A-SODEM-based iterative receiver. (d) A-SISO-based iterative receiver (the demodulator is part of the A-SISO).

(for the particular application in the perfect CSI case), averaging over the output symbols of CC2 seems to be a preferable choice. A reasonable choice for the probabilities needed in the averaging process, is to use the most recent soft-metrics produced by A-SISO2. This is identical to the operation of soft mapper (SOMAP) [BeDiMoPo98] in the case of perfect CSI. The only difference is that the demodulator and the SOMAP are now integrated with the A-SISO1, since a phase estimate is required for this operation. This solution is both simple to implement, and compatible with the notion that SISO blocks exchange information only in the form of soft metrics. A similar procedure can be followed for the evaluation of the transition metrics of A-SISO2.

Parameter Estimate Update   Several options are considered for updating the phase estimate in A-SISO1.

Starting from the simplest solution, the channel update in A-SISO1 is only performed for those time instants k, for which the symbol q_k is only a function of b_k and $c_k(1)$ (k is even). The resulting updates for this *punctured* DD-PLL become

$$\tilde{\theta}_k = \begin{cases} \tilde{\theta}_{k-1} + \lambda \Im \left\{ z_k \mathrm{QPSK}(b_k, c_k(1))^* e^{-j\tilde{\theta}_{k-1}} \right\} & k \text{ even} \\ \tilde{\theta}_{k-1} & k \text{ odd} \end{cases} \qquad (4.16)$$

where b_k and $c_k(1)$ are obtained from the state transition of A-SISO1. The immediate consequence of this sort of channel update is a loss of the full tracking ability of the estimator (*i.e.*, the effective loop bandwidth is halved). In addition, such an approach is not always applicable, since the symbol q_k may always be an explicit function of the symbol $d_k(1)$ as well, as in the case of non-punctured codes (this is also true in this case when considering phase estimation for A-SISO2).

In a more refined technique, the phase estimator – and in particular the DD-PLL (or DD-PLLs) – is updated for every time instant k. The symbols b_k and $c_k(1)$ are determined by the state transition of A-SISO1, while an estimate $\tilde{d}_k(1)$ of $d_k(1)$ – needed when k is odd – is determined by hard quantizing the most recent soft information of $d_k(1)$ available either from A-SISO2 or from any other soft block in the adaptive receiver. The resulting updates for this parallel DD-PLL become

$$\tilde{\theta}_k = \begin{cases} \tilde{\theta}_{k-1} + \lambda \Im \left\{ z_k \mathrm{QPSK}(b_k, c_k(1))^* e^{-j\tilde{\theta}_{k-1}} \right\} & k \text{ even} \\ \tilde{\theta}_{k-1} + \lambda \Im \left\{ z_k \mathrm{QPSK}(b_k, \tilde{d}_k(1))^* e^{-j\tilde{\theta}_{k-1}} \right\} & k \text{ odd} \end{cases} \qquad (4.17)$$

Finally, an even more sophisticated technique can be derived by utilizing a *mixed-mode* PLL. Such a PLL operates in a decision directed mode in terms of the symbols $(b_k, c_k(1))$, while it effectively averages out the symbol $d_k(1)$ (a simple PLL structure that operates by averaging equiprobable binary symbols has been proposed in [LiSi73]). Hybrid schemes that use a punctured PLL initially and switch to a parallel decision-directed operation are also possible.

Example 4.6. ──────────────────────────────

In this example, the first order DD-PLL and suboptimal binding term in (4.10) and (4.11) will be used. A hybrid approach for phase tracking is used. Specifically, A-SISO1 is run with the punctured DD-PLL of (4.16) on the initial iteration, and switches to the parallel decision-directed mode of (4.17) in the subsequent iterations. The rational behind this hybrid bootstrapping procedure is that in the first iteration, there are no soft (or hard) decisions available for the symbol $d_k(1)$. The activation schedule is described as follows: A-SISO1 (with internal SOMAP and de-modulator) $\rightarrow$ forward extrinsic information form A-SISO1 to A-SISO2 $\rightarrow$ A-SISO2 (with corresponding internal SOMAP and demodulator) $\rightarrow$ forward extrinsic information form A-SISO2 to A-SISO1 $\rightarrow$ A-SISO1, etc.

The generator polynomial of the two RSC codes is the one in Section 2.4.3.1 and the output symbol is formed exactly as described in (4.14). The interleaver size is $K = 16384$ symbols, and only APP-type soft information is exchanged between A-SISOs. In Fig-4.12, performance curves similar to those of Fig-4.10 are presented. The conclusions

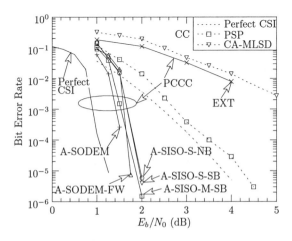

Figure 4.12. BER vs. E_b/N_0 for PCCC with phase dynamics and various A-SISO and A-SODEM configurations (the optimal performance for S receivers was achieved for $d = 0$). For comparison, the performance of CC with adaptive hard-decision detection is presented.

are similar to the SCCC case, with the only difference being the slight degradation of the A-SISO-S-SB and A-SISO-S-NB algorithms over the A-SISO-M-SB receivers. In addition, the A-SODEM-FW performance is very close to the performance of the A-SISO-based receivers, and the A-SODEM algorithm results in slightly better performance compared to A-SISOs. Also, as in the case of perfect CSI, the quantitative performance achieved using the SCCC and PCCC systems is very similar.

$\hspace{6cm}$ — *End Example*

A similar problem to that of phase tracking for PCCCs is decoding of a PCCC in the presence of a flat fading channel. The suggested transmission schemes utilize an additional *channel* interleaver in order to provide additional time diversity. Such a system is illustrated in the following example, which is based on the work in [HeCh00].

Example 4.7. ──

The standard PCCC code of the previous example is considered on an interleaved flat fading channel. A block of source symbols b_k is encoded, modulated and channel interleaved. After the channel interleaver, two training sequences (20 symbols) are attached at the head and tail of the interleaved symbol block (2040 symbols) and pilot symbols are inserted within the interleaved symbol block (1 every 30 symbols). The signal is transmitted through a flat fading channel and observed in AWGN.

The structure of the adaptive receiver, shown in Fig-4.13, is that of a decoupled estimator and decoder, which is similar to the A-SODEM-based receiver. The difference in this approach though is that the feed-

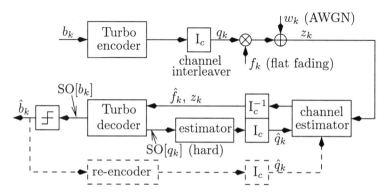

Figure 4.13.   Block diagram of the decoupled adaptive iterative detector for turbo codes on an interleaved flat fading channel.

back information from the decoder to the estimator is hard decisions, instead of soft decisions. For the first iteration, the channel is estimated by the known training sequences and pilot symbols. After the first iteration, the hard output of the turbo decoder is fed back to the channel estimator to refine the channel estimates. The refined channel estimates are passed back to the turbo decoder. In the simulations the channel taps are generated based on the model in (2.21). Two kinds of channel estimators are considered based on probabilistic channel models.

- An FI Kalman smoother based on a first order Gauss-Markov model $f_k = \alpha f_{k-1} + n_k$. The parameter α was selected such that the 10 dB bandwidth of the GM model is ν_d.

- A Wiener filter, which for M-PSK signaling can be designed a-priori [ViTa95, ChLe98] based on the channel correlation in (2.21). The estimate of f_k is

$$\hat{f}_k = \sum_{i=-N}^{N} \frac{z_{k-i}}{\hat{q}_{k-i}} g_k \qquad (4.18)$$

where $2N+1$ is the window size and $\{g_k\}$ are the pre-designed Wiener filter taps.

In Fig-4.14 the performance of these adaptive iterative detectors (labeled KF and WF for Kalman filter and Wiener filter, respectively) at the 10th iteration are shown for a normalized Doppler spread $\nu_d = 0.01$ and a window size $2N + 1 = 31$. For comparison, the performance of

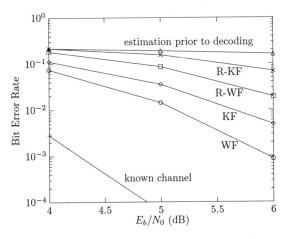

Figure 4.14. The performance of different decoupled adaptive iterative detection methods with punctured QPSK and $\nu_d = 0.01$ at the 10th iteration.

the receiver with parameter estimation prior to iterative decoding, and the known channel case is shown as well. In addition, the performance of a receiver that utilizes the final decoded symbols (after re-encoding) for the channel estimation is also presented (labels R-WF and R-KF are used to denote these two receivers). The performance of both R-WF and R-KF is much worse than the proposed adaptive receivers, while the WF-based receivers are better than the KF-based. This is most likely because the WF can better approximate the correlation in (2.21), whereas the KF applies a first order GM approximation. The accuracy of this approximation is more critical in the flat fading case than in the ISI case. Variations of these approaches, that provide minor performance improvements, are discussed in [HeCh00].

—————————————————————————————— *End Example*

4.5   Summary and Open Problems

The problem of identifying optimal and suboptimal (practical) algorithms that generate soft information in the presence of unknown parameters was the main topic of this Chapter. Several options were discussed for the formulation of the optimal A-SISO under the assumption of Gauss Markov, or deterministic parameters, and a linear observation model. Utilizing different simplifying assumptions, families of practical algorithms were derived, the basic characteristic of which is either forward adaptive processing, or forward-backward adaptive processing followed by binding. The application of these A-SISOs as part of an adaptive iterative receiver for a TCM system in frequency selective fading channel is straightforward and results in significant performance improvements. Similarly, the utilization of A-SISOs or A-SODEMs as part of the iterative decoder for SCCC and PCCC codes when carrier phase uncertainty is present, provides close to optimal performance even with significant channel dynamics.

Several theoretical and practical open problems arise from the material presented in this Chapter. On the theoretical front, the concept of forward-backward estimator correlator is fairly new and not fully understood, yet. In particular, only the linear observation model has been studied, for the special cases of GM and deterministic parameter model. Furthermore, the implications of the forward-backward EC to other detection and estimation problems (*e.g.*, MAP-SqD in flat fading channels, joint data detection and ISI channel acquisition) have not been explored. Another interesting theoretical topic is the derivation of optimal forward-backward soft decision structures that involve forward only estimation. In fact, the FA type algorithms developed herein can be thought of as suboptimal derivatives of a forward adaptive, backward non-adaptive EC optimal structure, similar to the one developed in [GeLo97, HaChAu00].

The problem of SCCC and PCCC receiver design for channels with high dynamics is very crucial for any practical application of turbo codes. Although the results presented herein show remarkable performance at significant phase dynamics, the problem of receiver design for turbo codes in the presence of unknown frequency shifts remains unsolved. A popular method for deriving such adaptive algorithms is to truncate the estimator memory. The linear predictive, or innovations based receiver of [MaMa98], as well as the "noncoherent" receivers in [CoFeRa00], are all instances of this approach, where open loop estimators are used to aid the detection process. A complete comparison between these algorithms is not available, but it is likely that the open loop methods are most robust to extreme dynamics when few pilots are available, and that the closed loop structures work better for a given complexity in less severe dynamics. Finally, the joint design of concatenated codes and adaptive iterative decoders may lead to more efficient and robust transmission schemes for the demanding wireless channels.

A-SISO-M-SB	Adaptive Soft Input Soft Output, Multiple estimators, Suboptimal Binding
A-SISO-S-NB	A-SISO, Single estimator, No Binding
A-SISO-S-SB	A-SISO, Single estimator, Suboptimal Binding
A-SODEM-FW	Adaptive Soft DEModulator, Forward only
AKF	Average Kalman Filter
APP-S-AKF-OB	A Posteriori Probability, Single estimator, Optimal Binding
EXT	EXTernal PLL
FA	Forward Adaptive
FABA	Forward Adaptive Backward Adaptive
FL-FB-FABA	Fixed Lag, Forward-Backward, Forward Adaptive Backward Adaptive
FL-FB-FA	Fixed Lag, Forward-Backward, Forward only adaptation
FL-L^{2}VS-CE	Fixed Lag, Lee and Li Vucetic Sato, Constraint Estimation
FL-L^{2}VS-UE	Fixed Lag, L^{2}VS, Unconstraint Estimation
MSM-M-KF-OB	Minimum Sequence Metric, Multiple estimators, Kalman Filter, Optimal Binding
MSM-M-LMS-NB	MSM, Multiple estimators, Least Mean Squares, No Binding
MSM-M-LMS-SB	MSM, Multiple estimators, LMS, Suboptimal Binding
MSM-M-LMS-SB	MSM, Multiple estimators, LMS, Suboptimal Binding
MSM-S-LMS-SB	MSM, Single estimator, LMS, Suboptimal Binding
PCKF	Partially Conditioned Kalman Filter
R-KF	Re-encoding, Kalman Filter
R-WF	Re-encoding, Wiener Filter

Table 4.1. Table of abbreviations specific to Chapter 4.

4.6 Problems

4.1. Consider the Gauss Markov channel and observation equation of (1.52) and (1.51), respectively. Derive a block Estimator Correlator for the above model, similar to the one obtained in Example 1.7 for the deterministic modeling case.

4.2. Find a closed-form expression for the integral

$$\int \frac{\mathcal{N}_n^{cc}(\mathbf{x}; \mathbf{m}_1; \mathbf{K}_1)\mathcal{N}_n^{cc}(\mathbf{x}; \mathbf{m}_2; \mathbf{K}_2)}{\mathcal{N}_n^{cc}(\mathbf{x}; \mathbf{m}_3; \mathbf{K}_3)} d\mathbf{x}$$

where $\mathcal{N}_n^{cc}(\mathbf{x}; \mathbf{m}; \mathbf{K})$ denotes the n-dimensional probability density function of a vector complex circular Gaussian random variable with mean $\mathbf{m}$, and non-singular covariance matrix $\mathbf{K}$.

4.3. [An99, AnCh00] Derive the forward-backward Estimator Correlator expressed in (1.87) for the Gauss Markov parameter model. Specifically, using the results of the previous problem, show that the binding factor $b_p(\cdot)$ is given by (1.88).

4.4. [An99, AnCh99] Derive a forward-backward Estimator-Correlator for the deterministic parameter model of (1.44). Work in the logarithmic domain and generalize the metric by introducing an exponentially decaying window relative to an arbitrary time k as follows

$$M_0^{K-1}(\mathbf{q}(\mathbf{a}), \mathbf{a}) = \min_{\mathbf{f}} \left[\sum_{m=0}^{K-1} \left(\frac{|z_m - \mathbf{q}_m^T \mathbf{f}|^2}{N_0} - \ln p(a_m) \right) \rho^{|k-m|} \right]$$

4.5. Derive a block, forward recursive, and forward-backward Estimator-Correlator for the non-linear observation model

$$z_k(\zeta) = q_k(\mathbf{a}(\zeta))e^{j\phi} + n_k(\zeta)$$

where $n_k(\zeta)$ is AWGN, defined as in (1.44), and the phase ϕ is modeled as an unknown deterministic constant.

4.6. [An99, AnCh00] Derive the forward-backward recursions (4.6) for the GM parameter model. Specifically, using a result similar to the one derived in Problem 4.2, give a closed-form expression for the binding term $b_p'(\cdot)$.

4.7. [ZhFiGe97] Starting from the Forward EC in (1.53), for an $(L+1)$-tap ISI/GM channel, and by folding the sequence tree into a trellis, defined by the trellis state $s_k^s = (a_{k-D}, \ldots, a_{k-1})$, derive a FL A-SISO that approximately evaluates $\text{APP}_p[a_{k-D}] \sim p(z_0^k, x_{k-D})$.

4.8. [IlShGi94] Consider the forward recursion in (4.6b) under the Gaussian assumption for the innovation term $p(z_k|t_k, \mathbf{z}_0^{k-1})$. Show that under this assumption, the recursion can be performed with the aid of partially conditioned Kalman filter (PCKF), as described in Section 4.2.2. In particular, derive forward recursions for the state-conditioned estimate $\tilde{\mathbf{f}}_{k|k-1}(s_k) = \mathbb{E}\left\{\mathbf{f}_k|s_k, \mathbf{z}_0^{k-1}\right\}$ under the above Gaussian assumption.

4.9. Show that under the Gaussian assumption for the innovation terms, the binding term $b_p'()$ in (4.6a) can be evaluated off-line, resulting in a function similar to (1.88)

$$b'_p(\tilde{\mathbf{f}}_{k|k}, \tilde{\mathbf{F}}_{k|k}, \tilde{\mathbf{f}}^b_{k|k+1}, \tilde{\mathbf{F}}^b_{k|k+1}) =$$
$$\frac{p(a_k)}{N_0} \frac{|\mathbf{K_f}||\mathbf{P}|}{|\tilde{\mathbf{F}}_{k|k}||\tilde{\mathbf{F}}^b_{k|k+1}|} \exp(\boldsymbol{\beta}^H \mathbf{P} \boldsymbol{\beta} - \gamma)$$

with

$$\mathbf{P}^{-1} = \tilde{\mathbf{F}}^{-1}_{k|k} + (\tilde{\mathbf{F}}^b_{k|k+1})^{-1} - \mathbf{K_f}^{-1} + \frac{\mathbf{q}^*_k \mathbf{q}^T_k}{N_0}$$

$$\boldsymbol{\beta} = \tilde{\mathbf{F}}^{-1}_{k|k} \tilde{\mathbf{f}}_{k|k} + (\tilde{\mathbf{F}}^b_{k|k+1})^{-1} \tilde{\mathbf{f}}^b_{k|k+1} + \frac{\mathbf{q}^*_k z_k}{N_0}$$

$$\gamma = \tilde{\mathbf{f}}^H_{k|k} \tilde{\mathbf{F}}^{-1}_{k|k} \tilde{\mathbf{f}}_{k|k} + (\tilde{\mathbf{f}}^b_{k|k+1})^H (\tilde{\mathbf{F}}^b_{k|k+1})^{-1} \tilde{\mathbf{f}}^b_{k|k+1} + \frac{|z_k|^2}{N_0}$$

where $\tilde{\mathbf{f}}_{k|k-1}$, $\tilde{\mathbf{f}}^b_{k|k+1}$ are the partially conditioned one-step forward and backward Kalman predictors and $\tilde{\mathbf{F}}_{k|k-1}$, $\tilde{\mathbf{F}}^b_{k|k+1}$ are the corresponding covariances, described in the previous problem.

4.10. [An99] Consider the forward recursion in (4.6b) under the Gaussian assumption for the innovation term $p(z_k|t_k, \mathbf{z}_0^{k-1})$. Under the additional assumption that the conditional means and covariances of the parameter $\mathbf{f}_k$ are not functions of the states, derive a forward update for the global estimate $\hat{\mathbf{f}}_{k|k-1}$. In doing so you need to assume that some sort of probabilistic description on t_{k-d}, $p(t_{k-d})$ is available at the receiver.

4.11. Compare the computational complexity and memory requirements for the FA type (*i.e.*, with forward only adaptation) forward-backward FL A-SISO algorithm, and the L^2VS-based FL A-SISO (with non-adaptive constrained ACS operations).

4.12. Derive single-estimator versions for the FA type generic algorithms discussed in Section 4.2.4 by inserting a tentative decision delay.

Chapter 5

APPLICATIONS IN TWO DIMENSIONAL SYSTEMS

We focus on detection problems associated with systems most naturally indexed in two dimensions in this chapter. The problem considered in most detail is a 2D ISI-AWGN channel, with page-access optical memories (POMs) providing the motivation. The algorithms and models developed, however, are applicable to any 2D combining and marginalization problem with local metric dependencies. We conclude this chapter by considering one such application, image halftoning.

5.1 Two Dimensional Detection Problem

5.1.1 System Model

Consider a generalization of the one-dimensional systems considered in Chapter 1. Specifically, let a digital page $a(\zeta; i, j)$ of independent random variables be "transmitted" and denote the associated noisy observation page by $z(\zeta; i, j)$. Analogous to the one-dimensional folding condition in (1.85), we focus on the case when the observation depends only locally on the conditional data page

$$P_{\mathbf{Z}(\zeta)|\mathbf{A}(\zeta)}(\mathbf{Z}|\mathbf{A}) = \prod_{(i,j)} p(z(i,j)|\mathbf{A}) = \prod_{(i,j)} p(z(i,j)|t(i,j)) \qquad (5.1)$$

where $\mathbf{A}$ represents the data page consisting of $a(i,j)$ and $t(i,j)$ is a local neighborhood description for $a(i,j)$. Specifically, $t(i,j) = \{a(i - m, j - n)\}_{(m,n) \in \mathcal{P}}$ where $\mathcal{P}$ is the *support region*. Note that since $p(\mathbf{A})$ factors into the product of marginal pmfs, whenever (5.1) holds we have

$$P_{\mathbf{Z}(\zeta),\mathbf{A}(\zeta)}(\mathbf{Z}, \mathbf{A}) = \prod_{(i,j)} p(z(i,j)|t(i,j))p(a(i,j)) \qquad (5.2)$$

where in both (5.1) and (5.2) each $t(i,j)$ is implicitly defined by the conditional value of $\mathbf{A}$. For concreteness, we focus on the special case of

$$\mathcal{P} = \{(m,n) : -L_\mathrm{c} \le m \le +L_\mathrm{c}, -L_\mathrm{r} \le n \le +L_\mathrm{r}\} \tag{5.3}$$

In other words, the dependency on the underlying page hypothesis has support region of size $(2L_\mathrm{c}+1) \times (2L_\mathrm{r}+1)$ with L_c providing a measure of the memory across columns (y-direction) and L_r doing the same for the memory down rows. Most applications that we consider may be viewed as a 2D system with inputs $a(i,j)$ and outputs $x(i,j;\mathbf{A})$ that depend on $\mathbf{A}$ via the conditional data in the support region, $t(i,j)$. Then $z(i,j)$ is the output of a memoryless stochastic channel for which $x(i,j)$ is the input $-$ *i.e.*, $p(z(i,j)|t(i,j)) = p(z(i,j)|x(i,j))$ where $x(i,j)$ is the output determined by $t(i,j)$.

5.1.2   Optimal 2D Data Detection

With the above assumption, MAP detection of either the symbol $a(\zeta;i,j)$ individually (*i.e.*, MAP symbol detection, MAP-SyD) or the page $\mathbf{A}(\zeta)$ collectively (*i.e.*, MAP page detection, MAP-PgD) can be formulated in terms of an additive metric function. Specifically, define

$$\mathrm{M}[t(i,j)] \triangleq -\ln[p(z(i,j)|t(i,j))p(a(i,j))] \tag{5.4a}$$
$$= -\ln[p(z(i,j)|x(t(i,j)))] - \ln[p(a(i,j))] \tag{5.4b}$$
$$= \mathrm{MI}[x(t(i,j))] + \mathrm{MI}[a(i,j)] \tag{5.4c}$$

where the specialization to a system with output $x(i,j)$ has been made. Thus, based on (5.2), MAP page and symbol detection are obtained, conceptually at least, by the following

$$\mathrm{M}[\mathbf{A}, \mathbf{X}(\mathbf{A})] = \sum_i \sum_j \mathrm{M}[t(i,j)] \tag{5.5a}$$

$$\hat{a}(i,j) = \arg\min_{a(i,j)} \left[\min_{\mathbf{A}:a(i,j)} \mathrm{M}[\mathbf{A}, \mathbf{X}(\mathbf{A})] \right] \quad \text{(MAP-PgD)} \tag{5.5b}$$

$$\hat{a}(i,j) = \arg\min_{a(i,j)} \left[\min^*_{\mathbf{A}:a(i,j)} \mathrm{M}[\mathbf{A}, \mathbf{X}(\mathbf{A})] \right] \quad \text{(MAP-SyD)} \tag{5.5c}$$

which is the special case of combining and marginalizing in the metric domain over the structure of a 2D system. Note that there is no reason that one must associate the input metric of $a(i,j)$ with the metric of $t(i,j)$ and we will consider models that associate $\mathrm{MI}[a(l,m)]$ with $\mathrm{M}[t(i,j)]$ for $(l,m) \neq (i,j)$. A related issue is how one handles edge information. In particular, for the one-dimensional case, this was obtained by $p(s_0)$. The analogous quantity in 2D depends on how one models the corresponding 2D system. Throughout this chapter we assume that $a(i,j)$ is defined for $i \in \{0, 1, \ldots I - 1\}$ and $j \in \{0, 1, \ldots J - 1\}$ and that the edges are terminated in a deterministic manner. For example, $x(0,0)$ depends on edge symbols, the values of which are assumed known at the detector with probability one.

Before the thresholding over $a(i,j)$, both (5.5b) and (5.5c) are special cases of the general combining and marginalization problem of Section 1.2.1.2. In fact, the only aspect that distinguishes the applications considered in this chapter is that the local dependencies are most naturally expressed using a 2D index set. Thus, problems of the same form as those in (5.5) are considered that are not data detection problems based on statistical source and channel models. These problems are data fitting or encoding problems, typically of the form (5.5b). This class has been referred to as the *2D-digital least metric (2D-DLM)* problem in [ChChOrCh98].

Example 5.1. ———————————————————————
Data detection for the 2D linear ISI-AWGN channel can be formulated as in (5.5). Specifically, the observation page $z(i,j)$ is

$$z(i,j) = x(i,j) + w(i,j) \tag{5.6a}$$

$$= \sum_{l=-L_c}^{L_c} \sum_{m=-L_r}^{L_r} f(l,m)a(i-l, j-m) + w(i,j) \tag{5.6b}$$

$$= f(i,j) \circledast a(i,j) + w(i,j) \tag{5.6c}$$

where the last equality defines a 2D convolution and the channel is assumed to have finite support. With this assumption, $t(i,j)$ is defined as above with $p(z(i,j)|x(i,j)) = \mathcal{N}_1(z(i,j); x(i,j); \sigma_w^2)$. The input data is assumed to be independent for different indices with a-priori probabilities $p(a(i,j))$. Thus, the MAP page and symbol detection problems are as defined in (5.5) with $\mathrm{MI}[x(i,j)] \equiv [z(i,j) - x(i,j)]^2/(2\sigma_w^2)$ and $\mathrm{MI}[a(i,j)] \equiv -\ln p(a(i,j))$.
——————————————————————————— *End Example*

While the 2D detection problem is analogous to problem of detecting the input to a one dimensional FSM corrupted by noise, the lack of a natural order for the 2D index set complicates the problem immensely. A straightforward realization of MAP-PgD is to build a look-up table (LUT) storing $\mathrm{M}[\mathbf{A}, \mathbf{X}(\mathbf{A})]$ for all possible input pages. However, the number of entries in this LUT is $|\mathcal{A}|^{I \times J}$. This is prohibitively complex even for a binary data page with relatively small page sizes. Another

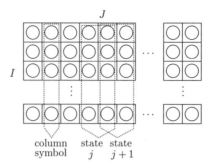

Figure 5.1. An algorithm for MAP-PgD using the Viterbi algorithm ($L_{\mathrm{r}} = 1$).

approach, which is illustrated in Fig-5.1, is to run the VA by treating each column in $\mathbf{A}$ as a single vector symbol. These vector symbols take on $|\mathcal{A}|^I$ different conditional values and the associated Viterbi algorithm is run on a trellis with $|\mathcal{A}|^{I(2L_{\mathrm{c}}+1)}$ transitions. Thus, while providing a huge complexity decrease relative to the exhaustive search, this is still impractical for the page sizes of interest. The problem of MAP-SyD is, similarly, prohibitively complex. Thus, the focus on this chapter will be on designing efficient approximate algorithms. In the following section we present bounds on the performance associated with MAP-SyD and MAP-PgD, which provide a way of gauging the performance of suboptimal detection algorithms against the best achievable.

5.2   Performance Bounds for Optimal 2D Detection

Both the upper and lower bounds in Sections 1.4.3 and 1.4.4 were derived in a manner that extends directly to the 2D case [Ch96]. In particular, for the upper bound, the notion of sufficient (*e.g.*, simple) sequences generalizes to sufficient 2D patterns. Specifically, conditioned on the transmitted page being $\mathbf{A}$, the set of simple patterns $\mathcal{S}_{(i,j)}(\mathbf{A})$ are those that differ from $\mathbf{A}$ at location (i,j) and have a connected pattern of disagreements. This is illustrated in Fig-5.2 for a (3×3) support region centered around the point of interest. More precisely, if $\mathbf{A}$ and $\tilde{\mathbf{A}}$ disagree at locations (i,j) and (k,l), then these disagreements are *connected* iff $t(i,j)$ and $t(k,l)$ share common elements – *i.e.*, recalling the dual variable/set interpretation introduced in Chapter 3, this requires that $t(i,j) \sqcap t(k,l) \neq \emptyset$. A simple pattern $\tilde{\mathbf{A}} \in \mathcal{S}_{(i,j)}(\mathbf{A})$ is one in which the set of all disagreements form a connected set in this sense. Thus, the pattern in Fig-5.2(c) is not simple because the two support regions do not intersect. The pattern in Fig-5.2(d) comprises two simple patterns that are isolated from each other, so the overall pattern is not simple.

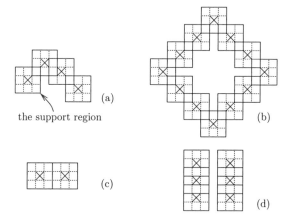

Figure 5.2.   Examples of difference patterns for $a(i,j)$ and $\tilde{a}(i,j)$ for an $L_c = L_r = 1$ support region. The disagreements are indicated by ×'s and other outlined points are in agreement. Examples of simple patterns are shown in (a)-(b) with non-simple patterns shown in (c)-(d). For the linear ISI case, these may be interpreted as (a)-(b) simple and (c)-(d) non-simple error patterns with the ×'s denoting the non-zero elements in the error pattern.

It is straightforward to verify that this definition of 2D simple patterns reduces to the standard simple sequence definition from Chapter 1 when the support region is 1D (*e.g.*, $L_r = 0$). Intuitively, the connected condition is that the conditional values of $a(\cdot, \cdot)$ at the two locations both affect at least one common system output $x(\cdot, \cdot)$. It is straightforward to extend the development of Section 1.4.3 to an upper bound on MAP-PgD detection using these simple patterns [Ch96, ChChThAn00]. Specifically, (1.130) remains valid for the 2D linear ISI-AWGN channel with $\mathcal{S}$ interpreted as 2D simple patterns with shift equivalences removed. These 2D simple error patterns are those for which the non-zero elements are connected in the above sense. The examples in Fig-5.2 may be interpreted as error patterns with the nonzero elements denoted by the "×-ed" boxes. For the linear ISI channel the connected condition for two nonzero terms means that the contributions to $e(i,j) \circledast f(i,j)$ from the two locations overlap.

An interesting and subtle point coming from this bound development is that a pairwise decisions can be made for portions of pages using only local information [Ch96]. For example, for the simple patterns in Fig-5.2(a)-(b), an optimal pairwise decision between $\mathbf{A}$ and $\tilde{\mathbf{A}}$ at the ×-locations can be made independently of the pairwise decision at other locations in the page. Thus, the assumption of a local region of agreement between different hypotheses plays the same role as a state agreement in 1D. In fact, this leads to a sort of 2D Viterbi algorithm which, unfortunately, has complexity that grows exponentially in the perimeter of the observation page used (see Problem 5.3).

Evaluation of this 2D bound can be approximated by exhaustive search over some finite region. However, the number of error patterns searched grows much more quickly in 2D than in 1D. Thus, in practice the search is limited to a relatively small region. The 2D version of the upper bound in (1.130) can, however, be shown to converge [ChChThAn00] for the linear ISI-AWGN channel.

Similarly, the lower bound in Section 1.4.4 generalizes directly to the 2D case. In the linear ISI-AWGN case, one can search all simple error patterns over a finite region to obtain different distance sets. A strictly valid lower bound for MAP-SyD of the form in (1.135) can then be formed using these sets. Specifically, one can purge sequences from $\mathcal{B}(d)$, until a pairwise partition exists, so that a bound of the form in (1.135) can be computed. This process is demonstrated in the following example.

Example 5.2. ────────────────────────────────────

Consider the observation model in (5.6a) with a (3×3) symmetric support region (*i.e.*, $L_{\mathrm{r}} = L_{\mathrm{c}} = 1$) and the channel taps are as follows

$$\mathbf{F} = \begin{pmatrix} f(-1,-1) & f(0,-1) & f(+1,-1) \\ f(-1,0) & f(0,0) & f(1,0) \\ f(-1,+1) & f(0,+1) & f(+1,+1) \end{pmatrix} = \begin{pmatrix} g & h & g \\ h & 1 & h \\ g & h & g \end{pmatrix} \quad (5.7)$$

where we have fixed the center tap to the value 1 so that the channel can be specified by two parameters, g and h. In this example we consider the coefficients $g = 0.0327$ and $h = 0.181$, which is referred to as Channel A in the following.

We consider the input alphabet $\mathcal{A} = \{0, 1\}$ so that the error alphabet is $\Delta\mathcal{A} = \{0, +1, -1\}$. All simple error patterns up to size (4×4) were searched exhaustively (*i.e.*, nonzero elements in the error patterns were limited to a (4×4) grid). This may be implemented by generating 3^{16} different patterns and discarding those that are not simple. A check should also be included to remove shift equivalences.

For each simple pattern $\mathbf{E}$, the normalized square distance

$$\bar{d}^2(\mathbf{E}) = \frac{1}{\|\mathbf{F}\|^2} \sum_{m,n} |e(m,n) \circledast f(m,n)|^2 = \frac{d^2(\mathbf{E})}{\sum_{i,j} f^2(i,j)} \quad (5.8)$$

was computed. After removing shift equivalences, the 400 patterns yielding smallest distances were retained to construct the bounds. This yielded 23 different values for $\bar{d}$, which we label in increasing order as $\bar{d}_1 < \bar{d}_2 < \bar{d}_3 \ldots$. The upper bound was approximated using all of these terms and the fact that $P_C(\mathbf{E}) = 2^{-w(\mathbf{E})}$ for the binary alphabet.

The smallest normalized square distance found was $\bar{d}_1^2 = 0.76$, which is generated by two different weight four error patterns. One pattern is

$$
\underbrace{\begin{matrix} 0 & 0 & 0 & 0 \\ 0 & +1 & -1 & 0 \\ 0 & -1 & +1 & 0 \\ 0 & 0 & 0 & 0 \end{matrix}}_{\tilde{e}(i,j)} = \underbrace{\begin{matrix} \star & \star & \star & \star \\ \star & 1 & 0 & \star \\ \star & 0 & 1 & \star \\ \star & \star & \star & \star \end{matrix}}_{\tilde{a}(i,j)} - \underbrace{\begin{matrix} \star & \star & \star & \star \\ \star & 0 & 1 & \star \\ \star & 1 & 0 & \star \\ \star & \star & \star & \star \end{matrix}}_{a(i,j)}
\tag{5.9}
$$

where the $\star$ indicates a *free position*. Specifically, recalling the convention that $\tilde{e}(i,j) = \tilde{a}(i,j) - a(i,j)$ where $a(i,j)$ is the true data, the above error pattern is expressed as such a difference and the transmitted sequence $a(i,j)$ is constrained to take the values 0 and 1 as shown, but the other locations are free to be selected as either 0 or 1 with the same values taken for $\tilde{a}(i,j)$ in the corresponding locations. Thus, only one transmitted page is consistent with the above error pattern.

The other error pattern is the opposite of that in (5.9) with the values of $\tilde{a}(i,j)$ and $a(i,j)$ exchanged. Thus, based on the finite search, the set of data pages that are consistent error sequences have distance $\bar{d}_1$ is

$$
\mathcal{B}(\bar{d}_1) = \left\{ \begin{matrix} \star & \star & \star & \star \\ \star & 1 & 0 & \star \\ \star & 0 & 1 & \star \\ \star & \star & \star & \star \end{matrix} , \begin{matrix} \star & \star & \star & \star \\ \star & 0 & 1 & \star \\ \star & 1 & 0 & \star \\ \star & \star & \star & \star \end{matrix} \right\}
\tag{5.10}
$$

Since in the four restricted locations, only 2 of the possible $2^4 = 16$ sequences are in $\mathcal{B}(d_1)$, $P(\mathcal{B}(d_1)) = 2/16 = 1/8$. This yields the lower bound

$$
P_{Sy}(\mathcal{Y}) \geq \frac{1}{8} Q \left(\sqrt{\frac{\|\mathbf{F}\|^2 \bar{d}_1^2}{4\sigma_w^2}} \right) = \frac{1}{8} Q \left(\sqrt{\frac{\gamma \bar{d}_1^2}{2}} \right)
\tag{5.11}
$$

where $\gamma = \|\mathbf{F}\|^2/(2\sigma_w^2)$ is a measure of the SNR.

A second lower bound is obtained by the weight-one error patterns which achieve $\bar{d}_6 = 1$ - *i.e.*, the 6th smallest distance found. This is a special case of the ISI-free lower bound of Example 1.21, and results in

$$
P_{Sy}(\mathcal{Y}) \geq Q \left(\sqrt{\frac{\gamma}{2}} \right)
\tag{5.12}
$$

A third lower bound can be obtained considering the distance $\bar{d}_3^2 = 0.873$ arising from the two distinct error patters

$$
\begin{matrix} 0 & 0 & 0 & 0 \\ 0 & +1 & -1 & 0 \\ 0 & 0 & 0 & 0 \\ 0 & 0 & 0 & 0 \end{matrix} \quad \text{and} \quad \begin{matrix} 0 & 0 & 0 & 0 \\ 0 & +1 & 0 & 0 \\ 0 & -1 & 0 & 0 \\ 0 & 0 & 0 & 0 \end{matrix}
\tag{5.13}
$$

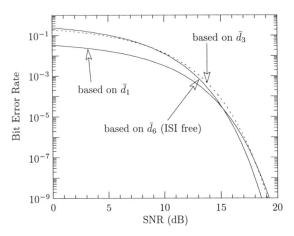

Figure 5.3. The best lower bound for Channel A from the finite error pattern search is achieved by combining the three component bounds shown.

along with their opposites. The corresponding $\mathcal{B}(\bar{d}_3)$ is

$$
\mathcal{B}(\bar{d}_3) = \left\{ \begin{matrix} \star & \star & \star & \star \\ \star & 0 & 1 & \star \\ \star & \star & \star & \star \\ \star & \star & \star & \star \end{matrix}, \begin{matrix} \star & \star & \star & \star \\ \star & 1 & 0 & \star \\ \star & \star & \star & \star \\ \star & \star & \star & \star \end{matrix}, \begin{matrix} \star & \star & \star & \star \\ \star & 0 & \star & \star \\ \star & 1 & \star & \star \\ \star & \star & \star & \star \end{matrix}, \begin{matrix} \star & \star & \star & \star \\ \star & 1 & \star & \star \\ \star & 0 & \star & \star \\ \star & \star & \star & \star \end{matrix} \right\}
$$

$$
= \left\{ \begin{matrix} 1 & 0 \\ 0 & \star \end{matrix}, \begin{matrix} 0 & 1 \\ 1 & \star \end{matrix} \right\} \cup \left\{ \begin{matrix} 1 & 1 \\ 0 & \star \end{matrix}, \begin{matrix} 0 & 0 \\ 1 & \star \end{matrix} \right\} \cup \left\{ \begin{matrix} 1 & 0 \\ 1 & \star \end{matrix}, \begin{matrix} 0 & 1 \\ 0 & \star \end{matrix} \right\}
$$
(5.14)

where only the restricted 2×2 part of the patterns is shown in the last equation. The partition in (5.14) yields the pairwise USI scheme. Since 6 of the 8 possible patterns in the three restricted positions are in $\mathcal{B}(\bar{d}_3)$, the associated lower bound is

$$
P_{Sy}(\mathcal{Y}) \geq \frac{6}{8} Q\left(\sqrt{\frac{\gamma \bar{d}_3^2}{2}} \right)
$$
(5.15)

In the evaluation of the lower bound in (1.135) (*i.e.*, as a maximum over all single-d lower pairwise USI lower bounds) over the SNR range of interest, it was found that only the three different distances discussed above were used, resulting in an overall lower bound of the form

$$
P_{Sy}(\mathcal{Y}) \geq \max \left\{ \frac{1}{8} Q\left(\sqrt{\frac{\gamma(0.76)}{2}} \right), \frac{3}{4} Q\left(\sqrt{\frac{\gamma(0.873)}{2}} \right), Q\left(\sqrt{\frac{\gamma}{2}} \right) \right\}
$$
(5.16)

Each of these three bounds are plotted in Fig-5.3. The bound based on $\bar{d}_6$ dominates at low SNR up to an SNR of 9 dB. The bound associated with $\bar{d}_3$ dominates from 9 dB up to 18.2 dB, above which the $\bar{d}_1$ bound dominates. The composite lower bound results from the maximization in (5.16).

—————————————————————————————— *End Example*

In Example 5.2, we can not conclude that the minimum distance is $\bar{d}_1$, due to the finite search. However, the lower bound stated is strictly valid. Thus, we can conclude that the ISI associated with Channel A causes an effective degradation of at least $-10\log_{10}(\bar{d}_1^2) = 1.18$ dB in SNR at sufficiently high SNR. We note that, while many of the tests in [AnFo75] have analogies in the 2D case, it does not appear that the method described therein extends to 2D. Specifically, there is no known subset of 2D simple error patterns which can be searched to find the minimum distance for any 2D channel of a given support size. While the upper bound is only an approximation, the same numerical result is obtained using a search of only (3×3) error patterns and including fewer terms in the sum. Conversely, using a 2D extension of the stack algorithm in [AnFo75], patterns up to size (4×12) have been searched without altering the subsequent bound (or finding a smaller distance). In the following, we use similar truncated bounds as upper bounds without further qualification. Specifically, upper and lower bounds obtained by this procedure are presented in Section 5.4 as a gauge for the performance of the suboptimal algorithms introduced.

5.2.1 Finding Small Distances

Recall from Section 1.4.3 that the problem of finding the minimum distance is a shortest path problem similar to the underlying data detection problem. The minimum distance problem differs in that the underlying alphabet is $\Delta\mathcal{A}$ and the minimizing input cannot be all zeros. Similarly, for the 2D linear ISI channel, minimization of (5.8) is a 2D-DLM problem with similar constraints. This means that, since 2D detection appears to be prohibitively complex, finding the minimum distance is likely to have similar complexity. This may not have severe practical consequences since, as demonstrated above, useful lower bounds can be obtained without an assurance that $d_{\min}$ has been found. Furthermore, the similarity of the data detection problem and the associated minimum distance problem allows one to apply an effective (suboptimal) detection algorithm to search for small distances.

Example 5.3. ——————————————————————————————
In Section 5.3, several iterative detection algorithms, or SISOs, are devel-

oped which are applicable to the 2D ISI channel. One of these, referred to as a "composite 2D SISO" in Section 5.3.1.3, was used to search for the error pattern that minimizes (5.8) for a given channel of the form in (5.7). The requirement of a nonzero error pattern was enforced by setting the input metrics for the error value at the center of the page appropriately. More precisely, let $e(i_0, j_0)$ be at or near the center of the page considered, then the value of $e(i_0, j_0) = +1$ is enforced by setting $\mathrm{MI}[e(i_0, j_0) = +1] = 0$ and the metrics associated with $e(i_0, j_0) = 0$ and $e(i_0, i_0) = -1$ to infinity. This also exploits the fact that $d(\mathbf{E}) = d(-\mathbf{E})$, so one need only consider $e(i_0, i_0) = +1$. The 2D SISO returns the nonzero error pattern minimizing the distance with the associated distance also noted. We denote the smallest distance found by this search as d_s.

This algorithm was run for a variety of 3×3 symmetric channels in (5.7) on a 30×30 page with the results shown in Fig-5.4. The normalized smallest distance found – $\bar{d}_\mathrm{s}$, along with the associated pattern is shown over the (g, h) plane. The results obtained by this suboptimal search

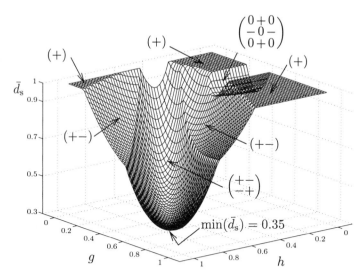

Figure 5.4. Normalized d_s and corresponding error patterns. The underlying 2D channels are defined in (5.7).

coincide with those found by exhaustive search and the stack algorithm over small pattern sizes. Only four minimizing error patterns were found for all channels considered. Moreover, all the $\bar{d}_\mathrm{s}$ error patterns found are either symmetric or asymmetric around the center in both the x and y directions. This is analogous to the symmetry found for the vast majority of minimizing sequences in 1D ISI channels [AnFo75, VaFo91].

We also observed that $\bar{d}_s$ reaches a minimum value of 0.35 when $g \cong 1/2$ and $h \cong 1/\sqrt{2}$. It is interesting to note that this "worst" 2D ISI channel is the outer product of the worst 3-tap 1D ISI channel – $(1/\sqrt{2}, 1, 1/\sqrt{2})$ [Pr95, pg.601] – with itself.

———————————————————————— *End Example*

5.3   Iterative 2D Data Detection Algorithms

5.3.1   Iterative Concatenated Detectors

As described in Chapter 2, once a subsystem decomposition or a graphical model for a system has been established, an iterative detection or message passing algorithm has been specified except for the activation schedule. Furthermore, these algorithms accept soft-in information on the 2D system inputs and outputs and update this information to produce soft-out information on the same. Thus, these algorithms may be viewed as suboptimal SISOs (*i.e.*, approximations to the marginal soft inverse of the 2D system). We use the term SISO to describe these algorithms in the following with the notion that they may be used as part of a larger iterative detector (*e.g.*, see Problem 5.6).

The 2D SISOs developed are, therefore, based largely on obtaining models for the underlying 2D system (or equivalently, the CM problem implied in (5.5)). We also apply some of the methods of complexity reduction developed in Chapter 3 to simplify the resulting algorithms.

5.3.1.1   A Concatenated Channel Model and 2D SISO

A general 2D system underlying (5.5) may be viewed as a mapping from the input page $a(i,j)$ to the page of transitions $t(i,j)$. In this section we demonstrate that this may be viewed as the serial concatenation of two 1D FSMs separated by a block interleaver [ChCh98]. For concreteness, we develop this for the linear ISI channel, but it will be apparent that this applies to a general 2D system as described in Section 5.1.1.

Consider the ISI channel with rectangular support defined in (5.3) and define the *row* vector[1]

$$\mathbf{v}(i,j) = [a(i, j - L_\mathrm{r}) \, a(i, j - L_\mathrm{r} + 1) \, \cdots \, a(i, j + L_\mathrm{r})] \qquad (5.17)$$

———————————————

[1]In order to make the development most intuitive, we momentarily depart from the convention of using column vectors exclusively.

as an inner vector symbol. The 2D convolution operation in (5.6a) can then be reformulated as

$$x(i,j) = \sum_{m=-L_c}^{L_c} \mathbf{v}(i-m,j)\mathbf{f}_m^T = \mathbf{v}(i,j) \circledast \mathbf{f}_i^T \qquad (5.18)$$

where the row vector $\mathbf{f}_m = [f(m,L_r)\ f(m,L_r-1)\ \cdots\ f(m,-L_r)]$ is the coefficient of the vector-based one-dimensional convolutional operation with the m-th row of $\mathbf{F}$. Consequently, (5.6a) is accomplished in two stages: the collapsing of the row memory in (5.17) and the 1D convolution of vectors in (5.18).

In fact, for a fixed i (row), the mapping from $a(i,j)$ to $\mathbf{v}(i,j)$ is an FSM with $\mathbf{v}(i,j)$ defining the transition. Similarly, the convolution in (5.18) is a 1D FSM in the column index variable i. Thus, this simple modeling trick yields the serially concatenated model for 2D ISI as illustrated in Fig-5.5. Specifically, the memory is first collapsed across each row, then the output can be computed independently running down each column. This is illustrated further in the following example.

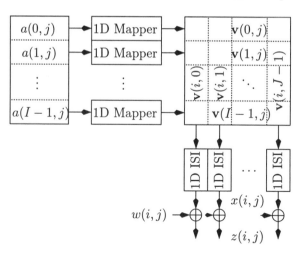

Figure 5.5. Block diagram for the (COL-row) concatenated channel model for a 2D ISI channel.

Example 5.4. ───────────────────────
An example of the above mapping for a 3×3 2D ISI is given in Fig-5.6 for a single transition (*i.e.*, $t(7,5)$). First, the binary input pixel array $a(i,j)$ is mapped *row-wise* into a 8-ary vector symbol array $\mathbf{v}(i,j)$. For instance, $\mathbf{v}(6,5) = [a(6,4),a(6,5),a(6,6)]$ and $\mathbf{v}(6,6) = [a(6,5),a(6,6), a(6,7)]$ (not shown). Notice that these vector symbols overlap and may

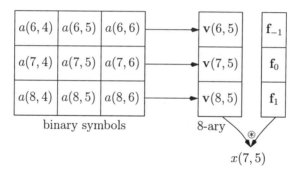

Figure 5.6. Illustration of the concatenated mapping process for a 3 × 3 2D ISI channel.

be viewed as the output of an FSM. Specifically, $\mathbf{v}(6,5)$ is the transition of the FSM with state $[a(6,4), a(6,5)]$ and input $a(6,6)$.[2] After the mapping in (5.17) has been applied to each row, the sequence $\mathbf{v}(i,j)$ is read out *column-wise* and fed into the second FSM, the vector-based 1D convolutional operation (5.18) which yields $x(i,j)$ in (5.6a). Specifically, note that $t(7,5)$ is defined by the values of $\mathbf{v}(6,5)$, $\mathbf{v}(7,5)$, and $\mathbf{v}(8,5)$.

—————————————————————————————— *End Example*

Clearly, the concatenated model applies to a general 2D system of the form described in Section 5.1. This follows because, as shown in Example 5.4, specification of $\{\mathbf{v}(i, j - m)\}_{m=-L_r}^{+L_r}$ determines the transition $t(i,j)$. It is worth stressing that there is *no approximation* in this model. Specifically, the concatenated model provides exactly the same input-output relation as the original 2D system.

A key observation is that the model in Fig-5.5 is equivalent to a serial concatenation of two 1D FSMs separated by a block interleaver as illustrated in Fig-5.7. For the serial concatenation in Fig-5.7, the inner FSM

Figure 5.7. Equivalent block diagram of the concatenated channel model in Fig-5.5 in terms of 1D subsystems.

corresponds to the column (vector ISI) FSM in Fig-5.5, the outer FSM corresponds to the row-wise FSM in Fig-5.5 and the block interleaver corresponds change between row-wise and column-wise ordering. The

---

[2]Notice that the channel is noncausal in both i and j. Thus, in the following, the notation of the previous chapters should be modified slightly to describe the underlying FSM.

system in Fig-5.7 takes a row-wise raster scan of the data page $a(i,j)$ yielding a_k. Specifically, $k = i \cdot J + j$, with i incremented once for every J times that j is. This is processed by the 1D FSM to produce $\mathbf{v}(i,j)$ ordered by row-wise scanning – *i.e.*, $\mathbf{v}_k$. The block interleaver reorders $\mathbf{v}(i,j)$ to correspond to a column-wise scan. Specifically, we use the index variable $n = i + j \cdot I$ with j incremented once for each I times that i is incremented. The inner FSM performs the vector ISI convolution and produces $x(i,j)$ in column-scan order. Note that our assumption of a known pattern around the edges of the page ensures that the state of the row (column) FSM are known before and after each row (column) is scanned.

In this model, the outer FSM is a row-wise simple FSM with memory $2L_r$. The column-wise FSM (5.18) is a vector-based 1D ISI (simple FSM) with memory $2L_c$ and inputs $\mathbf{v}(i,j)$ that are drawn from an alphabet of size $|\mathcal{A}|^{2L_r+1}$. We refer to the structure illustrated in Fig-5.7 (and, equivalently Fig-5.5) as the "COL-row" version of the concatenated model. The analogous ROW-col version of this model is defined by associating the outer FSM with column-wise processing. We capitalize the COL (ROW) in the COL-row (ROW-col) version as a reminder that the FSM associated with the column processing has many more states.

5.3.1.2   Concatenated 2D SISO

Based on Fig-5.7 and the development of Chapter 2 an effective 2D SISO algorithm can be obtained by the same processing used for serially concatenated FSMs with interleaving. The resulting algorithm is illustrated in Fig-5.8 for the system notation in Fig-5.7 and, equivalently,

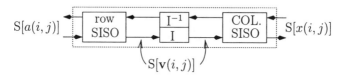

Figure 5.8.   The 2D SISO algorithm (COL-row version).

in Fig-5.9, for the diagram in Fig-5.5. This *concatenated 2D SISO* consists of two types of processors: the column SISOs and the row SISOs which correspond to the inner stage and outer stage, respectively. The parameters of these two SISOs, which may be implemented using the forward-backward algorithm, are listed in Table 5.1.

The 2D SISO processing is the same as that of the iterative detector for a serially concatenated system. As suggested by Fig-5.9, however, one has the option of executing all of the row SISOs in parallel since the processing is decoupled. Similarly, the column SISOs may be exe-

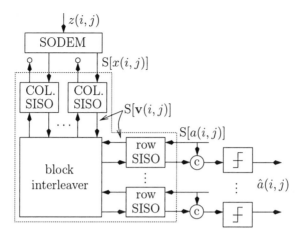

Figure 5.9. A simple 2D SISO: the iterative concatenated detector (COL-row version).

Table 5.1. The parameters of column SISOs and row SISOs in the iterative concatenated detection algorithm (COL-row version).

Processor	Transition		Input symbol	
	Def.	Size	Def.	Size
Column SISO	$\{\mathbf{v}(i-l,j)\}_{l=-L_c}^{L_c}$	$\|\mathcal{A}\|^{(2L_r+1)\times(2L_c+1)}$	$\mathbf{v}(i,j)$	$\|\mathcal{A}\|^{(2L_r+1)}$
Row SISO	$\mathbf{v}(i,j)$	$\|\mathcal{A}\|^{(2L_r+1)}$	$a(i,j)$	$\|\mathcal{A}\|$

cuted in parallel. Note, however, that the row SISOs cannot be activated until all column SISOs have completed their processing for a given iteration. Also, as implied by Fig-5.8, the processing can be done serially for columns, then rows, etc.

One iteration of the COL-row SISO is defined by activation of all column SISOs, then all row SISOs. For the first activation of the COL (inner) SISO, the soft-in information on the hidden variables $\mathbf{v}(i,j)$ is set to uniform. The COL SISOs and row SISOs exchange and update soft information on the vector symbols $\mathbf{v}(i,j)$. Note that since extrinsic soft-out information on the 2D system inputs $a(i,j)$ and outputs $x(i,j)$ can be produced using this algorithm, this may be viewed as a 2D SISO which approximates the marginal soft inverse of the 2D system.

5.3.1.3 Modified Concatenated 2D SISOs

The COL-row SISO described above is inherently asymmetric with respect to the row and column processing. For example, the column (inner) SISO is far more complex for a square support region. It is rea-

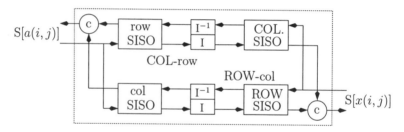

Figure 5.10.   The composite 2D SISO algorithm consisting of ROW-col and col-ROW 2D SISO algorithms operating independently.

sonable, therefore, to consider combining a COL-row and ROW-col 2D SISO to obtain a more symmetric algorithm. We refer to this as the *composite* 2D (concatenated) SISO algorithm. Exactly how one combines the operation of the COL-row and ROW-col SISOs is not initially clear. However, our experiments suggest that an effective way to do so is as illustrated in Fig-5.10. Since these two submodules operate on different soft information, they can operate independently and thus simultaneously. The soft information from each is combined only after the final iteration to produce the final soft output via

$$\mathrm{SO}[u(i,j)] = \mathrm{SO}_{\mathrm{Rc}}[u(i,j)] \copyright \mathrm{SO}_{\mathrm{Cr}}[u(i,j)] \qquad (5.19)$$

where $u(i,j)$ can be either $x(i,j)$ or $a(i,j)$, and the subscript represents the corresponding version of concatenated 2D SISO used.

For a square support region, complexity is increased by a factor of two when a composite SISO is used in place of the COL-row or ROW-col SISO. Alternatives exist to trade increased complexity for (possibly) better performance. For example, in the COL-row model, the definition of $\mathbf{v}(i,j)$ can be modified to

$$\mathbf{v}(i,j) \triangleq [a(i,j-L_{\mathrm{r}}-1), a(i,j-L_{\mathrm{r}}), \dots a(i,j+L_{\mathrm{r}})] \qquad (5.20)$$

Note that, compared to (5.17), an extra input $a(i,j-L_{\mathrm{r}}-1)$ is included. Consequently, each inner (column) SISO can incorporate two columns of $\mathrm{SI}[x(i,j)]$ instead of one. This is expected to improve performance since it decreases the degree of marginalization (*i.e.*, heuristic rule 2 in Section 2.7.1). Carrying this notion to the extreme, one obtains an optimal detector of the form shown in Fig-5.1 (with the VA running in the y-direction).

When a suboptimal SISO is used for the row or column SISO, soft information filtering (see Chapter 3) may provide an improvement in performance. One such configuration of interest in the following is shown in Fig-5.11. Furthermore, since the 2D SISO itself is a suboptimal approx-

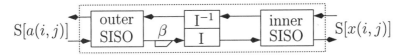

Figure 5.11. A modified 2D SISO algorithm with soft information filtering.

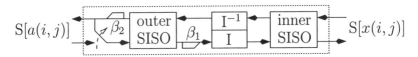

Figure 5.12. A modified 2D SISO algorithm using self iteration and soft information filtering.

imation to the marginal soft inverse of the 2D system (even if the row and column SISOs are "full-state" forward-backward algorithms), the concept of self-iteration is applicable as well. A configuration of interest in the following, show in Fig-5.12, uses soft-in information filtering and self-iteration.

Complexity Reduction   The complexity of the concatenated 2D SISO algorithm grows exponentially in the area of the support region. This may be prohibitive even for moderate values of L_r and L_c. For example, when $L_r = L_c = 2$ (*i.e.*, a 5×5 2D ISI channel), the inner SISO has $|\mathcal{A}|^{20}$ states, which is prohibitive, even for a binary inputs. The decision feedback concepts introduced in Chapter 3 can be applied to reduce this complexity. For example, many 2D ISI channels are symmetric in nature and have energy that falls off quickly from the center tap. Such a 2D SISO can be based on a modeled memory of $L_1 < L = L_r = L_c$ with decision feedback used for the taps outside of this region. Specifically, the original channel can be partitioned into two parts: the taps with index $-L_1 \leq i, j \leq L_1$ and the rest. One may then feedback the hard decisions $\hat{a}(i,j)$ or soft estimates $\tilde{a}(i,j)$ for all positions related to the taps in the second part. Consequently, the complexity is comparable to a 2D SISO operating on a $(2L_1 + 1) \times (2L_1 + 1)$ channel. This concept is illustrated in Fig-5.13.

In particular, this can be applied to the COL-row SISO by enforcing this hard information at the $SI[x(i,j)]$ port at each iteration. For the case when hard decisions are fedback, $\hat{a}(i,j)$ can be obtained from the last iteration and used in each of the shaded locations in Fig-5.13 to recompute $SI[x(i,j)]$ at each iteration. Similar processing could be done by averaging over soft information at each iteration – *i.e.*, using $\tilde{a}(i,j)$, the average value of $a(i,j)$ according to the beliefs from the last iteration. Alternatively, this averaging could be done once, before the first

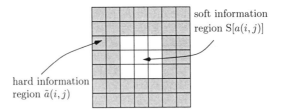

Figure 5.13.   Truncation of a 2D ISI: from 7×7 to 3×3.

iteration, and these values could be used for all subsequent iteration. The last approach has the advantage that the values of $\mathrm{SI}[x(i,j)]$ are the same at each iteration.

5.3.2   Distributed 2D SISO Algorithms

The concatenated channel model obtained in Section 5.3.1.1 is based on a model which is essentially a reordering in a 1D fashion. Models that are more naturally 2D can be established by considering the dependencies of local variables directly. Once a graphical model is established, the corresponding SISO follows from the principles developed in Chapter 2. One characteristic of these resulting SISOs is that they tend to naturally admit parallel activation schedules and architectures. This is a desired property for many 2D applications since the signal format and system hardware is naturally 2D and locally connected. We consider two graphical models that differ from the concatenated models presented thus far and suggest fully parallel algorithms based on parallel activation schedules. It is important to note, however, that other schedules are possible for message passing on the same graphical models.

5.3.2.1   Graphical Model 1 and Associated 2D SISOs

The graphical model for the 1D ISI channel shown in Fig-2.58 extends directly to the 2D case. This is illustrated in Fig-5.14, where only one mapper from $t(i,j)$ to $x(i,j)$ is shown. Analogous to the model in Fig-2.58, there is a broadcaster node for each input variable $a(i,j)$ that sends the value to each $t(i,j)$ affected by its value (*i.e.*, for the square support, this is $t(i+l,j+m)$ for $|l| \leq L_{\mathrm{c}}$ and $|m| \leq L_{\mathrm{r}}$). While it is cumbersome to show the complete diagram, this may be though of as a set of connections from the input plane (made up of $a(i,j)$) to the output plane (made up of $x(t(i,j))$). The node that maps the input symbols to $x(i,j)$ is denoted by the subsystem $\mathsf{T}_{(i,j)}$. This graph is referred to as the graphical model 1 for the 2D system (2D-GM1).

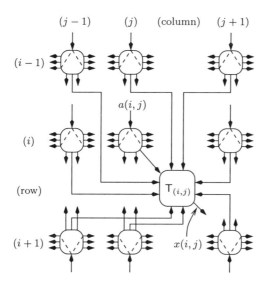

Figure 5.14. Graphical model 1 for the 2D system (illustrated by a 3×3 2D ISI channel). Note that the only output subsystem shown is $\mathsf{T}_{(i,j)}$ with all its connections shown.

Clearly, 2D-GM1 has cycles so that applying iterative detection (message passing) is, in general, suboptimal. Nonetheless, one can apply the standard message passing techniques to this 2D model in exactly the same manner used in the 1D case described in Example 2.15. We refer to this algorithm with parallel activation of all $\mathsf{T}_{(i,j)}^{-s}$ nodes, followed by parallel activation of all SOBC nodes, as the *fully-parallel (2D SISO) algorithm 1* (FPA1).

For many applications considered, the energy in the 2D channel impulse response decays quickly away from the center. Thus, the connection cutting concept introduced in Chapter 3 can be considered for complexity reduction of FPA1. Since the center tap typically contains the most energy in the applications considered (*i.e.*, $f^2(i,j)$ is maximized by $i = j = 0$), we consider cutting all message passing connections from $\mathsf{T}_{(i,j)}^{-s}$ back to the SOBCs except for the SOBC corresponding to $a(i,j)$. Note that this corresponds precisely to the algorithm in Example 3.4 (and the distributed sparse SISO algorithm in Section 3.4) with a single pivot selected to be the center tap. The corresponding simplified version of FPA1 is shown in Fig-5.15.

To compensate for the performance degradation associated with this connection cutting, one may apply 2D cross initial combining (see Problem 5.12) and soft information filtering. Specifically, the soft information passed back to the SOBC for $a(i,j)$ can be filtered. This reduced

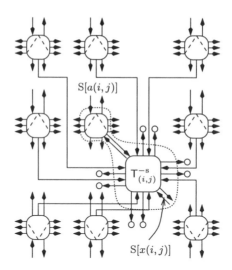

Figure 5.15. The SISO associated with the graphical model in Fig-5.14 using connection cutting.

complexity version of the FPA1 (*i.e.*, connection cutting, filtering, and cross-combining), was suggested in [ChChNe98] where it was referred to as the *2D distributed detection algorithm* ($2D^4$).

5.3.2.2   Graphical Model 2 and Associated 2D SISOs

While the 2D-GM1 model does not implicitly raster the data page, it does result in an algorithm (FPA1) that marginalizes the soft information on $t(i,j)$ all the way down to the input symbol level before combining it at the next iteration. According to heuristic rule 2 of Section 2.7.1, one should avoid marginalization whenever possible to improve performance. An algorithm [Th00] which does not marginalize all information to the input symbol plane is described by defining two auxiliary variables $r(i,j)$ and $c(i,j)$ which, together with one input variable specify $t(i,j)$. For example, for the case of $L_r = L_c = 1$ these are defined as

$$r(i,j) = \begin{pmatrix} a(i-1,j-1) & a(i-1,j) & a(i-1,j+1) \\ a(i,j-1) & a(i,j) & a(i,j+1) \end{pmatrix} \qquad (5.21a)$$

$$c(i,j) = \begin{pmatrix} a(i-1,j-1) & a(i-1,j) \\ a(i,j-1) & a(i,j) \\ a(i+1,j-1) & a(i+1,j) \end{pmatrix} \qquad (5.21b)$$

In general, the auxiliary variables are

$$r(i,j) \triangleq t(i-1,j) \sqcap t(i,j) \qquad (5.22a)$$

$$c(i,j) \triangleq t(i,j-1) \sqcap t(i,j) \qquad (5.22b)$$

$$t(i,j) \sqsubset r(i,j) \sqcup c(i,j) \sqcup a(i + L_\mathrm{c}, j + L_\mathrm{r}) \qquad (5.22\mathrm{c})$$

Thus, these auxiliary quantities establish direct connections among the $t(i,j)$ variables instead of through $a(l,m)$ plane as in 2D-GM1 (see Fig-5.14). Using the relationship (5.22c), the graphical model in Fig-5.16 is established. We refer to this as the graphical model 2 for 2D systems (2D-GM2). It is interesting to note that this directed graph is

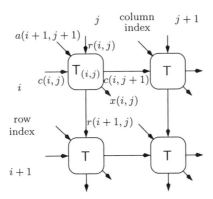

Figure 5.16. Graphical model 2 for 2D systems (shown for $L_\mathrm{r} = L_\mathrm{c} = 1$). Note that only the inputs and outputs of subsystem $\mathsf{T}_{(i,j)}$ are labeled. The inputs of $\mathsf{T}_{(i,j)}$ are $r(i,j)$ fed from $\mathsf{T}_{(i-1,j)}$, $c(i,j)$ fed from $\mathsf{T}_{(i,j-1)}$ and $a(i+1,j+1)$. The output is $x(i,j)$

"causal" in the "southeast" direction – *i.e.*, all connections in the graph are southeast directed so that a give subsystem receives its inputs only after the outputs of all northwest systems have produced outputs. Due to the auxiliary variables introduced, $\mathsf{T}_{(i,j)}$ is connected only to a single 2D system input variable $a(i+1,j+1)$. Nevertheless, 2D-GM2 has numerous cycles, the shortest of which is length 4 as may be observed from Fig-5.16.

Given the 2D-GM2, the corresponding SISO follows by applying the standard rules described in Chapter 2. In addition, since there are short cycles in this model, we also introduce the option of soft information filtering with the resulting algorithm diagrammed in Fig-5.17. Similar to the FPA1 and 2D^4 2D SISOs, this SISO can be implemented in a fully distributed manner by parallel activation of the $\mathsf{T}_k^{-\mathrm{s}}$ nodes. We refer the algorithm shown in Fig-5.17 (*i.e.*, with filtering) and parallel activation as the *fully-parallel (2D SISO) algorithm 2* (FPA2). Notice that FPA2 has lower computational complexity than the FPA1 because the marginalization to the symbol plane and subsequent recombining is avoided. The storage requirements for the FPA2, however, are greater than that of the FPA1 due to this same characteristic.

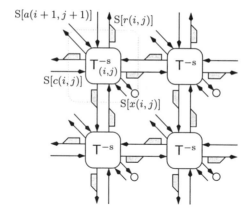

Figure 5.17.   The FPA2 2D SISO developed from the 2D-GM2 in Fig-5.16. Soft information filters have been added as illustrated.

5.4   Data Detection in POM Systems

5.4.1   POM System Model

There is a great interest in parallel recording and retrieval techniques for the future high capacity storage systems. The page-oriented optical memory (POM) system is a promising candidate because of its potential for simultaneously achieving high capacity, fast data transfer and 2D parallel access (*e.g.*, see [ChChNe99] and references therein). However, POM systems operating near their capacity are subject to numerous sources of noise and interference arising from the optical system itself. Specifically, the blur resulting from the low-pass nature [Go68] of the optical system causes neighboring bits in a data page to overlap during the retrieval procedure. Thus this pre-detection retrieval procedure can be modeled by a 2D ISI (space continuous) channel

$$s(x, y) = \sum_{(i,j)} a(i,j) h(x - i\Delta, y - j\Delta) \qquad (5.23)$$

where $a(i,j)$ is the stored binary data, and Δ is the vertical and horizontal detector (*e.g.*, CCD pixel) spacing. The effect of interference is captured by $h(x,y)$, the point spread function of the optical system, which is truncated to the region $\{(x,y) : |x| < (L_r + 0.5)\Delta, |y| < (L_c + 0.5)\Delta\}$ in our system model. The optical-to-electrical conversion process is assumed to take place by means of an array of detectors (*e.g.*, CCD) that integrate the intensity of the signal over both time and space.[3] Further-

---

[3] As shown in (5.23), the integration over time (*e.g.*, temporal matched filtering) has already taken place.

more, it is assumed that a detector is centered on each pixel location and the fill factor is unity (*i.e.*, no spatial guard bands between detectors exist). For a POM system using incoherent signaling (*i.e.*, optical intensity modulation and detection), the function $s(x,y)$ in (5.23) characterizes the intensity function directly. A discrete-space model for the output of the (i,j)-th detector, $x(i,j)$ is the linear ISI model in (5.6a) with channel coefficients determined by spatial integration over the optoelectronic detector region

$$f(l,m) = \int_{-\Delta/2}^{\Delta/2} \int_{-\Delta/2}^{\Delta/2} h(x + l\Delta, y + m\Delta) dx dy \qquad (5.24)$$

The blurred output from the 2D ISI channel is then corrupted by various sources of post-detection noise, which may be modeled as AWGN. Other signal and noise models (*e.g.*, coherent field noise, shot noise, etc.) are possible, especially in coherent POM systems [ChChNe99]. In the current context, these result in different soft information metrics and/or nonlinear ISI channels.

Thus, the data retrieval problem for POM systems may be modeled as a 2D detection problem of the form discussed in Section 5.1 and the iterative detection algorithms (SISOs) developed in Section 5.3 can be applied. Effective ISI mitigation techniques translate to increased POM storage capacity by possibly reducing the required detector spacing. Furthermore, the noise level is often determined by the number of pages that are written to a particular medium. Therefore, an SNR gain associated with an effective algorithm translates to an increased storage capacity (*i.e.*, see [ChChNe99] for a detailed description of these trade-offs).

Two channels inspired from POM systems are used in the following investigation. The Gaussian blur, $h(x,y) = \mathcal{N}_2((x,y); \mathbf{0}; \sigma_b^2 \mathbf{I})$ is a typical point spread function in a POM system. Channel A is a 3×3 truncated Gaussian blur channel with $\sigma_b = 0.623$ [4] which results in a discrete-space channel in the form of (5.7) with $a = 0.1235$ and $b = 0.3515$. Channel B is a 5×5 truncated Gaussian blur with $\sigma_b = 0.8$. Specifically, the upper-left portion of this symmetric channel is

$$\begin{pmatrix} 0.0040 & 0.0317 & 0.0639 \\ 0.0317 & 0.2534 & 0.5034 \\ 0.0639 & 0.5034 & 1.000 \end{pmatrix} \qquad (5.25)$$

We assume iid-uniform binary inputs of intensity 0 and 1, and AWGN with variance σ_w^2. The SNR is defined by $\gamma = \|\mathbf{F}\|^2 / 2\sigma_w^2$. Due to the

---

[4]In the following, σ_b is expressed relative to Δ (*i.e.*, equivalently assume that $\Delta = 1$).

operational SNR, only min-sum processing is considered. For the iterative algorithms, the iteration is terminated once convergence is observed,
i.e., after N_c iterations.

5.4.2   Existing Detection Algorithms

Various detectors have been developed to solve the 2D data detection problem. For comparison, the following detectors are tested on the
channels A and B.

Thresholding Detector   Letting $m_z \triangleq \mathbb{E}\{z(i,j)\}$, the threshold
decision (TH) rule is: $\hat{a}(i,j) = 1$ if $z(i,j) > m_z$; otherwise, $\hat{a}(i,j) = 0$.
For a binary input alphabet $\{0,1\}$, AWGN noise, and the channel in
(5.7) with $g \geq 0$ and $h \geq 0$, it can be shown that the performance of TH
does not have a BER floor if and only if $(g+h) < 0.25$. Therefore, when
the 2D ISI is severe, it can be expected that the threshold detector will
fail.

MMSE Equalizer   Denoting $v_0(\zeta) = v(\zeta) - \mathbb{E}\{v(\zeta)\}$, the minimum
mean square error (MMSE) equalizer is defined by a $(2Q+1) \times (2Q+1)$
linear filter $g(i,j)$ and the operation

$$\check{a}_0(i,j) = g(i,j) \circledast z_0(i,j) \tag{5.26}$$

The MMSE filter coefficients $g(l,m)$ are obtained by solving the Wiener-
Hopf equations, and the corresponding decision rule is: $\hat{a}(i,j) = 1$ if
$\check{a}_0(i,j) > 0$, otherwise $\hat{a}(i,j) = 0$. The details of the MMSE equalizer
design for 2D ISI channels and the impact of their use on POM storage
capacity are discussed in detail in [ChChNe99].

Decision Feedback Equalizer   The decision rule in the MMSE
equalizer can be replaced with a simple iterative decision making process [NeChKi96]. After obtaining the tentative decisions $\hat{a}(i,j)$ by the
MMSE equalizer, we calculate the quantity $x_0(i,j)$, which is the 2D
ISI output at location (i,j) assuming $a(i,j) = 0$. Similarly, $x_1(i,j)$ is
obtained by setting $a(i,j) = 1$. An updated decision is obtained as
$\hat{a}(i,j) = \arg\min_l |z(i,j) - x_l(i,j)|$. This iteration can be performed
a fixed number of times, or until no further improvement/change occurs. This algorithm is referred to as (2D) decision feedback equalization
(DFE).

Decision Feedback VA   Although the standard VA [Fo73] is not
directly applicable to MAP-PgD for the 2D ISI channel, the so-called
decision-feedback VA (DFVA) [HeGuHe96, KeMa98] adapts the VA to
provide a reasonable 2D detection algorithm. As illustrated in Fig-5.18
for a 3×3 2D ISI channel, the DFVA runs a VA row-wise, with a state

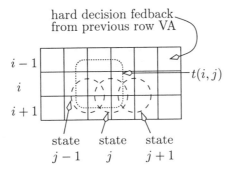

Figure 5.18.   The decision feedback VA (row-wise version) for a 3×3 2D ISI channel.

consisting of a 2×2 symbol matrix. The transition metrics are computed using hard decisions fed-down from the previous row. After reaching the end of each row, the VA traces back on the best path and makes decisions only on the row i. This process is continued down rows until the entire data page has been detected. A scheme combining the row-wise and column-wise versions along with hard-decision iterations has been developed and successfully applied to image processing problems in [Mi99]. Like other decision feedback algorithms, the DFVA may suffer significant error propagation effects at low SNR.

5.4.3   The Performance of Iterative Detection Algorithms

The performance of various algorithms on Channel A is shown in Fig-5.19. In addition, the performance bounds derived in Section 5.2 are shown. Channel A is a relatively severe channel 2D ISI channel for practical POM systems. The thresholding detector completely fails for this channel. Both the DFE and DFVA detectors perform 6 dB worse than the MAP-PgD upper bound. The composite 2D SISO slightly outperforms the concatenated COL-row 2D SISO with both converging relatively quickly (4 iterations). Based on the performance bounds, it is unlikely that a practical algorithm can be designed that will perform better than the composite 2D SISO. Several COL-row 2D SISOs using SW-SISO modules were applied to Channel A with fairly good performance. These solutions provide another algorithm with a fairly high degree of parallelism. As expected, the larger the window size in the SW-SISOs, the better the performance. Note that soft information filtering improves the performance of the COL-row detectors based on SW-SISO with $D = 2$. In summary, the detectors based on iterative detection can tolerate up to 6 dB more noise which translates to a storage capacity increase [ChChNe99].

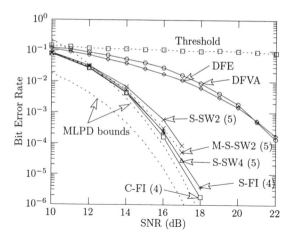

Figure 5.19. Performance of various iterative concatenated detectors on Channel A. Each curve is labeled by: algorithm used (iteration number used). S(C)-FI(SW) is the "single" (composite) concatenated 2D SISO using FI(SW)-SISO modules. The number following "SW" is the parameter D used with SW-SISO. The M-S-SW is the algorithm in Fig-5.12, with parameters $\beta_1 = 0.7$ and $\beta_2 = 0.15$ and SW-SISOs.

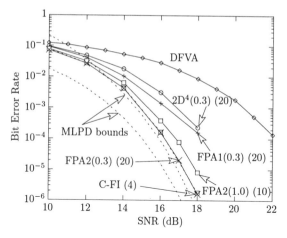

Figure 5.20. Performance of various iterative distributed detectors on Channel A. Each curve is labeled by: algorithm used (number of iterations). The number following FPA1 (or FPA2) is the value of the soft information filter parameter β which was optimized empirically.

Fully Parallel Algorithms   The distributed algorithms developed in Section 5.3.2 were applied to Channel A with the performance is shown in Fig-5.20. One iteration of these distributed algorithms corresponds to a parallel activation of the $\mathsf{T}_{(i,j)}^{-\mathsf{s}}$ nodes and the SOBC nodes (when appropriate). Note that the $2\mathrm{D}^4$ algorithm performs almost as well as the FPA1 despite a substantial complexity reduction and the severity of Channel A. As suggested by heuristic rule 2 in Section 2.7.1, the performance of the FPA2 is better than FPA1. Since both 2D-GM1 and 2D-GM2 have short cycles and a parallel activation schedule is used, both of the associated algorithms benefit from soft information filtering. Specifically, with $\beta = 0.3$, the performance of the FPA2 is comparable to the composite 2D SISO (using FI-SISO modules). It is worth noting that if a row-column activation schedule for message passing is used on 2D-GM2, filtering is not required to achieve the best performance shown in Fig-5.20 [Th00b].

Reduced Complexity 2D SISOs For Channel B, the 2D SISO algorithms are prohibitively complex. Since the energy in the ISI taps decays quickly away from the center, reduced-complexity versions based on decision feedback are expected to perform reasonable well. Specifically, we chose to truncate Channel B by keeping the central 3×3 portion as in Fig-5.13. A COL-row concatenated SISO with this complexity reduction was simulated for Channel B (*i.e.*, this has complexity comparable to the corresponding 2D SISO for Channel A) with the results shown in Fig-5.21. According to Fig-5.21, substituting $\tilde{a}(i,j) = 0$

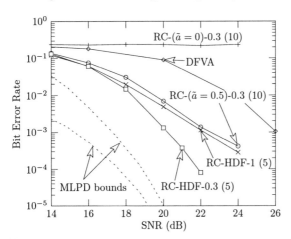

Figure 5.21. Performance of various reduced-complexity simple 2D SISO (Fig-5.11) detectors on Channel B. Each curve is labeled by: RC-hard information feedback scheme-β (number of iterations).

in the positions corresponding to the truncated taps is not viable. Using $\tilde{a}(i,j) = \mathbb{E}\{a(\zeta; i,j)\} = 0.5$ (*i.e.*, averaging over the a-priori probabilities) and using this value for each iteration results good performance. Using hard decision feedback with $\text{MI}[x(i,j)]$ adjusted before each activation of the COL SISO provides similar performance. With filtering and β optimized, as shown in Fig-5.11, the performance of the hard decision feedback technique is substantially improved. The performance of this algorithm is approximately 6 dB better than the DFVA and 3 dB from the lower bound. The DFVA, however, has $2^{3\times4} = 2^{12}$ states which is substantially more complex that 5 iterations of the reduced complexity COL-row SISO. Finally, using the soft information on $a(i,j)$ available at each iteration to feedback a soft estimate was found to provide no significant gain relative to the hard decision feedback algorithm (curve is not shown).

5.5   Digital Image Halftoning

Digital image halftoning refers to approximately rendering a gray level image as a binary (black and white) image. When a binary image is displayed or printed, the human visual system [Gr81] and/or the imperfections of the printing device [PaNe95] result in a blurring or low pass filtering of the image. While the blurring effect is undesired in a storage system (*e.g.*, POM) or many imaging systems, it enables effective image halftoning by averaging the black spots and white spaces to approximate the original gray level image.

Several halftoning methods have been developed [Ul87]. Since the "receiver" of halftone images is a human observer, good quantitative measures of quality for halftoning are difficult to specify. While well-defined metrics for quality such as least squared error (LSE) roughly capture the halftone quality, other aspects such as regular patterns, edge degradation, poor contrast can degrade the perceptual quality of the halftone. Image halftoning is an area with considerable (and impressive) results that are too numerous to summarize here. Our objective is to show that the halftoning problem may be formulated as a 2D-DLM problem and that the iterative algorithms presented in Section 5.3 can provide high quality results. This application also provides an opportunity to display the characteristics of the 2D detection algorithms visually which provides additional insight into their respective characteristics. We make no claim, however, to the optimality of these approaches in terms of quality or complexity due to the above perceptual quality issue and the vast body of research available (*e.g.*, . [Ul87, AnAl92, MuAh92, GeReSu93, ScPa94, NePa94, PaNe95, LiAl96, Wo97, PaNe99, LaArGa98, Ul00]). In Fig-5.22, a 512×512, 256 gray-level version of "Lena" is shown, which is a benchmark digital image selected as the test image for the experiments that follow.

We consider the LSE halftoning problem which is analogous to data detection in a linear ISI-AWGN channel. Specifically, let $z(i,j)$ represent the gray scale image and $a(i,j)$ a halftone. The least squares image halftoning (LSIH) problem is the 2D-DLM problem in (5.5b) with

Figure 5.22. The Lena image with 256 gray-levels.

$\mathrm{MI}[x(i,j)] = [z(i,j) - x(i,j)]^2$ and $x(i,j) = f(i,j) \circledast a(i,j)$ with $f(i,j)$ selected to capture the imaging system characteristics. Thus, with this model and objective, the LSIH is precisely equivalent to the ML-PgD problem for the 2D ISI-AWGN channel.

5.5.1 Baseline Results

A straightforward way to obtain a halftone image is by thresholding the gray-scale pixel: $a(i,j) = 1$ if $z(i,j) > 0.5$; otherwise, $a(i,j) = 0$. Thresholding is a special case of the more general LSIH problem presented earlier, derived under the assumption that $x(i,j) = a(i,j)$. Applied to Fig-5.22 this simple halftoning algorithm yields the image in Fig-5.23. This image is of poor quality, since most important details of the original image have been lost or severely distorted.

To obtain better halftoning results, a nontrivial filter $f(i,j)$ is required. One choice, inspired by the work in [Wo97], is the 3×3 2D linear filter with coefficients

$$\begin{pmatrix} 0.2219 & 0.1439 & 0.0355 \\ 0.1439 & 0.0980 & 0.0306 \\ 0.0355 & 0.0306 & 0.0174 \end{pmatrix} \qquad (5.27)$$

where the taps are normalized to sum to one. We emphasize that this filter only represents one *reasonable* choice for the given filter size and has not been carefully optimized.

The DFVA, presented in Section 5.4.2, can be used as an approximate solution to the LSIH problem. In Fig-5.24, the result of DFVA is shown using the filter in (5.27). Compared to (Fig-5.23), this image provides much better quality. However, there exist many low frequency, undesired patterns, which reflect the row-by-row processing of the DFVA.

One property of the 2D SISOs developed in Section 5.3 is that they handle "noncausal" filter patterns very naturally (*i.e.*, filters with $f(i,j) \neq 0$ in all directions). Specifically, hard decision feedback on a rastered model is not performed. Thus, it is expected that the undesirable patterns in the DFVA halftone will not occur with a 2D SISO solution. Using the (min-sum) COL-row 2D SISO algorithm with zero soft-in metrics for $a(i,j)$, the image shown in Fig-5.25 is obtained after three iterations. In terms of the objective criterion, the LSE of Fig-5.25 is approximately 40% less than that of Fig-5.24. In addition to having a smaller LSE, Fig-5.25 has much better perceptual quality. However, this halftone image still contains undesirable patterns, especially on the face and right portion in the image corresponding to portions in the original image in Fig-5.22 that have nearly constant gray level. A good halftone output of a constant gray level image should have the properties of random noise with the proportion of white to black pixels determined by the average gray level. This perceptual quality measure is difficult to capture with the LSE objective function. However, the iterative algorithms can be biased toward a random image to alleviate these regular patterns.

5.5.2   Random Biasing

The soft-in metrics for the 2D SISO allow one to naturally bias the halftoning process toward a particular image. For example, an initial bias toward a gray pattern suitable for a particular imaging device may be appropriate. Alternatively, a random biasing scheme can be used to alleviate the undesirable patterns in Fig-5.25. In the following we explore this option.

At the inner SISO of the COL-row 2D SISO, the soft-in metrics for $\mathbf{v}(i,j)$ are *initialized* using

$$\text{MI}[\mathbf{v}(i,j)] = \sum_{l=-L_{\mathrm{r}}}^{L_{\mathrm{r}}} [a(i,j-l)d(i,j-l) + (1-a(i,j-l))(r-d(i,j-l))]$$

$$(5.28)$$

Figure 5.23.   Halftone Lena image by thresholding.

Figure 5.24.   Halftone Lena image by DFVA.

Figure 5.25. Halftone Lena image using the COL-row 2D SISO with the uniform initialization.

Figure 5.26. Halftone Lena image using the 2D SISO with a random initialization.

Figure 5.27.   Halftone Lena image using a modified 2D SISO with a 7×7 filter.

Figure 5.28.   Halftone Lena image using error diffusion.

where $d(i,j)$ is a random number uniformly distributed in $[0,r]$ and r is a design parameter. At the outer SISO, uniform soft-in metrics (*i.e.*, $\mathrm{MI}[a(i,j) = 0] = \mathrm{MI}[a(i,j) = 1] = 0$) are used. After the first iteration, the soft information exchanged on $\mathbf{v}(i,j)$ is computed and exchanged in the standard fashion. This random initialization has the effect of biasing the image toward a random halftone. This provides a "clouding" of the halftone which is partially removed with each iteration. A halftone version of Lena using three iterations and the random biasing with $r = 30000$ is shown in Fig-5.26. Compared to Fig-5.25, the undesirable patterns are reduced significantly. In addition, the edges in the original image are well preserved and the image contrast is good.

5.5.3   Larger Filter Support Regions

Better performance is expected when the support of $f(i,j)$ is larger. However, this results in a significant increase in complexity for the 2D SISO algorithms. In this section, we use a 7×7 symmetric filter adopted from [Wo97] and a reduced complexity 2D SISO. Specifically, the filter $f(i,j)$ is (only the left upper part (4×4) given for compactness)

$$\begin{pmatrix} -0.0029 & 0.0030 & 0.0091 & 0.0116 \\ 0.0030 & 0.0174 & 0.0306 & 0.0355 \\ 0.0091 & 0.0306 & 0.0980 & 0.1439 \\ 0.0116 & 0.0355 & 0.1439 & 0.2219 \end{pmatrix} \tag{5.29}$$

Decision feedback is used as described in Section 5.12 with $L_1 = 1$. Specifically, for the first iteration $\tilde{a}(i,j) = z(i,j)$ is used and $\hat{a}(i,j)$ is used for subsequent iterations (*i.e.*, the tentative halftone image from the previous iteration). With a uniform initialization it was found that this halftoning scheme works well in most regions in the image, but provides poor perceptual quality in those regions that are nearly all black or all white. The modified 2D SISO with self-iteration and filtering as shown in Fig-5.12 (using FI-SISO modules) was used to improve the quality in these regions. The results are shown in Fig-5.27 using $\beta_1 = 0.3$ and $\beta_2 = 0.65$ and 10 iterations.

A simple and effective technique for halftoning is error diffusion [PaNe95]. The result of this algorithm applied to the Lena image is shown in Fig-5.28. It can be observed that, although Fig-5.27 is perceptually superior to Fig-5.23 and Fig-5.24, a comparison between Fig-5.27 and Fig-5.28 is difficult. The error diffusion version is smoother, while the 2D SISO version is sharper (*i.e.*, better edge preservation) and has better contrast and fewer artifacts.

5.5.4 High Quality and Low Complexity using 2D-GM2

An algorithm which produces a halftone which has the best attributes of the error diffusion method and the method that yields the image in Fig-5.27 is obtained using message passing on 2D-GM2 in Section 5.3.2.2 with decision feedback. The result of this algorithm is the halftone shown in Fig-5.29. Specifically, the 7×7 filter in (5.29) is used with hard de-

Figure 5.29.   Halftone Lena using the 2D-GM2 model with decision feedback to create an effective 3-pixel L-shaped support region.

cision feedback used in all locations of the support except for three: $f(0,0)$, $f(0,1)$, and $f(1,0)$ (*i.e.*, the effective $\mathcal{P}(i,j)$ consists of $a(i,j)$, $a(i-1,j)$ and $a(i,j-1)$). Thus, the support region used by the algorithm is a small "L" shaped region. The hard decisions $\hat{a}(i,j)$ fedback in the other locations are the best hard decisions currently available. For the initial activation, the hard decisions are randomly selected to be 0 or 1, each with probability $1/2$. In contrast to FPA2, the activation schedule used is a column-row serial schedule as described in the following. First, forward and backward messages are passed up and down the entire first column. Next, this same complete forward-backward schedule is executed up and down the second column. This proceeds sequentially to the right for each column. After the last column has been processed, a full forward-backward schedule is run on the first row. This continues sequentially for each row and, after processing the last row, one iteration is complete. Note that this schedule enables updated hard decisions to be fedback early – *i.e.*, these hard decisions are updated after every node activation, not each iteration as defined above.

The quality of the image in Fig-5.29 is quite good, exhibiting the smoothness of the error diffusion halftone in Fig-5.28 and the sharpness of that in Fig-5.27. Also, the associated processing complexity is substantially lower than that used to produce Fig-5.27. Specifically, the algorithm producing the image in Fig-5.29 passes messages on binary variables (see Problem 5.14) and has an effective $t(i,j)$ that takes only 8 values. In contrast, the algorithm used for Fig-5.27 has an effective $t(i,j)$ that takes on 2^9 values making it roughly 64 times as complex.

5.6   Summary and Open Problems

Two dimensional systems provide an interesting framework to apply the concepts developed in Chapters 1-3. In particular, there are several reasonable system models that lead to iterative detection algorithms with substantially different characteristics – *i.e.*, trading performance, computational complexity, storage requirements, and parallelism. Furthermore, performance bounds are tractable for 2D systems and provide an opportunity to compare the performance of iterative detection to that of the optimal receiver for a nontrivial problem. Most reasonable iterative detectors were found to perform near optimally.

We believe that the methods presented in this chapter may prove to be applicable to a number of relevant problems. Any system which is most naturally described in terms of a multidimensional index set with local dependencies is a candidate for these methods. This may include, for example, multi-track magnetic recording channels, space-time channels (*e.g.*, multi-carrier systems with frequency and temporal distortion/coding), and various problems in image processing and compression. For some of these areas, the adaptive methods of Chapter 4 may also be applicable in concert with the 2D methods introduced in this chapter. There is considerable room for application and modification of these techniques in the data encoding/fitting problems. In fact, we have applied the iterative algorithms in place of the DFVA [KeMa98] for near-lossless image compression with significant improvements [ChChOrCh98]. In many of these potential applications, however, complexity reduction is imperative since the complexity grows exponentially in the support region and, in many of these applications, $|\mathcal{A}|$ is quite large. The reduced complexity approaches discussed herein represent only a first step in the development of iterative algorithms that could be useful for gray level image deblurring, for example. While a rectangular support region may be a reasonable assumption for data detection problems, it may be an overkill for many data encoding/fitting problems, as suggested by the L-shaped region used to create the best halftone.

Several theoretical topics related to the performance bounds and the theoretical complexity of the 2D detection problem are interesting future directions. In the former, the searching algorithms used to approximate the minimum distance and the upper bound can certainly be improved. Application of the concepts in [AnFo75, Ve86] and their extension to 2D is an interesting and difficult open problem. Similarly, a formal analysis of the complexity for the data detection problem with a finite support size has not been performed. For an arbitrary support, it can be shown that the problem is NP-Hard [Th00] (see [Co90, DaLu93] for other considerations of the computational complexity of general inference problems). Translation of these results to a finite support channel is difficult and has not been done. This subclass of 2D detection problems based on a finite support region, however, may be the most relevant question regarding complexity. For example, in the 1D case, the general class of sequence detection problems is NP-Hard, yet for the finite support channel, the Viterbi algorithm provides an efficient solution. Based largely on intuitive arguments, we conjecture that optimal detection for a relatively simple 2D channel (*e.g.*, a 3×3 support region), requires complexity that is exponential in the perimeter of the *observation region* used (see Problem 5.3). Similarly, the theoretical complexity of the $d_{\min}$ search, even relative to that of the MAP-PgD problem, is not clear.

1D	One Dimensional
2D	Two Dimensional
2D^4	2D Distributed Data Detection (algorithm)
2D-GM1(2)	Graphical Model 1(2) for 2D systems
CCD	Charge Coupled Device
COL-row	Column-Row (concatenated)
DFVA	Decision Feedback VA
FPA1(2)	Full-Parallel Algorithm 1 (2)
LSE	Least Squared Error
LSIH	Least Squared (Error) Image Halftoning
LUT	Look Up Table
MMSE	Minimum Mean Squared Error
NP	Non-deterministic Polynomial

Table 5.2.   Table of abbreviations specific to Chapter 5 (continued on next page)

POM	Page-access Optical Memory
ROW-col	Row-Column (concatenated)
TH	Threshold detector

Table 5.2.   (Continuation) Table of abbreviations specific to Chapter 5

5.7   Problems

5.1. Consider the support region $\mathcal{P} = \{(m,n) : m = 0, 0 \leq n \leq L\}$ which corresponds to the 1D simple FSM relation. Show that simple 2D patterns for this case are the simple sequences from Chapter 1.

5.2. Discuss how the Viterbi algorithm applied as shown in Fig-5.1 could be used to search for small distances for an ISI channel.

5.3. Consider making a pairwise decision between any two data hypotheses which agree on a certain region.

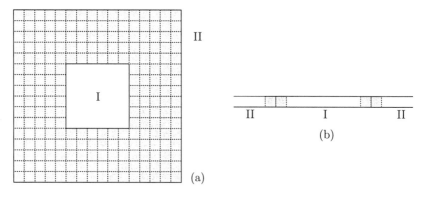

Figure 5.30.   Moats of agreements for (a) a 3×3 ($L_r = L_c = 1$) support region and (b) a 1D ISI channel with $L = 2$. Agreements are shown as shaded pixels.

   (a) Consider the case shown in Fig-5.30(a) with a 3×3 support region. Show that this pairwise decision can be made separately on the interior region I and the exterior region II.

   (b) Repeat this argument for the support region corresponding to a 1D channel with $L = 2$ and the regions shown in Fig-5.30(b).

   (c) Explain how the decoupling in the case of the above 1D example enables the Viterbi algorithm. What is the relation between the "moat" of agreements and the states used in the trellis?

(d) Describe the analogous "2D Viterbi algorithm" that arises from the applying the above reasoning to the case shown in Fig-5.30(a).

(e) Compare the growth in "moat-conditions" as the size of the interior region I grows in each of the two cases in Fig-5.30. What is the reduction in complexity relative to exhaustive search for the 1D and 2D algorithms?

5.4. Consider finding the worst 1D ISI channel with memory length L. Specifically, the worst channel is the one with smallest normalized minimum distance. Show that, given L, there exists at least one worst channel which has either even or odd symmetry around its center tap.

5.5. Show that the conclusion from Problem 5.4 extends to 2D ISI channels. Specifically, show that for a given $L = L_r = L_c$, there is a worst channel that has either even or odd symmetry around its center tap.

5.6. Consider the ISI-AWGN channel where the input $a(i,j)$ is actually the output of a 2D error correction code. Specifically, suppose that the page $a(i,j)$ is produced by an uncoded binary page $b(m,n)$ such that each 3×3 "transition" on the uncoded data page defines four coded bits (*i.e.*, a rate 1/4, 2D convolutional code) so that there are four times as many binary variables in the coded page **A** than in the uncoded page **B**. Describe how this serial concatenation of 2D systems may be decoded iteratively using two 2D SISOs.

5.7. Compare the complexity of the COL-row and ROW-col concatenated 2D SISOs when $L_r = 3$ and $L_c = 1$. In general, when $L_r \neq L_c$, which model is simpler?

5.8. Repeat Example 5.4 for the ROW-col version of the concatenated 2D SISO algorithm.

5.9. Consider *separable* 2D ISI channels of the form $\mathbf{F} = \mathbf{g}\mathbf{h}^{\mathrm{T}}$ – *i.e.*, $\mathbf{F}$ is the outer produce of two 1D ISI channels. Can all 2D ISI channels be written in this form? If not, describe the restriction on g and h for the channel in (5.7) to be separable. Show that the cardinality of the inner (hidden) symbols in the concatenated model can be reduced in the case of a separable channel.

5.10. [Th00] Draw the implicit index diagram for the COL-row concatenated model.

5.11. Based on the correspondence between Fig-5.5 and Fig-5.7 describe how the iterative detector for a serially concatenated TCM-

ISI system with block interleaving can be implemented in a parallel fashion. Is this the case if the initial and/or final states of the TCM and ISI FSMs is unknown?

5.12. [ChChNe98] In the 2D⁴ algorithm, given $\{z(i,j), z(i\pm1,j), z(i,j\pm1)\}$, develop a combining scheme at $t(i,j)$ node and simplify as much as possible. Consider some larger observation combining regions. Compare the resulting complexities of the combining scheme.

5.13. Consider the column-row activation schedule for the message passing algorithm on 2D-GM2 described in Section 5.5.4 for data detection. Specifically, simulate this under the conditions used to produce Fig-5.20. Does filtering help with this activation schedule?

5.14. Consider the 2D-GM2 model with the non-rectangular support regions shown in Fig-5.31. Determine the auxiliary variables $r(i,j)$ and $c(i,j)$ in each case. Discuss the applicability and relative complexity of the concatenated model and the 2D-GM1 model to these support regions.

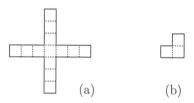

(a)   (b)

Figure 5.31.   Non-rectangular 2D support regions.

5.15. For the threshold (TH) detector show that for the channel in (5.7) there exists an error floor when $(g+h) > 0.25$ and this error floor is

$$\frac{1}{2^8} \sum_{i=0}^{4} \sum_{j=j_i}^{4} \binom{4}{i}\binom{4}{j}$$

where j_i is the minimum value of j such that $ig + jh \geq m_z$.

5.16. The noise-free output signal of a coherent POM channel may be modeled by [ChChNe99]

$$x(i,j) = \sum_{k,l,m,n=-L}^{L} R((k,l);(m,n))a(i-k,j-l)a(i-m,j-n)$$

$$(5.30)$$

where $R(\cdot)$ is a known function characterizing the incoherent detection of the coherent signal. Consider the case where this signal is distorted by some form of pixel-wise independent noise. Which, if any, of the algorithms presented in this chapter applicable to this channel?

Chapter 6

IMPLEMENTATION ISSUES: A TURBO DECODER DESIGN CASE STUDY

with

PETER A. BEEREL
University of Southern California

An overview of the key steps associated with implementing a turbo decoder is presented in this chapter. We focus on the specific design of a decoder for the PCCC system of the example in Section 2.4.3.1. Specifically, for that rate $1/2$ PCCC with $K = 1024$ interleaver and 4-state RSC constituent codes, we outline the design of a decoder using fixed-interval, MSM-based SISOs. Our motivation for selecting min-sum processing for the PCCC decoder is primarily pedagogical. More precisely, as shown in Fig-2.30, the APP-based decoder yields an improvement of up to approximately 0.5 dB in coding gain, and betters the MSM-based decoder performance in only 6 iterations. Bitwidth analysis and the like, however, are most easily explained for the min-sum version and then extended to the min*-sum version as required. Furthermore, our development does not exploit the specific structure of the constituent codes, except for the purposes of concrete examples. Thus, we address general issues in hardware design that should be applicable to the implementation of many iterative detection algorithms. In all applications except the (low-SNR) decoding of turbo-like codes, the min-sum version is preferred over the min*-sum version because of negligible performance differences and simpler implementation. All the design issues and considerations assume a traditional design methodology for synchronous circuits as described in a variety of good textbooks (*e.g.*, [WeEs93]).

6.1   Quantization Effects and Bitwidth Analysis

All simulation results in the previous chapters demonstrating iterative detection techniques were based on high precision floating-point computer simulations (*e.g.*, at least 32-bit floating-point precision). When implementing these algorithms in hardware, however, fixed-point arithmetic requiring fewer bits is often used. The meaning of *fixed-point* arithmetic is that the number of bits used to represent a number (*e.g.*, 7 bits) and the location of the decimal (binary) point within the number (*e.g.*, between the 3rd and 4th bit) are fixed. Representing numbers as fixed-point is equivalent to using integers to approximate real numbers after scaling. Fixed-point representation often simultaneously leads to significant reductions in hardware area and energy consumption, while increasing the hardware speed. Care must be taken to ensure that using fixed-point representations and arithmetic does not induce quantization error large enough to cause a substantial degradation in coding performance. Thus, there is an important tradeoff between hardware complexity and coding performance that requires careful bitwidth analysis.

Bitwidth analysis is typically first performed on the inputs and outputs of a chip or block. The bitwidths of internal variables may then be derived using subsequent analysis. If the operations of the block are reasonably simple, then one can perform a worst-case analysis to determine the internal bitwidths required to ensure that the only quantization effects are those associated with the quantization of the block inputs. That is, using the fixed-point representation of these internal quantities mimics the infinite precision processing of the quantized inputs. However, when the operations of a block are more complicated or when the worst-case analysis yields too many bits to be practical, more careful analysis/simulation is needed.

6.1.1   Quantization of Channel Metrics

For many decoding algorithms, the inputs to the decoding block are often obtained from the channel through an analog-to-digital (A/D) converter. In this section, we are not concerned with the details of an A/D implementation; rather, we focus on the number of bits required at the output of the converter.

Example 6.1. ―――――――――――――――――――――――

For the turbo decoder in the example of Section 2.4.3.1, the inputs are the soft-in metrics from the channel – *i.e.*, $\{MI[c_k(0)]\}$, $\{MI[c_k(1)]\}$, and $\{MI[d_k(1)]\}$ which are derived from $M[x_k(i)]$. Thus, a first step in understanding the quantization effects is to characterize different fixed-point representations of these quantities while representing all other quantities using infinite precision. For example, the soft information on the uncoded symbols, as well as all quantities internal to the two SISOs, will be considered unquantized for this example.

As in the example of Section 2.4.3.1, we would like to use only one number (*i.e.*, the negative log-likelihood ratio) for soft information on each binary variable. Since this information will take positive and negative values, a 2's complement binary representation is convenient for performing all of the internal integer arithmetic. The quantizer is specified by the quantization regions, the reconstruction levels, and the bit labels of the reconstruction levels. In addition, there may be some preprocessing of the input to the quantizer to adjust the range. It is desirable that the reconstruction levels correspond directly to the value represented by the bit label in 2's complement. To simplify the exposition, we assume that the quantizer reconstruction levels are integers, but this discussion generalizes to fixed-point approaches with the binary-point at any location. For example, if 3 bits are used, then the integers $-4, -3, \ldots 2, 3$ can be represented. The asymmetry around the origin is not desired, but is a result of the 2's complement representation.

Recall that the associated channel model for the PCCC system is

$$z_k(i) = \sqrt{E_s} x_k(i) + w_k(i) \tag{6.1}$$

where $x_k(i) = \pm 1$ according to the value of the underlying coded symbol and $w_k(i)$ is a real AWGN sequence with variance $N_0/2$. Also, recall that, for min-sum processing, this was the input metric associated with a $+1$ on the channel and that it could be multiplied by an arbitrary positive constant for all time indices. For a given value of E_b/N_0, this scaling is equivalent to specifying the value of E_s in (6.1) – *i.e.*, we may consider (6.1) to model the output of a gain control circuit that sets the signal level to some known value. Since the reconstruction values are fixed, the selection of this gain will affect performance. An example of this is shown in Fig-6.1, where a uniform quantizer is shown with $\sqrt{E_s} = 2$ and $E_b/N_0 = 2.0$ dB.

In Fig-6.2, we show simulation results for 3 and 4 bit quantizers designed as described above for the system in the example of Section 2.4.3.1. In each curve, the choice of $\sqrt{E_s}$ has been optimized empirically. The curves labeled "5 or more" and "6 or more" in Fig-6.2 show the performance of a decoder that uses infinite precision internal data representation and therefore shows the degradation associated only with channel metric quantization. The meaning of this terminology and the other curves shown in Fig-6.2 are discussed in Section 6.1.3. Similar simulation results can be found in the literature [HsWa99, WuWo99, AuMc99].

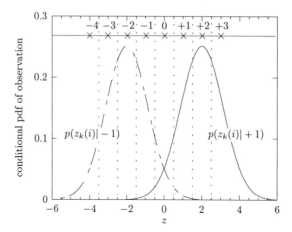

Figure 6.1. A three bit quantizer having integer reconstruction levels corresponding to the 2's complement encoding. Also shown is the pdf of the associated channel metrics under the conditional data values for $E_b/N_0 = 2.0$ dB and $\sqrt{E_s} = 2$.

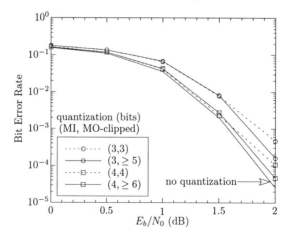

Figure 6.2. The effect of quantization of the channel metrics on the performance for the MSM-based iterative decoder. The clipping effects shown are discussed in Section 6.1.3.

From these results, we conclude that representing the channel soft-in metrics with 4 bits is sufficient to achieve coding performance close to that obtained with infinite precision. Note that this is a factor of 8 smaller than the 32-bit floating-point representations of a naive microprocessor-based implementation.

The effect of scaling the observation is also shown for this example in Fig-6.3. For a larger number of quantization levels, the performance is less sensitive to the scaling factor used. Also note that Fig-6.1 represents the best choice for scaling for the 3-bit quantizer at the value of E_b/N_0

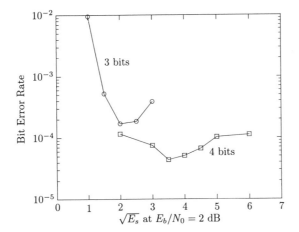

Figure 6.3. The effect of observation (channel metric) scaling on the performance for the MSM-based iterative decoder.

shown.

—————————————————————————— *End Example*

The conclusion drawn in Example 6.1 is dependent upon a number of factors. For example, the normalization strategy requires that only one number be stored for soft information on each binary variable. However, this implicitly makes this soft information take on both positive and negative values. If two numbers were stored (*e.g.*, using $(z_k - \sqrt{E_s})^2$ and $(z_k + \sqrt{E_s})^2$), one would need only unsigned arithmetic and the required bit-widths for these soft-in values may change (see Problem 6.1). In hardware, however, handling signed arithmetic is not more complex than handling unsigned arithmetic. Moreover, because the area of the design considered turns out to be dominated by memory, using the normalization strategy significantly reduces the required chip area.

The result of Example 6.1 is intuitive given knowledge of soft-in quantization requirements for a Viterbi decoder of a convolutional code in isolation. Specifically, most of the gain associated with soft-in Viterbi decoding over hard-in Viterbi decoding can be achieved with 3-bit (8 level) quantization [LiCo83]. However, the turbo decoder operation differs significantly in several ways. For example, the soft-in metrics from the channel are used once each iteration. The range and resolution requirements for each iteration could differ. For example, in subsequent iterations the soft-in on the uncoded bits (*i.e.*, MI[b_k] and MI[a_k]) will change as the data beliefs become more reliable. It is expected that the most demanding of these quantization requirements will determine the overall quantization requirements.

The simulations in Fig-6.2 use floating-point representations of all soft measures other than the channel soft-in information. When the channel metrics are quantized, all internal soft information can be represented in fixed-point without any additional performance degradation so long as the bitwidths of these variables are sufficiently large (*i.e.*, to prevent overflow). In the next sections we consider the required bitwidths for the forward and backward state metrics and the soft-information on the uncoded bits.

6.1.2   Bitwidth Analysis of the Forward/Backward State Metrics

The first step in analyzing the required bitwidths of internal computations is to determine the dynamic range of the quantities involved in the algorithm. To represent a quantity in 2's complement notation with dynamic range $[-\Delta, \Delta)$ requires $n_\Delta = \lceil \log_2 2\Delta \rceil$ bits, where $\lceil x \rceil$ represents the smallest integer larger than x.

At first glance, the key quantities in the forward-backward algorithm are the forward and backward state metrics. These values grow with each ACS step which suggests a large dynamic range requiring many bits to represent. One possible way to avoid using a large number of bits is to normalize the state metrics either every step or intermittently. For example, the smallest state metric can be subtracted off of all state metrics. This can be done without changing the results of the algorithm, but it incurs significant costs in hardware area, latency, and energy consumption.

With more careful analysis, however, one may observe that the algorithm refers to state metrics only in the form of differences between pairs of state metrics. More precisely, the algorithm uses differences of path metrics to perform the forward and backward ACS operations as well as slightly modified path metric differences to compute soft-outputs during the completion operation. This is important because this characteristic of the forward-backward algorithm limits the dynamic range of these differences.

In hardware, one can take advantage of the limited dynamic range using 2's complement representation and modulo-2 arithmetic [He89]. The underlying principle behind this approach is that when the dynamic range of the result of a sequence of 2's complement additions and subtractions lies within the range $[-\Delta, \Delta)$, the computation can be safely performed using module 2^{n_Δ} arithmetic. For example, if two numbers a, b are to be compared using subtraction and $|a - b| < \Delta$, which implies the result is within the range $[-\Delta, \Delta)$, then the subtraction can be evaluated as $(a - b)$ mod 2Δ [He89]. This implies that n_Δ bits are required and that any under or over flow of the various adders/subtracters in the design can be ignored. For our design this implies that the state metrics may under and over flow but the final results will be accurately represented.

Example 6.2. ————————————————————————————
Consider the case where $\Delta = 3$ which means that $\lceil \log_2(2 \cdot 3) \rceil = 3$ bits are required to represent the dynamic range. Thus, from the above discussion, using 3 bits to represent the quantities allows over and under flow conditions to be ignored in ACS operations. These three bits can represent the numbers $-4, -3, \ldots, 2, 3$. Consider the ACS operations illustrated in Fig-6.4. The comparison is implemented by examin-

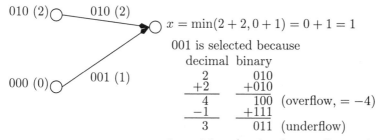

$$x = \min(2 + 2, 0 + 1) = 0 + 1 = 1$$

001 is selected because

decimal   binary

2	010
$+2$	$+010$
4	100 (overflow, $= -4$)
-1	$+111$
3	011 (underflow)

is positive, despite the overflow and underflow

Figure 6.4.   Example of modulo arithmetic applied to an ACS operation.

ing the sign of the result of subtracting the two quantities. Thus, the ACS operation comparing $(2 + 2)$ and $(1 + 0)$ is carried out by computing $(2 + 2) - (1 + 0)$ and comparing against zero. First, when the additions are performed (*i.e.*, $1 + 0$ and $2 + 2$), the $2 + 2$ operation causes an overflow: $010 + 010 = 100$, yielding a somewhat surprising intermediate answer of -4. Continuing with the subtraction, we have $-4 + (-1) = 100 + 111 = 011$, which represents an underflow.[1] Because the carry bit is discarded, this yields the correct answer, 3. Since this is a positive number we correctly declare $2 + 2 > 1 + 0$ and select the lower branch.
———————————————————————————— *End Example*

This simple modulo-2 technique for ACS-based (*i.e.*, min-sum) algorithms may be applied to determine the required bitwidth for the forward

———————————————————

[1] Recall that -1 in 2's complement arithmetic with 3 bits is 111

and backward state metrics given the bitwidths of the soft-in metrics. This is illustrated in the following detailed example.

Example 6.3. ──

Continuing with Example 6.1, we can determine the bitwidths required for the forward and backward state metrics for each SISO given the bitwidths selected for the soft-in metrics. Specifically, based on Example 6.1, we use 4 bits to represent the quantized channel metrics. Furthermore, assume that the current soft-in on the uncoded bits (*i.e.*, b_k or a_k) is also represented by 4 bits. Without the modulo arithmetic technique of Example 6.2, the bitwidth of the forward metrics could be conservatively taken to be 16 bits. This follows from the fact that every subpath through two steps of the trellis introduces at most five 4-bit numbers (three from the even index and two from the odd index) to the path metric. For an interleaver of length K this means that paths consist of the sum of $5K/2$ 4-bit numbers. This requires $\log_2(5K/2) + 4$ bits to represent, which for $K = 1024$, amounts to 16 bits.

Using the modulo arithmetic technique we can reduce the number of required bits dramatically. The first step is to determine the dynamic range of the differences between forward metrics at any given time. Specifically, one must determine how large $|F_{k-1}[s_k] - F_{k-1}[s'_k]|$ for $s_k \neq s'_k$ may be. It can be shown by a simple argument that, for the constituent codes described in the example of Section 2.4.3.1, the difference between state metrics lies within the range $[-5 \cdot 2^{w-1}, 5 \cdot 2^{w-1}]$, where w is the bitwidth of the 2's complement soft-in information on both the FSM input and output (see Problem 6.2). Thus, for this example, 4-bit soft-in metrics imply that the dynamic range of differences between forward state metrics lies within $[-40, 40]$, implying that the state metrics can be represented in 7 bits. However, one must be careful to realize that the numbers which will be input to the comparators are partial path metrics equal to the state metrics plus one set of branch metrics. It is the dynamic range of the difference between inputs that determines the required bit-widths of the ACS units because the comparison of these numbers is done with subtraction and the result must be accurate for the correct selection to take place. In particular, the dynamic range of the result of the subtractor is $[-64, 64]$, which is $3 \cdot 2^{w-1}$ larger than that of the state metrics. This range is slightly larger than that which can be represented with 7 bits. Consequently, to be conservative, the ACS units should use 8 bits. Alternatively, however, 7 bits can be used, in which case the overflow that occurs when the difference is exactly 64 will cause some degradation in coding performance. If this performance degradation is sufficiently small, this tradeoff may be worthwhile.

In general, it can be shown that modulo arithmetic requires at most one extra bit compared to the method of normalizing to the smallest state metric. For example, the resulting dynamic range using such normalization would be $[0, 64]$, which in unsigned arithmetic can be represented in 7 bits.

——————————————————————————— *End Example*

In Example 6.3, using modulo arithmetic saved 8 bits for the representation of the state metrics. If a min*-sum algorithm was used instead of the min-sum version considered, this analysis becomes somewhat more complicated. In particular, the bitwidth analysis depends on the details of how the min*$(\cdot)$ function is implemented. For example, the $-\ln(1 + \exp(-|x - y|))$ component may be implemented using a look-up table that stores m-bit 2's complement values. Then, each state metric would have an additional m-bit number added to it which must be considered during bitwidth analysis. In particular, the dynamic range of the inputs to the comparators would increase by 4 times 2^{m-1}, because the maximum absolute difference between partial path metrics would increase by no more than 4 table look-up terms (*i.e.*, see Problem 6.6). If these terms were represented with 4 bits, then the dynamic range would increase by 32, requiring the state metrics to be represented by only one extra bit.

The value of m selected for the bit-width of the min*$(\cdot)$ look-up table typically cannot be found using worst case analysis because the value of $-\ln(1 + \exp(-|x - y|))$ generally does not have a finite binary expansion when x and y do (*e.g.*, see Problem 6.5). Therefore, implementation of a min*-sum processor inherently has internal quantization effects that must be evaluated via further fixed-point simulations. Furthermore, because a min*-sum algorithm is not invariant to metric multiplication by a global constant, care must be exercised in considering the quantization levels and the definition of the min*$(\cdot)$ look-up table. Specifically, one cannot scale up the observation to fit a specific set of quantizer reconstruction levels (as done in Fig-6.1 for the min-sum case) and still use the standard definition of min*$(\cdot)$ (*i.e.*, see Problem 6.4).

6.1.3   Soft-Out Metric Bitwidths

The soft-outputs produced by one SISO will be used as soft-input information for the next SISO operation. Typically the soft-inputs are assumed to be represented with fewer bits than the soft-outputs. Consequently, some form of mapping of soft-output to soft-input information is required. This is illustrated by the following example.

Example 6.4. ───────────────────────────────────────

The completion operation is also an ACS operation, so that the discussion in Section 6.1.2 is applicable. Continuing with Example 6.3, we can determine the required bitwidths for the soft-outputs based on the soft-in bitwidths. Specifically, the inputs to the comparators in the ACS operation in the completion operation are the sums of one forward metric, one backward metric, and one branch metric. The maximum absolute difference between these two inputs is thus $2 \cdot 40 + 24 = 104$, implying the ACS units for the completion step can safely be represented in 8 bits. The soft-output is the difference between these two quantities summed with one additional soft-in metric and can also be accurately represented with 8 bits.

─────────────────────────────────────── *End Example*

Example 6.4 illustrates a fundamental issue with iterative detection hardware: worst-case analysis on the bitwidths of the internal reliability metrics suggest that after each SISO activation, additional bits be allocated to the soft-information. Intuitively, this is satisfying for the theoretician, since, as beliefs become more reliable, the negative log likelihoods will diverge. Carrying through this worst-case allocation of bits, however, is not practical in most hardware designs of interest. For example, in the scenario of Example 6.4, after each SISO activation, 4 more bits should be allocated for the soft information on the uncoded bits. After the 20 such activations associated with ten iterations, the 4-bits allocated to the channel metrics has grown to approximately 80 bits for the final soft outputs which is overly conservative in practice.

It is reasonable, therefore, to re-map the soft-out metrics into a smaller dynamic range by some method. One way to do this is to *clip* the soft-out metrics to a certain bitwidth. For example, if one were to clip 5-bit 2's complement integers to the corresponding 3-bit representation, the values $\{-16, -15 \cdots -5\}$ and $\{+4, +5, \ldots +15\}$ would be mapped to -4 and $+3$, respectively (*e.g.*, see Fig-6.1). In the example design this would mean that the 8-bit soft-out metrics would be clipped down to smaller bitwidth m by forcing any soft-output value that is outside the range of m-bits to that of the closest value that can be represented in m bits. A particularly attractive version of clipping is to clip the soft-out metrics to the same bitwidth as the soft-in metrics. Note that this clipping process changes the turbo decoding algorithm. Specifically, it is no longer guaranteed that the performance will be the same as that of the turbo decoder using quantized channel metrics but infinite precision internal representations.

Example 6.5. ─────────────────────────────────────

Consider the extension of Example 6.4 where the 8-bit soft-output metrics are clipped back to 4 bits. A simple circuit to perform this 8-to-4 bit clipping is shown in Fig-6.5. Since this clipping affects the performance

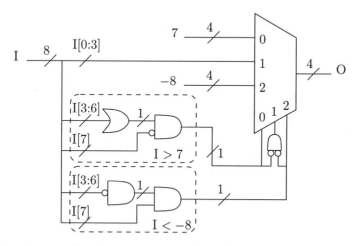

Figure 6.5. An example of clipping circuitry (after [Pa00]). The gates shown with input I[3:6] are short-hand for the corresponding gates with multiple inputs: I[3],I[4],I[5] and I[6].

in general, simulations have been run to assess the impact of clipping with the results shown in Fig-6.2. Specifically, results are shown for 3 and 4 bit quantization of the channel metrics with clipping to various bit widths. For 3-bit quantization, clipping to 5 or more bits does not affect the performance in practice. For 4-bit quantization of the channel metrics, clipping to 6 or more bits does not affect the observed performance. The 8-to-4 bit clipping with 4-bit input metric quantization suffers only a small performance degradation and, thus, this clipping is adopted in the reference design. Also, in general, the best choice of the quantizer parameters may be different with and without clipping.

───────────────────────────────── *End Example*

Clipping may be viewed as a re-mapping of the soft-out metrics to the quantization region supported by the channel metric quantizer. It reduces the dynamic range in favor of precision and is best suited to conditions where the soft-out information has low reliability (*i.e.*, small metric differences). This concept may be generalized to allow for other methods of *requantization* or *warping* of soft-outputs and channel metrics that are the inputs to a SISO. For example, it is also possible to premultiply all soft-inputs (soft-outputs from the previous iteration and channel metrics) by a negative power of two by means of a simple right shift before quantization[MaPiRoZa99]. This effectively reduces the precision of the soft-inputs in favor of a larger dynamic range.[2] This may yield improved results for conditions where the soft-out information has high reliability (*i.e.*, large metric differences).

Slightly different methods apply when the soft inputs are represented with positive numbers (recall that this can be done with a different normalization but requires more storage). For example, it is possible to identify the smallest soft-input (channel metric and/or soft-output from the previous iteration) and subtract it from all soft-inputs, making the soft-inputs range from 0 to a modified maximum value. At the same time, the most significant bits of the soft-inputs that are all the same can be identified and dropped [MaPiRoZa99]. After both of these operations are done, the result can be forced into the allocated number of bits by dropping the necessary number of least significant bits. This warping maximizes the dynamic range of the SISO inputs.

Warping and requantization alters the turbo algorithm from its fixed-point implementation that assumes floating-point internal precision. Consequently, the coding performance of the modified algorithm must be evaluated, usually using refined fixed-point simulations.

6.2   Initialization of State Metrics

In theory, the value of infinity can be assigned to a state metric or transition metric when there is certainty about this quantity according to some known initialization information. Specifically, in the example of Section 2.4.3.1, the forward and backward state metrics were initialized according to the fact that each edge state is zero. Thus, the forward and backward metrics associated with the zero state are set to zero, with all other state metrics at these times set to infinity. In a software implementation, a very large floating-point number can be used in place of infinity. In fixed-point implementations, however, more care is required in this initialization process. Directly implementing this special value convention in hardware is inefficient at best. Alternatively, setting the infinite-valued metrics to the maximum value of the finite-precision representation may also not work because it violates the expected maximum dynamic range of path metric differences necessary for the modulo arithmetic to work. For the min-sum algorithm the goal is to initialize

---

[2]This is equivalent to scaling all metrics by a single positive constant. Thus, from the discussion in Section 2.4.1, it follows that this is strictly valid for min-sum based algorithms, but not for min*-sum algorithms. In the latter, such scaling alters the algorithm and the effects of this modification must be assessed by further fixed-point simulations.

the other state metrics to numbers which guarantee they do not violate the bounds on path metric differences and also do not become part of survivor paths. More precisely, they should not affect the state metric values. For the four-state trellis of the example of Section 2.4.3.1, these constraints can be easily hand generated, as the following example illustrates.

Example 6.6.
Continuing with the design example, let $F_{-1}[i]$, for $i = 0, \ldots, 3$ represent the 7-bit initial forward state metrics and let $M_k[i, j]$ refer to the metric of the transition t_k between states $s_k = i$ and $s_{k+1} = j$. To guarantee state $s_0 = 2$ does not effect the metric of state $s_1 = 0$ we have

$$F_{-1}[0] + M_0[0, 0] \leq F_{-1}[2] + M_0[2, 0] \tag{6.2}$$

Similarly, to guarantee that $s_0 = 2$ does not effect the metric of state $s_1 = 1$, we have

$$F_{-1}[0] + M_0[0, 1] \leq F_{-1}[2] + M_0[2, 1] \tag{6.3}$$

Combined with the constraints that $F_{-1}[0] = 0$ and no two state metrics should differ by more than 40 to ensure the module-2 arithmetic works properly, this yields the two-sided constraint

$$\max \left\{ (M[0, 0] - M[2, 0]), (M[0, 1] - M[2, 1]) \right\} \leq F_{-1}[2] \leq 40. \tag{6.4}$$

Initializing $F_{-1}[2]$ and $F_{-1}[3]$ to ensure that $s_0 = 2$ and $s_0 = 3$ do not effect any state metrics is slightly more complicated because they are not directly compared with the initial value $F_{-1}[0]$. For an M^L state standard trellis, the general rule for state $s_0 = i, i \neq 0$ not to effect any state metric is as follows.

- The (unique) path metric from $s_0 = i$ to any state s reachable from $s_0 = i$ in the first L steps of the trellis should be no smaller than the (unique) path metric from state $s_0 = 0$ to state s.

- $F_{-1}[i] \leq 40$.

For our four-state trellis, it can easily be shown that choosing $F_{-1}[i] = 40$ for $i \neq 0$ satisfies the above two-sided constraints.

End Example

Notice that the same analysis does not apply for implementations of the min*-sum algorithm because in finite precision it is not possible to initialize the state metrics at time 0 such that they do not influence state metrics at later times. Consequently, one must initialize the state metrics to the largest value that does not violate the modulo arithmetic and analyze the effect on coding performance using further refined simulation. In fact, it may be possible that the effect of this on initialization motivates using more bits (*i.e.*, see Problem 6.7).

6.3   Interleaver Design and State Metric Memory

The forward-backward algorithm requires numerous quantities to be stored. The interleaver/de-interleaver stores soft-outputs and an input-buffer is needed to store the soft-inputs from the channel. In addition, for activation schedules in which the forward and backward state metrics are computed before needed, storing these is also required.

One basic means of storing these quantities is to use several all-purpose static random-access-memories (RAMs). These are well-known structures that densely store many bits of data. The basic RAM consists, of an address decoder, a 2-D array of memory cells, and row/column multiplexors, as illustrated in Fig-6.6. For the case of the interleaver design

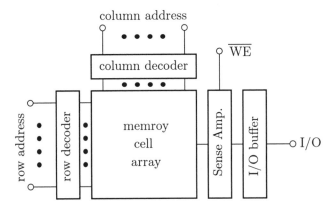

Figure 6.6.   Basic architecture of a random access memory (RAM).

an additional read-only-memory (ROM) is typically used to store the pseudo-random, but fixed, interleaving pattern. A ROM has much the same structure as that of a RAM but has simpler and smaller array of memory cells whose contents are hard-coded.

The latency of these memories depends largely on their size; the larger the memory the more challenging it is to make them fast. Consequently, to ensure that the memories do not overly constrain the clock cycle time, breaking them up into smaller memories, where possible, is advisable. When this is not possible, the speed of these memories may be raised using analog sense-amplifiers that quickly react to changes in voltages of internal bit lines, at the cost of some hardware complexity. In addition, the energy consumption can be reduced using self-timed circuitry which reduces the internal voltage-swing of the high-capacitive internal wires.

One interesting feature of the forward-backward algorithm is that the memory accesses for the state and soft-inputs are sequential. Consequently, one can use a sequential access memory (SAM) instead of a RAM. Note that the sequential nature of accessing enables pairs of quantities needed in successive steps to be grouped together in the memory and read/written as one quantity with twice the bit-width. Thus, reading/writing can take up to two clock cycles without impacting the clock cycle time. Larger groupings schemes are also possible if the memory access times are particularly slow. One implementation of a SAM involves replacing the address decoder of a RAM with a shift register that contains a single '1' bit that identifies which row of memory cells should be read from or written to. This reduces the required area and latency to some degree.

In the activation schedule with concurrent forward and backward operation (*i.e.*, see Fig-2.40(b)), the type of SAM needed is more precisely a last-in first-output (LIFO) buffer also known as a stack. These units can be implemented with a bank of registers with multiplexed inputs, as shown in Fig-6.7. They are substantially less dense than RAM-based

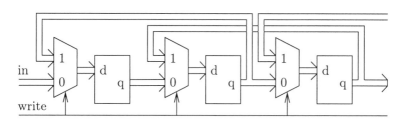

Figure 6.7.   Circuitry for a last-in-first-out (LIFO) memory.

memories, but can be made with extremely low latency. They may also consume more energy because of high switching activity on all internal wires. To reduce energy consumption, a LIFO buffer can be implemented with a tree based configuration of storage elements instead of the linear configuration shown. Additional multiplexors and control are used to choose which branches of the tree should be updated and in what direction. As a consequence, the number of storage elements that must be updated during one step can be significantly reduced, thereby saving energy. In fact, it may be possible for this type of design to consume lower energy than traditional RAM-based designs.

Example 6.7. ─────────────────────────────────────

An estimate of the area required for both RAM-based and LIFO memories is given in [Pa00]. A 1-bit LIFO unit incorporating a mux and a 1-bit register required $134 \times 256 \; \lambda^2$, where λ represents the basic dimensions of our layout. This is in contrast to a 1-bit RAM cell that requires $34 \times 56\lambda$. To be fair, we should also add approximately 10% additional area to the RAM cell to consider the additional area of the decoder and multiplexor amortized over all cells. This yields an estimate that the LIFO requires 15 times more area than the RAM-based memory.

─────────────────────────────────────── *End Example*

6.4   Determination of Clock Cycle Time and Throughput

The basic activation schedules described in Fig-2.40(b) provide a basis for which a more refined schedule should be designed. This refined schedule should describe in detail all the operations to occur in each clock cycle. The clock cycle time is then dictated by the longest latency of any sequence of operations scheduled to operate in one clock cycle.

It is well-known that the performance bottleneck of forward-backward algorithm lies in the ACS operation. The reason is that the algorithm requires one step of ACS to be completed before the next step begins. Consequently, the clock cycle time of high-throughput designs should target the latency of an ACS unit.

However, the peripheral parts of the ACS operation that involve branch metric calculation, memory accesses of state metrics, or calculation of control signals, may or may not be scheduled in the same clock cycle as the central add-compare-select operation. If they are, the clock cycle time may have to be much larger than the core ACS operation alone requires. In addition to the forward and backward recursions, the completion operation is also an ACS operation. If the ACS unit has enough bits to support both operations, it can perform both the state metric recursions and the completion operations. In this case, however, the clock cycle time may be determined by the logic that implements the more complex ACS operation required for the completion operation. While in many cases designs with longer clock cycle time often are simpler to understand and build, they are not necessarily significantly smaller nor do they necessarily consume significantly less energy per decoded bit than more aggressive designs with higher clock frequencies. In other words, careful architecture design can yield much higher clock rates with little cost in area and energy consumption.

The key architectural feature that we are alluding to above is called
pipelining. Pipelining is the decomposition of the algorithm into stages of
logic, each stage concurrently operating in a single clock cycle. The idea
is that stages earlier in the pipeline pre-compute quantities needed later
in the pipeline and are stored in pipeline registers (storage elements)
until needed. For example, an early pipeline stage may access memory to
retrieve quantities that are needed in the ACS operation in subsequent
pipeline stages, thereby helping ensure that the memory access time
does not impact the clock cycle time. We now describe one pipelined
architecture designed to ensure that the clock cycle time is determined
by the latency of an ACS unit.

Example 6.8. ————————————————————————————————
Continuing with the reference design for the decoder in the example of
Section 2.4.3.1, we consider a high-performance SISO architecture for
the fixed-interval forward-backward algorithm with parallel forward and
backward ACS recursions (*i.e.*, see Fig-2.40(b)) as illustrated in Fig-6.8.
This architecture is decomposed into five pipeline stages and has two

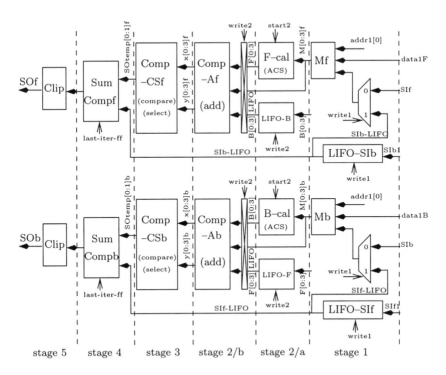

Figure 6.8.   A pipelined SISO architecture (after [Pa00]).

halves, the top half for computing the forward operation path and the bottom for simultaneously computing the backward operation path.

Stage 1 is responsible for generating branch metrics. Stage 2 is responsible for performing 7-bit ACS operations to compute the forward and backward state metrics. Notice that the results of these ACS operations are stored in a LIFO buffer to be used later for the completion operation. In particular, notice that only the first two stages are active for the first half of the total number of clock cycles. Once the forward and backward pointers cross, the next two stages are responsible for the completion operation, and the last stage is responsible for output clipping.

The goal of the architecture is to ensure that the critical path that determines the cycle time is dictated by the ACS unit in Stage 2/1. The detailed architecture of this block is illustrated in Fig-6.9. The two

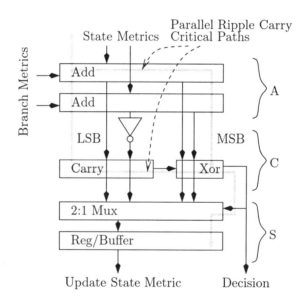

Figure 6.9.   ACS architecture and critical path (after [Bl93]).

adders are implemented with a basic adder architecture called a *ripple carry architecture* which requires relatively low area. The comparison can be implemented using subtraction implemented by a stripped-down version of an adder with one of the inputs inverted and the sum logic for all but the most significant bit removed. In particular, the most significant sum bit of the result indicates the result of the comparison. One advantage of this choice is that all three units operate in parallel. That is, the comparator begins computing its result as soon as the least significant output bit of the adders are available. Consequently, the critical path is not much longer than that of a single ripple carry adder.

Simulations of this component laid-out using the Hewlett-Packard 0.25μ CMOS process at 3.3V yielded an achievable clock frequency of approximately 75 MHz. The area of the SISO design, not including the LIFO buffers is $3300 \times 8000\ \lambda^2$. For our 0.25μ process, $\lambda = 0.15\mu m$, meaning the SISO requires area of only $0.5 \times 1.2 = 0.6\text{mm}^2$.

The area of the memory, on the other hand, is substantially more for this $K = 1024$ example. Using LIFOs for the soft-input and state metrics and a RAM-based interleaver, the total area required for a decoder can be estimated at $2.2 \times 2 = 4.4\text{mm}^2$.

In retrospect, it seems wise to replace the LIFOs with small, fast RAMs and thereby save significant area. In particular, this would yield an estimated area of 2.0mm^2, or a savings of more than a factor of two. If designed carefully, we believe that the RAMs will have little or no impact on clock cycle time.

The forward-backward algorithm for a block-size of 1024 with concurrent forward and backward processing requires approximately 1030 clock cycles, with pipelining similar to the above example. For 10 iterations, this means the algorithm requires approximately 20,600 total clocks to complete. We saw in the above design, that an estimated 75 MHz is achievable. This means that the above design can achieve a throughput of

$$\frac{1024 \text{ bits}}{(13.3 \text{ ns/clock cycle} \times 20,600 \text{ clock cycles})} = 3.7\text{Mb/sec} \qquad (6.5)$$

———————————————————————————— *End Example*

6.5 Advanced Design Methods

The above basic design can be improved in a number of ways to improve throughput, latency, and power.

6.5.1 Block-level Parallelism

One means of increasing the throughput of turbo decoders is to use multiple SISOs and interleaver/deinterleavers and pipeline across blocks of data [MaPiRoZa99]. Masera et al. suggested using 20 SISOs, 20 interleavers, and 20 de-interleavers in which case 20 different blocks can be processed in parallel, yielding an increase in throughput of 20 with a similar increase in area. For example, applying this method to the reference design in Example 6.8 would provide a throughput of 74 Mbps. It is also possible to use fewer than 20 (*e.g.*, a design with 2 is proposed in [Pa00]) with the obvious linear scaling in both throughput and area.

6.5.2   Radix-4 SISO Architectures

Another well-known idea to increase throughput is to process two steps of the trellis at a time. This is often called a "radix-4" architecture [Bl93]. The advantage of radix-4 architectures is that the SISO operations requires approximately half as many clock cycles to compute. The disadvantage is that the ACS operations in the forward/backward recursions and in the completion operations are more complex. The result is either a significantly larger design or a slightly slower clock cycle.

Example 6.9. ───

The ACS operation required in the forward/backward recursion operations of radix-4, 4-state trellis involves 4 incoming paths. This can be done with a bank of four adders and a tree of 2-way comparators. The tree of 2-way comparators increases the latency of the ACS operation demanding a larger clock cycle time. This extra latency can be avoided, however, by using 6 (*i.e.*, 4 choose 2) comparators and simple select logic [Bl93], as illustrated in Fig-6.10.

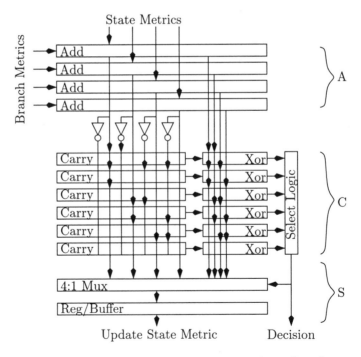

Figure 6.10.   Fast 4-way ACS architecture (after [Bl93]).

The ACS operation required in the completion step of a radix-4 4-state trellis involves comparing 8 different quantities. A similar parallel

8-way ACS can be implemented but it would require 8 choose 2, or 28 comparators. This large increase in area can be avoided by using a more traditional tree of comparators requiring 7 comparators. Scheduling this tree of comparators in one clock may require a significantly larger clock cycle time than otherwise would be needed. Consequently, it makes sense to break-up the tree of comparators into two halves, each half scheduled in one clock cycle.

—————————————————————————————— *End Example*

6.5.3   Fixed and Minimum Lag SISOs

Recall that one attractive alternative to fixed-interval SISOs is the fixed-lag SISOs described in Section 2.5. Masera et al. in [MaPiRoZa99] propose an efficient implementation of FL-SISOs that uses D banks of ACS units in a pipelined fashion to implement the backward recursions as illustrated in Fig-2.44. This implementation requires many more banks of ACS units than our fixed-interval approach but the number of forward and backward state metrics that must be stored is dramatically reduced.

A detailed implementation of a minimum-lag SISO with pipelined backward recursions is also described in [MaPiRoZa99]. This architecture is the result combining of the minimum-lag concept described in Section 2.5.6 and the pipelining method in Fig-2.44 for the specific choice of $H = D$. This architecture, which was initially suggested by Viterbi [Vi98], performs D completion operations for each backward recursion of length $2D - 1$. This requires only 2 ACS banks to implement the pipelined backward recursion with a modest amount of additional memory compared to the fixed-lag implementation. The architecture can be generalized to any value of H (see Problem 6.8).

Note that the overall latency of decoding a block of data for these designs is limited by the calculation of the forward state metric and is thus linear in the interleaver block size. Consequently, the throughput is no different than that of the fixed-interval SISOs described earlier.

6.5.4   Minimum Half Window (Tiled) SISOs

As described in Section 2.5.6, the minimum-half window SISOs with overlapping combining windows, or tiles, can also be activated in parallel, thereby yielding even lower latency and higher throughput than the fixed and minimum-lag approaches.

The latency of the minimum-half window implementations is approximately linear in the size of each tile (*i.e.*, $2D + H$ in Fig-2.50). Consequently, the increase in decoding throughput is the block size divided by the tile size. This increase is upper bounded by the sliding-window SISO special case (*i.e.*, $H = 1$), where the latency is $2D + 1$ ACS clock times. For acceptable performance, the minimum value of D should be several times the constraint length of the code to ensure reasonable coding gain (*e.g.*, see Example 2.11).

6.5.5 Sliding Window SISOs

An important property of the sliding-window SISO is that it completely decouples the soft-output computations. Each soft-output computation can be viewed as simply two sequences of ACS units, one to compute the forward state metrics and one to compute the backward state metrics, followed by an ACS completion step (see Fig-2.45). This effectively breaks the ACS feedback loop which limits the throughput of other SISO algorithms [ViMaPiRo00].

In particular, the sliding-window SISO promotes the use of internally-pipelined ACS units that can operate with much higher clock frequencies. For example, it is possible to fully-pipeline the ACS unit down to the bit-level where the clock cycle time is now dictated approximately by the latency of the computation of a single bit (*e.g.*, a 1-bit addition) [ViMaPiRo00]. The only tricky part of this bit-level pipelining is that the adder outputs must be appropriately delayed using internal shift-registers to be aligned with the results of the subtracter-based comparator which may not be ready until several clock cycles after the additions are completed.

Analog simulations of such ACS units in a conventional 0.5μ CMOS process suggest that a 1GHz clock frequency can be achieved [ViMaPiRo00]. Note that this is over 5 times faster than the estimated latency of a complete non-internally-pipelined ACS unit, thereby significantly reducing overall decoding latency and increasing throughput.

6.5.6 Tree SISOs

The latency of computing each tile[3] is the size of the tile because of the inherent sequential nature of the ACS recursions. While radix-4 and sliding-window algorithms alleviate this bottleneck, more aggressive tree SISOs (*e.g.*, see Section 2.5.5) architectures completely eliminate it. They are the highest performance architectures known to date but also require the most hardware resources.

---

[3]Recall that the fixed-interval SISO can be viewed as the special case of a tiled SISO with one tile.

In particular, tree SISOs allow tiles to be computed in time that is logarithmic in the tile size. Furthermore, there is no "overlap" penalty with tiled tree-SISOs [BeCh00]. Specifically, note that in the tiled forward-backward SISOs in Fig-2.50, there is overlap between the recursions performed in each sub-window. This overlap is not required for tree-SISOs when implemented in the form of Fig-2.48. Rough area and performance estimates of a tile-based architecture appear in [BeCh00]. It is estimated that, for the same PCCC code considered here with $K = 1024$, the proposed tiled tree SISO ($D = 16$) would require roughly 40 million transistors, but have a latency of only 14 clock cycles. The clock cycle time was estimated to be 5 ns yielding a total block latency for 10 iterations of just 700 ns. This translates to a throughput of almost 1.5 Gb/second. The tree-SISO in Example 2.19 and [ThCh00], which has much lower complexity with roughly twice the latency, is also a potential solution to extremely high throughput or low-latency processing.

For even higher throughput needs, it is also possible to pipeline the tree-SISO architectures across interleaver blocks (as described in Section 6.5.1). In particular, 20 such tiled tree-SISOs and associated interleavers can be used to achieve a factor of 20 in increased throughput, yielding a throughput of 30 Gb/second.

Moreover, unlike architectures based on the forward-backward algorithm, the tree-SISOs can easily be internally pipelined, yielding even higher throughputs with linear hardware scaling. In particular, if dedicated hardware is used for each stage of the tree-SISO, pipelining the tree-SISO internally may yield another factor of 4 to 5 in throughput, with no increase in latency. The tree-based architecture could thus support over 120 Gb/second.

While the hardware costs of such an aggressive architecture may be beyond practical limits today, given the continued increasing densities of VLSI technology, such systems may become cost-effective in the future for very high-speed applications.

6.5.7 Low-Power Turbo Decoding

A few other turbo coding works have focused on activation schedules and architectures that reduce power consumption. We refer the interested reader to [ScCaEn99, LiTsCh97, GaSt98].

CMOS	Complementary Metal Oxide Semiconductor
LIFO	Last In First Out
RAM	Random Access Memory
ROM	Read Only Memory
SAM	Sequential Access Memory
VLSI	Very Large Scale Integration

Table 6.1. Table of abbreviations specific to Chapter 6

6.6 Problems

6.1. Simulate the $K = 1024$, rate $1/2$ PCCC considered using soft-in information on the channel of the form $\alpha(z_k - \sqrt{E_s})^2$ and $\alpha(z_k + \sqrt{E_s})^2$ using different quantization methods. Specifically, by varying $\alpha > 0$, find a good quantizer for the region $[0, \infty)$ which has the reproduction labels naturally labeled as integers. For example, the label 010 should correspond to the real value of 2. Reproduce the Fig-6.2 and Fig-6.3 (including the pdfs). What bit width is required for each of the channel soft-in values for good performance?

6.2. Using the metrics defined in Fig-2.29, show that if the input metrics for all quantities are represented by w bits, the dynamic range of state metrics is upper bounded by $5 \cdot 2^{w-1}$. **Hint:** Consider the case where $s_k = i$ is part of a surviving path entering $s_{k+2} = j$ where $s_{k+2} = j$ has the minimum metric among states at time $k + 2$. What is an upper bound of the metrics $F_{k+1}[s_{k+2}]$ for the other values of s_{k+2}.

6.3. Repeat the hand computation of the PCCC decoder for a fixed-point implementation as described in Problem 2.11. Specifically, based on Fig-6.3 and the fact that the data in Table 2.5 was generated with $\sqrt{E_s} = 1$, multiply the observations by 3.5 and quantize to 4-bit 2's complement integers using a uniform quantizer and symmetric quantization bins (*i.e.*, analogous to the quantizer shown in Fig-6.1). Produce the tables requested in Problem 2.11 first assuming no clipping. Repeat this exercise with clipping of the output metrics to 4-bits. Do the decisions after two iterations vary in the three cases?

6.4. Recall that from (2.26), that the min*-sum processing uses the input metrics $\frac{4\sqrt{E_s}}{N_0} z_k(i)$.

(a) Show that based on the model in (6.1), $v_k(i) = \frac{4\sqrt{E_s}}{N_0} z_k(i)$ has the model

$$v_k(i) = 2\frac{E_b}{N_0}x_k(i) + n_k(i) \tag{6.6}$$

where $x_k(i) \in \{-1, +1\}$ (as before) and $n_k(i)$ is an AWGN sequence with zero mean and variance $4E_b/N_0$. Note that $E_b = E_s/2$.

(b) Explain why, in order to use the standard definition of

$$\min{}^*(x, y) = \min(x, y) - \ln(1 + e^{-|x-y|})$$

these metrics cannot be scaled by a multiplicative constant in order to meet a desired quantizer dynamic range.

(c) For the example shown in Fig-6.1, since $E_b/N_0 = 2$ dB it follows that $E_b/N_0 = 1.58$ (not in dB). Plot the conditional pdfs for $v_k(i)$ analogous to those shown in Fig-6.1. Consider using fixed-point (not necessarily integer), 2's complement representation of $v_k(i)$ that is approximately equivalent to quantization of $z_k(i)$ shown in Fig-6.1. Where would you place the quantizer reconstruction levels to accomplish this?

(d) Suppose, that instead of moving the reconstruction levels, $v_k(i)$ was multiplied by $1/1.58$ and the integer reconstruction levels shown in Fig-6.1 were used. Describe how the "min*(·)" look-up table must be modified. **Hint:** see Problem 2.8.

6.5. Suppose that x and y are represented using fixed-point, 2's complement with 3 bits as determined in Problem 6.4c. What values does $-\ln(1+\exp(-|x-y|))$ take on? What does this imply about representing this term using 2's complement?

6.6. Consider a min*-sum implementation where $-\ln(1 + \exp(-|x - y|))$ is represented by m bits in 2's complement fixed-point notation. Show that ΔF for the PCCC considered is increased by no more than $4 \cdot 2^{m-1}$ relative to ΔF for the min-sum case. Assuming $m = 4$ and all other soft-inputs are also represented using 4 bits, what is ΔF. What is the minimum required bitwidth of the ACS units?

6.7. Consider the implementation of the min*-sum algorithm described in Problem 6.6. What is the maximum value that state metrics can be initialized to without violating the state metric constraints required by modulo arithmetic. Simulate the effect of this initialization strategy and determine the degradation (if any) on coding performance compared with that of using infinite-precision initialization.

6.8. Consider the minimum-lag SISO described in Section 2.5.6 with $H = 2$. Draw the architecture for this minimum-lag SISO with a parallel backward recursion. How many ACS units are needed for the backward recursion? How much total memory (in bits) is required?

6.9. Compute ΔF for a min-sum forward-backward algorithm run on standard radix-4 trellis for 4-bit input metric quantization.

References

[Ag99]   D. Agrawal. *GMD decoding of Euclidean-space codes and itera-tive decoding of turbo codes.* PhD thesis, University of Illinois at Urbana-Champaign, Urbana, IL, 1999.

[Aj99]   S. M. Aji. *Graphical Models and Iterative Decoding.* PhD thesis, California Institute of Tech., 1999.

[AjMc00]   S. M. Aji and R. J. McEliece. The generalized distributive law. *IEEE Trans. Inform. Theory,* 46(2):325–343, March 2000.

[AlReAsSc99] P. D. Alexander, M. C. Reed, J. A. Asenstorfer, and C. B. Schlegel. Iterative multiuser interference reduction: Turbo CDMA. *IEEE Trans. Commun.,* 47:1008–1014, July 1999.

[An99]   A. Anastasopoulos. *Adaptive Soft-Input Soft-Output Algorithms for Iterative Detection.* PhD thesis, University of Southern California, Los Angeles, CA, August 1999.

[AnAl92]   M. Analoui and J. P. Allebach. Model based halftoning using direct binary search. In *Human Vision, Visual Proc., and Digital Display III,* volume 1666 of *Proc. SPIE,* pages 96–108. SPIE, February 1992.

[AnCh97]   A. Anastasopoulos and K. M. Chugg. An efficient method for simulation of frequency selective isotropic Rayleigh fading. In *Proc. Vehicular Tech. Conf.,* pages 2084–2088, Phoenix, AZ, May 1997.

[AnCh97b]   A. Anastasopoulos and K. M. Chugg. Iterative equaliza-tion/decoding of TCM for frequency-selective fading channels. In *Proc. Asilomar Conf. Signals, Systems, Comp.,* pages 177–181, November 1997.

[AnCh98]   A. Anastasopoulos and K. M. Chugg. TCM for frequency-selective, interleaved fading channels using joint diversity com-bining. In *Proc. International Conf. Communications,* Atlanta, GA, June 1998.

[AnCh99]   A. Anastasopoulos and K. M. Chugg. Adaptive iterative detection for turbo codes with carrier-phase uncertainty. In *Proc. Globecom Conf.,* pages 2369–2374, Rio de Janeiro, Brazil, December 1999.

[AnCh00]   A. Anastasopoulos and K. M. Chugg. Adaptive Soft-Input Soft-Output algorithms for iterative detection with parametric uncertainty. *IEEE Trans. Commun.*, 48(10), October 2000.

[AnFo75]   R. R. Anderson and G. J. Foschini. The minimum distance for MLSE digital data systems of limited complexity. *IEEE Trans. Inform. Theory*, 21:544–551, September 1975.

[AnMo79]   B. D. O. Anderson and J. B. Moore. *Optimal Filtering.* Prentice Hall, Englewood Cliffs, NJ, 1979.

[AnMo84]   J. B. Anderson and S. Mohan. Sequential coding algorithms: A survey cost analysis. *IEEE Trans. Commun.*, COM-32:169–176, February 1984.

[AnPo98]   A. Anastasopoulos and A. Polydoros. Adaptive soft-decision algorithms for mobile fading channels. *European Trans. Commun.*, 9(2):183–190, March/April 1998.

[AuMc99]   J. Au and P. J. McLane. Performance of turbo codes with quantizaed channel measurements. In *Proc. Globecom Conf.*, Dallas, TX, December 1999. (Comm. Theory Symp.).

[Ba87]   G. Battail. Pondération des symboles décodé par l'algorithme de Viterbi. *Ann. Télecommun.*, 42:31–38, January 1987. (in French).

[BaCu98]   E. Baccarelli and R. Cusani. Combined channel estimation and data detection using soft statistics for frequency-selective fast-fading digital links. *IEEE Trans. Commun.*, 46:424–427, April 1998.

[BaCoJeRa74]   L. R. Bahl, J. Cocke, F. Jelinek, and J. Raviv. Optimal decoding of linear codes for minimizing symbol error rate. *IEEE Trans. Inform. Theory*, IT-20:284–287, March 1974.

[Be63]   P. A. Bello. Characterization of randomly time-variant linear channels. *IEEE Trans. Commun. Systems*, 11:360–393, 1963.

[BeAdAnFa93]   C. Berrou, P. Adde, E. Angui, and S. Faudeil. A low complexity soft-output Viterbi decoder architecture. In *Proc. International Conf. Communications*, pages 737–740, June 1993.

[BeBeMa98]   N. Benvenuto, L. Bettella, and R. Marchesani. Performance of the Viterbi algorithm for interleaved convolutional codes. *IEEE Trans. Vehic. Technol.*, 47(3):919–923, August 1998.

[BeCh00]   P. A. Beerel and K. M. Chugg. An $O(\log_2 N)$-latency SISO with application to broadband turbo decoding. In *Proc. IEEE Military Comm. Conf.*, Los Angeles, CA, October 2000. (see also the USC, Comm. Sciences Inst. report CSI-00-05-01).

[BeDiMoPo98]   S. Benedetto, D. Divsalar, G. Montorsi, and F. Pollara. Soft-input soft-output modules for the construction and distributed iterative decoding of code networks. *European Trans. Teleommun.*, 9(2), March/April 1998.

[BeDiMoPo98b]   S. Benedetto, D. Divsalar, G. Montorsi, and F. Pollara. Serial concatenation of interleaved codes: performance analysis, design, and iterative decoding. *IEEE Trans. Inform. Theory*, 44(3):909–926, May 1998.

[BeGl96]   C. Berrou and A. Glavieux. Near optimum error correcting coding and decoding: turbo-codes. *IEEE Trans. Commun.*, 44(10):1261–1271, October 1996.

[BeGlTh93]   C. Berrou, A. Glavieux, and P. Thitmajshima. Near shannon limit error-correcting coding and decoding: turbo-codes. In *International Conference on Communications*, pages 1064–1070, Geneva, Switzerland, May 1993.

[BeLuMa93]   N. Benvenuto, G. Lubello, and R. Marchesani. Multitrellis decmposition of the Viterbi algorithm for multipath channels. In *Proc. ICC'93*, pages 746–750, Geneva, Switzerland, May 1993.

[BeMa96]   N. Benvenuto and R. Marchesani. The Viterbi algorithm for sparse channels. *IEEE Trans. Commun.*, 44:287–289, March 1996.

[BeMo96]   S. Benedetto and G. Montorsi. Unveiling turbo codes: some results on parallel concatenated coding schemes. *IEEE Trans. Inform. Theory*, 42(2):408–428, March 1996.

[BeMoDiPo96]   S. Benedetto, G. Montorsi, D. Divsalar, and F. Pollara. A soft-input soft-output maximum a posteriori (MAP) module to decode parallel and serial concatenated codes. Technical report, JPL-TDA, November 1996. 42–127.

[BePa79]   C. A. Belfiore and Parks. Decision feedback equalization. *Proc. IEEE*, 67, August 1979.

[BeSa94]   N. Benvenuto and A. Salloum. Performance of the multitrellise Viterbi algorithm for sparse channels. In *Proc. GLOBECOM'93*, pages 1493–1497, San Francisco, CA, November 1994.

[BiDiMcSi91]   E. Biglieri, D. Divsalar, P. J. McLane, and M. K. Simon. *Introduction to Trellis-Coded Modulation with Applications*. Macmillian, New York, 1991.

[Bl93]   P. J. Black. *Algorithms and Architectures for High Speed Viterbi Decoding*. PhD thesis, Stanford University, California, March 1993.

[Ca74]   C. R. Cahn. Phase tracking and demodulation with delay. *IEEE Trans. Inform. Theory*, IT-20:50–58, January 1974.

[CeMaSa79]   E. Cerny, D. Mange, and E. Sanchez. Synthesis of minimal binary decision trees. *IEEE Transactions on Computers*, 28:472–482, 1979.

[ChHa66]   R. W. Chang and J. C. Hancock. On receiver structures for channels having memory. *IEEE Trans. Inform. Theory*, IT-12:463–468, October 1966.

[Ch95]   K. M. Chugg. *Sequence Estimation in the Presence of Parametric Uncertainty*. PhD thesis, University of Southern California, Los Angeles, CA, August 1995.

[Ch96]   K. M. Chugg. Performance of optimal digital page detection in a two-dimensional ISI/AWGN channel. In *Proc. Asilomar Conf. Signals, Systems, Comp.*, November 1996.

[Ch98]   K. M. Chugg. The condition for the applicability of the Viterbi algorithm with implications for fading channel MLSD. *IEEE Trans. Commun.*, 46(9):1112–1116, September 1998.

[Ch99]   X. Chen. *Iterative Data Detection: Complexity Reduction and Applications*. PhD thesis, University of Southern California, Los Angeles, CA, December 1999.

[ChAn00]   K. M. Chugg and A. Anastasopoulos. On symbol error probability bounds for ISI-like channels. submitted to *IEEE Trans. Commun.* (see also: Technical Report CSI-00-04-01, Communication Sciences Institute, USC), April 2000.

[ChCh98]   X. Chen and K. M. Chugg. Near-optimal page detection for two-dimensional ISI/AWGN channels using concatenated modeling and iterative detection. In *Proc. International Conf. Communications*, Atlanta, GA, 1998.

[ChCh98b]   K. M. Chugg and X. Chen. Efficient architectures for soft-output algorithms. In *Proc. International Conf. Communications*, Atlanta, GA, 1998.

[ChCh00]   X. Chen and K. M. Chugg. Reduced state soft-in/soft-out for complexity reduction in iterative and non-iterative data detection. In *Proc. International Conf. Communications*, New Orleans, LA, 2000.

[ChCh00b]   X. Chen and K. M. Chugg. Iterative soft-in/soft-out algorithms for arbitrary sparse channels. *IEEE Trans. Commun.* (submitted for publication, Jan. 2000).

[ChChNe98]   X. Chen, K. M. Chugg, and M. A. Neifeld. Near-optimal parallel distributed data detection for page-oriented optical memories (Special issue: Advanced optical storage technologies). *IEEE J. Select. Topics Quantum Electron.*, 4(5):866–879, Sept./Oct. 1998.

[ChChNe99]   K. M. Chugg, X. Chen, , and M. A. Neifeld. Two-dimensional equalization in coherent and incoherent page oriented optical memory. *J. Opt. Soc. Amer. A*, 16(3):549–562, March 1999.

[ChChOrCh98]   K. M. Chugg, X. Chen, A. Ortega, and C-W. Chang. An iterative algorithm for two-dimensional digital least metric problems with applications to digital image compression. In *Proc. Intl. Conf. Image Processing*, Chicago, IL, 1998.

[ChChThAn00]   K.M. Chugg, X. Chen, P. Thiennviboon, and A. Anastasopoulos. Two-dimensional data detection: Achievable performance and near-optimal algorithms. Technical Report CSI-00-11-01, Communication Sciences Institute, USC, Los Angeles, CA, November 2000.

[ChDeOr99]   K. M. Chugg, K. Demirciler, and A. Ortega. Soft information for source channel coding and decoding. In *Proc. Asilomar Conf. Signals, Systems, Comp.*, volume Sess. MP4-3, October 1999.

[ChLe98]   K. M. Chugg and K. Lerdsuwanakij. Fading channel sequence detection based on rational approximations to the Clarke doppler spectrum. In *Proc. International Conf. on Telecommunications*, Chalkidiki, Greece, 1998.

[ChPo96]   K. M. Chugg and A. Polydoros. MLSE for an unknown channel – part I: Optimality considerations. *IEEE Trans. Commun.*, 44:836–846, July 1996.

[ChPo96b]   K. M. Chugg and A. Polydoros. MLSE for an unknown channel
              – part II: Tracking performance. *IEEE Trans. Commun.*, pages
              949–958, August 1996.

[Cl68]   R. Clark. A statistical theory of mobile radio reception. *Bell
              Sys. Tech. J.*, 47:97–1000, 1968.

[Co90]   G. F. Cooper. The computational complexity of probabilistic in-
              ference using Bayesian belief networks. *Artif. Intell.*, 42(2-3):393–
              405, March 1990.

[CoFeRa00]   G. Colavolpe, G. Ferrari, and R. Raheli. Noncoherent itera-
              tive (turbo) detection. *IEEE Trans. Commun.*, 48(9):1488–1498,
              September 2000.

[CoXiXi99]   Ling Cong, Wu Xiaofu, and Yi Xiaoxin. On SOVA for non-
              binary codes. *IEEE Communications Letters*, 3:335–337, Decem-
              ber 1999. (authors' family names listed first on paper).

[CoLeRi90]   T. H. Cormen, C.E. Leiserson, and R. L. Rivest. *Introduction to
              Algorithms*. The MIT Press, Cambridge, Mass., 1990.

[CuMa99]   R. Cusani and J. Mattila. Equalization of digital radio channels
              with large multipath delay for cellular land mobile applications.
              IEEE Trans. Commun., 47(3):348–351, March 1999.

[DaLu93]   P. Dagum and M. Luby. Approximating probabilistic inference in
              Bayesian belief networks is NP-hard. *Artif. Intell.*, 60(1):141–153,
              March 1993.

[DaMeVi94]   A. N. D'Andrea, U. Mengali, and G. M. Vitetta. Approximate ML
              decoding of coded PSK with no explicit carrier phase reference.
              IEEE Trans. Commun., 42:1033–1040, Feb./Mar./April 1994.

[DaMo99]   F. Daneshgaran and M. Mondin. Design of interleavers for turbo
              codes: Iterative interleaver growth algorithms of polynomial com-
              plexity. *IEEE Trans. Inform. Theory*, 45:1845–1859, September
              1999.

[Di97]   D. Divsalar. Serial and hybrid concatenations of codes with in-
              terleavers. In *IEEE Communcation Theory Workshop*, Tucson,
              AZ, April 1997.

[Di99]   D. Divsalar. A simple tight bound on error probability of block
              codes with application to turbo codes. Technical Report TDA
              Progress Report 42-139, Jet Propulsion Labs., Pasadena, CA,
              November 1999.

[DiPo97]   D. Divsalar and F. Pollara. Serial and hybrid concatenated codes
              with applications. In *Intern. Symposium on Turbo Codes and
              related topics*, Brest, France, September 1997.

[DoDi95]   S. Dolinar and D. Divsalar. Weight distributions for turbo codes
              using random and nonrandom permutations. Technical Report
              TDA Progress Report 42-122, Jet Propulsion Labs., Pasadena,
              CA, August 1995.

[DuHe89]   A. Duel-Hallen and C. Heegard. Delayed decision feedback esti-
              mation. *IEEE Trans. Commun.*, 37:428–436, May 1989.

[EyQu88]   M. V. Eyuboğlu and S. U. Qureshi. Reduced-state sequence
              estimation with set partitioning and decision feedback. *IEEE
              Trans. Commun.*, COM-38:13–20, January 1988.

[FeGeFi98]   I. J. Fevrier, S. B. Gelfand, and M. P. Fitz. Fast computation of efficient decision feedback equalizers for high speed wireless communications. In *Proc. IEEE ICASSP'98*, volume 6, pages 3493–3496, Seattle, WA, May 1998.

[FeMe89]   G. Fettweis and H. Meyr. Parallel Viterbi algorithm implementation: Breaking the ACS-bottleneck. *IEEE Trans. Commun.*, 37:785–790, August 1989.

[Fo72]   G. D. Forney, Jr. Lower bounds on error probability in the presence of large intersymbol interference. *IEEE Trans. Commun.*, COM-20:76–77, February 1972.

[Fo72b]   G. D. Forney, Jr. Maximum-likelihood sequence estimation of digital sequences in the presence of intersymbol interference. *IEEE Trans. Inform. Theory*, IT-18:284–287, May 1972.

[Fo73]   G. D. Forney, Jr. The Viterbi algorithm. *Proc. IEEE*, 61:268–278, March 1973.

[Fo75]   G. J. Foschini. Performance bound for maximum-likelihood reception of digital data. *IEEE Trans. Inform. Theory*, 21:47–50, January 1975.

[Fo97]   G. D. Forney, Jr. Iterative decoding. In *26th Annual IEEE Communication Theory Workshop*, Tuscon, AZ, April 1997.

[Fo00]   G. D. Forney, Jr. Codes on graphs: Normal realizations. *IEEE Trans. Inform. Theory*. (submitted for publication, April 2000).

[FoBuLiHa98]   M. P. C. Fossorier, F. Burkert, S. Lin, and J. Hagenauer. On the equivalence between SOVA and max-log-MAP decodings. *IEEE Communications Letters*, 5:137–139, May 1998.

[Fr98]   B. J. Frey. *Graphical Models for Machine Learning and Digital Communications*. MIT Press, Cambridge, MA, 1998.

[FrAn98]   V. Franz and J. B. Anderson. Concatenated decoding with a reduced-search BCJR algorithm. *IEEE J. Select. Areas Commun.*, 16(2):186–195, February 1998.

[FrKs98]   B.J. Frey and F.R. Kschischang. Early detection and trellis splicing: Reduced complexity iterative decoding. *IEEE J. Select. Areas Commun.*, pages 153–159, February 1998.

[FrMc98]   B. J. Frey and D. J. C. MacKay. A revolution: Belief propagation in graphs with cycles. In M. I. Jordan, M. I. Kearns, and S. A. Solla, editors, *Advances in Neural Information Processing Systems*, pages 470–485. MIT Press, 1998.

[FrWe99]   C. Fragouli and R. Wessel. Symbol interleaved parallel concatenated trellis coded modulation. In *Proc. International Conf. Communications*, pages 42–46, Vancouver, B.C., Canada, June 1999. (Comm. Theory Mini-Conf.).

[Ga62]   R. G. Gallager. Low density parity check codes. *IEEE Trans. Inform. Theory*, 8:21–28, January 1962.

[Ga63]   R. G. Gallager. *Low-Density Parity-Check Codes*. MIT Press, Cambridge, MA, 1963.

[GaSt98]   D. Garret and M. Stan. Low power architecture for the soft-output Viterbi algorithm. In *Proc. International Symposium on Low-Power Electronics and Design*, 1998.

[GeLo97]   M. Gertsman and J. H. Lodge. Symbol-by-symbol MAP demodulation of CPM and PSK on Rayleigh flat-fading channels. *IEEE Trans. Commun.*, 45(7):788–799, July 1997.

[GeReSu93]   R. Geist, R. Reynolds, and D. Suggs. A Markovian framework for digital halftoning. *ACM Trans. Graph.*, 12:136–159, April 1993.

[Gi62]   A. Gill. *Finite-State Machines*. McGraw-Hill, 1962.

[Go66]   B. Goldberg. 300 kHz–30 MHz MF/HF. *IEEE Trans. Commun. Technol.*, COM-14(6):767–784, December 1966.

[Go68]   J. W. Goodman. *Introduction to Fourier Optics*. McGraw-Hill, Inc., 1968.

[GoCh00]   R. Golshan and K. M. Chugg. Iterative coded multiuser detection with a Verdú soft demodulator. In *Proc. Asilomar Conf. Signals, Systems, Comp.*, October 2000.

[GoVa89]   G. H. Golub and C. F. Van Loan. *Matrix Computations*. Johns Hopkins Press, 2nd edition, 1989.

[Gr81]   D. J. Granrath. The role of human visual models in image processing. *Proc. IEEE*, 69(5):552–561, May 1981.

[Ha65]   M. A. Harrison. *Introduction to Switching and Automata Theory*. McGraw-Hill, 1965.

[Ha96]   S. Haykin. *Adaptive Filter Theory*. Prentice-Hall, Englewood Cliffs, NJ, 3rd edition, 1996.

[HaChAu00]   A. Hansson, K. M. Chugg, and T. Aulin. A forward-backward algorithm for fading channels using forward-only estimation. Technical Report CSI-00-11-02, Communication Sciences Institute, USC, Los Angeles, CA, November 2000.

[HaEl00]   A. R. Hammons, Jr. and H. El Gamal. On the convergence of the turbo decoder. In *Conference on Information Sciences and Systems (CISS)*, March 2000. (WA6-3).

[HaHo89]   J. Hagenauer and P. Hoeher. A Viterbi algorithm with soft-decision outputs and its applications. In *Proc. Globecom Conf.*, pages 1680–1686, Dallas, TX, November 1989.

[HaOfPa96]   J. Hagenauer, E. Offer, and L. Papke. Iterative decoding of binary block and convolutional codes. *IEEE Trans. Inform. Theory*, 42(2):429–445, March 1996.

[HaRu76]   C. R. P. Hartmann and L. D.Rudolph. An optimum symbol-by-symbol decoding rule for linear codes. *IEEE Trans. Inform. Theory*, IT-22(5):514–517, September 1976.

[HaSt97]   A. Hafeez and W. E. Stark. Soft-output multiuser estimation for asynchronous CDMA channels. In *Proc. Vehicular Tech. Conf.*, pages 465–469, Phoenix, AZ, May 1997.

[He60]   C. W. Helstrom. *Statistical Theory of Signal Detection*. Pergamon Press, 1960.

[He89]   A. P. Hekstra. An alternative to metric rescaling in Viterbi decoders. *IEEE Trans. Commun.*, 37(11), November 1989.

[He00]   J. Heo. Adaptive iterative detection based on belief propaga-
              tion. Technical Report CSI-00-05-03, Communication Sciences
              Institute, USC, Los Angeles, CA, May 2000. (Ph. D. Disertation
              proposal).

[HeCh00]   J. Heo and K. M. Chugg. Adaptive iterative detection for turbo
              codes on flat fading channels. In *WCNC*, Chicago, IL, September
              2000.

[HeChAn00]   J. Heo, K. M. Chugg, and A. Anastasopoulos. A compari-
              son of forward-only and bi-directional fixed-lag adaptive SISOs.
              In *Proc. International Conf. Communications*, pages 1660–1664,
              New Orleans, LA, June 2000.

[HeGuHe96]   J. F. Heanue, Gürkan, and L. Hesselink. Signal detection for page-
              access optical memories with intersymbol interference. *Applied
              Optics*, 35:2431–2438, May 1996.

[HeJa71]   J. A. Heller and I.M. Jacobs. Viterbi decoding for satellite and
              space communication. *IEEE Trans. Commun. Technol.*, COM-
              19:835–848, October 1971.

[HeWi98]   C. Heegard and S. Wicker. *Turbo Coding*. Kluwer Academic
              Publishers, 1998.

[HsWa99]   J.-M. Hsu and C.-L. Wang. On finite-precision implementation
              of a decoder for turbo-codes. In *IEEE International Symposium
              on Circuit and Applications*, volume 4, pages 423–426, 1999.

[Il92]   R. A. Iltis. A Bayesian maximum-likelihood sequence estima-
              tion algorithm for a priori unknown channels and symbol timing.
              IEEE J. Select. Areas Commun., 10:579–588, April 1992.

[IlShGi94]   R. A. Iltis, J. J. Shynk, and K. Giridhar. Bayesian algorithms
              for blind equalization using parallel adaptive filtering. *IEEE
              Trans. Commun.*, 42:1017–1032, Feb./Mar./Apr. 1994.

[Je96]   F. V. Jensen. *An Introduction to Bayesian Networks*. Springer-
              Verlag, 1996.

[JeJe94]   F. V. Jensen and F. Jensen. Optimal junction trees. In *Conf. Un-
              certainty in Artificial Intelligence*, pages 360–366, San Francisco,
              CA, March 1994.

[JeLaOl90]   F. V. Jensen, S. L. Lauritzen, and K. G. Olesen. Bayesian updat-
              ing in causal probabilistic networks by local computation. *Com-
              putational Statistics Quarterly*, 4:269–282, 1990.

[KeMa98]   L. Ke and M. W. Marcellin. Near-lossless image compression:
              Minimum-entropy, constrained-error DPCM. *IEEE Trans. Imag.
              Processing*, 7(2):225–228, February 1998.

[Ko71]   H. Kobayashi. Simultaneous adaptive estimation and decision al-
              gorithm for carrier modulated data transmission sysytems. *IEEE
              Trans. Commun.*, 19:268–280, June 1971.

[KoBa90]   W. Koch and A. Baier. Optimum and sub-optimum detection of
              coded data disturbed by time-varying intersymbol interference.
              In *Proc. Globecom Conf.*, pages 807.5.1–5, December 1990.

[KoWe99]   C. Komninakis and R. D. Wesel. Pilot-aided joint data and chan-
              nel estimation in flat correlated fading. In *Proc. Globecom Conf.*,

pages 2534–2539, Rio de Janeiro, Brazil, 1999. (Comm. Theory Symposium).

[KsFr98] F.R. Kschischang and B.J. Frey. Iterative decoding of compond codes by probability propagation in graphical models. *IEEE J. Select. Areas Commun.*, pages 219–231, February 1998.

[KsFrLo00] F. R. Kschischang, B. J. Frey, and H.-A. Loeliger. Factor graphs and the sum-product algorithm. *IEEE Trans. Inform. Theory*, 2000. submitted for publication.

[KwKa98] D. Kwan and S. Kallel. A truncated best-path algorithm. *IEEE Trans. Commun.*, 46:568–572, May 1998.

[LaArGa98] D. L. Lau, G. R. Arce, and N. C. Gallagher. Green-noise digital halftoning. *Proc. IEEE*, 86(12):2424–2444, December 1998.

[LaSp88] S. L. Lauritzen and D. J. Spiegelhalter. Local computation with probabilities on graphical structures and their application to expert systems. *J. Roy. Statist. Soc. B*, pages 157–224, 1988.

[Le74] L-N. Lee. Real-time minimal-bit-error probability decoding of convolutional codes. *IEEE Trans. Commun.*, 22:146–151, February 1974.

[LeChPo99] K. Lerdsuwanakij, K. M. Chugg, and A. Polydoros. Quanitzation-based estimation. In *Proc. Asilomar Conf. Signals, Systems, Comp.*, volume 1, pages 37–41, 1999.

[LiAl96] D. J. Lieberman and J. P. Allebach. Digital halftoning using the direct binary search algorithm. In *Proc. IST Int. Conf. High Technology*, pages 114–124, Chiba, Japan, September 1996.

[LiCo83] S. Lin and D. Costello, Jr. Error Control Coding: Fundamentals and Applications. Prentice-Hall, 1983.

[LiRi99] X. Li and J. A. Ritcey. Trellis-coded modulation with bit interleaving and iterative decoding. *IEEE J. Select. Areas Commun.*, 17:715–724, April 1999.

[LiSi73] W. C. Lindsey and M. K. Simon. *Telecommunication Systems Engineering.* Prentice-Hall, Englewood Cliffs, New Jersey, 1973.

[LiTsCh97] L. Lin, C. Y. Tsui, and S. R. Cheng. Low power soft output Viterbi decoder scheme for turbo decoding. In *IEEE Intern. Symp. on Circuits and Systems*, pages 1369–1372, June 1997.

[LiVuSa95] Y. Li, B. Vucetic, and Y. Sato. Optimum soft-output detection for channels with intersymbol interference. *IEEE Trans. Inform. Theory*, 41:704–713, May 1995.

[LoMo90] J. Lodge and M. Moher. Maximum likelihood estimation of CPM signals transmitted over Rayleigh flat fading channels. *IEEE Trans. Commun.*, 38:787–794, June 1990.

[LuWi98] L. Lu and S. W. Wilson. Synchronization of turbo coded modulation systems at low SNR. *Proc. International Conf. Communications*, Atlanta, GA, June 1998.

[Ma99] D. J. C. MacKay. Good error correcting codes based on very sparse matrices. *IEEE Trans. Inform. Theory*, 45:399–431, February 1999.

[MaMa98]   I. D. Marsland and P. T. Mathiopoulos. Multiple differential detection of parallel concatenated convolutional (turbo) codes in correlated fast Rayleigh fading. *IEEE J. Select. Areas Commun.*, 16(2):265–275, February 1998.

[MaNe96]   D. J. C. MacKay and R. M. Neal. Near Shannon limit performance of low density parity check codes. *Electronics Letters*, 32(18):1645–1646, August 1996.

[MaPiRoZa99] G. Masera, G. Piccinini, M. Ruo Roch, and M. Zamboni. VLSI architectures for turbo codes. *IEEE Transactions on VLSI*, 7(3), September 1999.

[MaPr73]   F. R. Magee and J. G. Proakis. Adaptive maximum-likelihood sequence estimation for digital signaling in the presence of intersymbol interference. *IEEE Trans. Inform. Theory*, 19:120–124, January 1973.

[Ma75]   J. E. Mazo. A geometric derivation of Forney's upper bound. *Bell Sys. Tech. J.*, 54:1087–1094, Jul.–Aug. 1975.

[Ma75b]   J. E. Mazo. Faster-than-Nyquist signaling. *Bell Sys. Tech. J.*, 54:1450–1462, October 1975.

[Mc74]   P. L. McAdam. *MAP bit decoding of convolutional codes.* PhD thesis, University of Southern California, Los Angeles, CA, 1974.

[Mc99]   R. McEliece. Iterative decoding of RA codes. In *1999 IEEE Comm. Theorey Workshop*, Aptos, CA, 1999.

[McKe97]   N. C. McGinty and R. A. Kennedy. Reduced-state sequence estimator with reverse-time structure. *IEEE Trans. Commun.*, 45(3):265–268, March 1997.

[McKeHo98]   N. C. McGinty, R. A. Kennedy, and P. Hoeher. Parallel trellis Viterbi algorithm for sparse channels. *IEEE Trans. Commun.*, pages 143–145, May 1998.

[McMaCh98]   R. J. McEliece, D. J. C. MacKay, and J. F. Cheng. Turbo decoding as an instance of Pearl's "belief propagation" algorithm. *IEEE J. Select. Areas Commun.*, 16:140–152, February 1998.

[McWeWe72]   P. L. McAdam, L. R. Welch, and C. L. Weber. M.A.P. bit decoding of convolutional codes. *Proc. IEEE Int. Symp. Info. Theory*, 1972.

[Me95]   J. M. Mendel. *Lessons in estimation theory for signal processing, communications, and control.* Prentice Hall, Englewood Cliffs, NJ, 1995.

[Mi60]   D. Middleton. *An Introduction to Statistic Communication Theory.* McGraw-Hill, New York, 1960.

[Mi99]   C. L. Miller. *Image Restoration Using Trellis-Search Methods.* PhD thesis, University of Arizona, 1999.

[Mo82]   B. M. E. Moret. Decision trees and diagrams. *Computing Surveys*, 14:593–623, 1982.

[Mo97]   G. Montorsi. Soft-input soft-output modules to iteratively decode networks of concatenated codes. In *26th Annual IEEE Communication Theory Workshop*, Tucson, AZ, April 1997.

[Mo97b]   M. Moher. *Cross-Entropy and Iterative Detection*. PhD thesis, Carelton University, Ottowa, Canada, May 1997. (OCIEE-97-05).

[Mo98]   M. Moher. An iterative multiuser decoder for near-capacity communications. *IEEE Trans. Commun.*, 46:870–880, July 1998.

[MoAu99]   P. Moqvist and T. Aulin. Improved lower bounds on the symbol error probability for ISI channels. Technical report, Dept. of Comp. Eng., Chalmers Univ. of Technology, Goteborg, Sweden, January 1999. http://www.ce.chalmers.se/staff/pmoqvist/.

[MoAu99b]   P. Moqvist and T. Aulin. Certain aspects on map algorithms for turbo codes. In *Radio Vetenskap och Kommunikation*, Karlskrona, Sweden, June 1999. (in English).

[MoAu00]   P. Moqvist and T. Aulin. Power and bandwidth efficient serially concatenated CPM with iterative decoding. In *Proc. Globecom Conf.*, San Francisco, CA, December 2000.

[MoGu98]   M. Moher and P. Guinand. An iterative algorithm for asynchronous coded multiuser detection. *IEEE Communications Letters*, pages 229–231, August 1998.

[MuAh92]   J. B. Mulligan and A. J. Ahumada, Jr. Principled halftoning based on models of human vision. In *Human Vision, Visual Proc., and Digital Display III*, volume 1666 of *Proc. SPIE*, pages 109–121. SPIE, February 1992.

[MuGeHu96]   S. H. Müller, W. H. Gerstacker, and J. B. Huber. Reduced-state soft-output trellis-equalization incorporating soft feedback. In *Proc. IEEE Globecom'96*, pages 95–100, Westminster, London, November 1996.

[MuWeJo99]   K. P. Murphy, Y. Weiss, and M. I. Jordan. Loopy belief propagation for approximate inference: an empirical study. In *Uncertainty in AI*, pages 737–740, June 1993.

[NeChKi96]   M. A. Neifeld, K. M. Chugg, and B. M. King. Parallel data detection in page-oriented optical memory. *Optic Letters*, 21:1481–1483, September 1996.

[NePa94]   D. L. Neuhoff and T. N. Pappas. Perceptual coding of images for halftone display. *IEEE Trans. Imag. Processing*, 3(4):341–354, July 1994.

[Pa00]   P. Pawawongsak. EE577b VLSI design project: A design of a turbo decoder chip. Technical Report CENG-00-005, Computer Engineering Division, University of Southern California, July 2000.

[PaNe95]   T. N. Pappas and D. L. Neuhoff. Printer models and error diffusion. *IEEE Trans. Imag. Processing*, 4(1):66–80, January 1995.

[PaNe99]   T. N. Pappas and D. L. Neuhoff. Least-square model-based halftoning. *IEEE Trans. Imag. Processing*, 8(8):1102–1116, August 1999.

[Pe86]   J. Pearl. Fusion, propagation, and structuring in belief networks. *Artif. Intell.*, 29(3):241–288, September 1986.

[Pe88]   J. Pearl. *Probabilistic Reasoning in Intelligent Systems: Networks of Plausible Inference*. Morgan Kaufmann, 1988.

[PeSeCo96]   L. Perez, J. Seghers, and D. Costello. A distance spectrum inter-
                pretation of turbo codes. *IEEE Trans. Inform. Theory*, 42:1698–
                1709, November 1996.

[PiDiGl97]   A. Picart, P. Didier, and A. Glavieux. Turbo-detection: A new
                approach to combat channel frequency selectivity. In *Proc. Inter-
                national Conf. Communications*, Montreal, Canada, 1997.

[Po94]   H. V. Poor. *An Introduction to Signal Detection and Estimation.*
                Springer-Verlag, 1994.

[Pr91]   J. Proakis. Adaptive equalization for TDMA digital mobile radio.
                IEEE Trans. Veh. Tech., pages 333–341, May 1991.

[Pr95]   J. G. Proakis. *Digital Communications.* McGraw-Hill, Inc., New
                York, 3rd edition, 1995.

[Ra89]   L. R. Rabiner. A tutorial on hidden Markov models and se-
                lected applications in speech recognition. *Proc. IEEE*, 77:257–
                286, February 1989.

[RaPoTz95]   R. Raheli, A. Polydoros, and C-K. Tzou. Per-survivor processing:
                A general approach to MLSE in uncertain environments. *IEEE
                Trans. Commun.*, 43:354–364, Feb–Apr. 1995.

[ReScAlAs98]   M. C. Reed, C. B. Schlegel, P. D. Alexander, and J. A. Asenstor-
                fer. Iterative multiuser detection for CDMA with FEC: Near-
                single-user performance. *IEEE Trans. Commun.*, 46:1693–1699,
                December 1998.

[Ri98]   S. Riedel. New symbol-by-symbol map decoding algorithm for
                high-rate convolutional codes that use reciprocal dual codes.
                IEEE J. Select. Areas Commun., pages 175–185, February 1998.

[RiShUr00]   T. Richardson, A. Shokrollahi, and R. Urbanke. Design of prov-
                ably good low density parity check codes. *IEEE Trans. Inform.
                Theory*, 2000. (submitted for publication).

[RiUr00]   T. Richardson and R. Urbanke. The capacity of good low den-
                sity parity check codes under message passing decoding. *IEEE
                Trans. Inform. Theory*, 2000. (submitted for publication).

[RoSi97]   M. E. Rollins and S. J. Simmons. Simplified per-survivor Kalman
                processing in fast-frequency-selective fading channels. *IEEE
                Trans. Commun.*, 45:544–553, May 1997.

[RoViHo95]   P. Robertson, E. Villebrum, and P. Hoeher. A comparison of
                optimal and suboptimal MAP decoding algorithms operating in
                the log domain. In *Proc. International Conf. Communications*,
                pages 1009–1013, Seattle, WA, 1995.

[ScCaEn99]   C. Schurgers, F. Catthoor, and M. Engles. Energy efficient data
                transfer and storage organization for a MAP turbo decoder mod-
                ule. In *Proc. International Symposium on Low-Power Electronics
                and Design*, 1999.

[ScPa94]   M. A. Schulze and T. N. Pappas. Blue noise and model-based
                halftoning. In *Human Vision, Visual Proc., and Digital Display
                V*, volume 2179 of *Proc. SPIE*, pages 182–194. SPIE, February
                1994.

[SeFi95]   J. P. Seymour and M. P. Fitz. Near-optimal symbol-by-symbol detection schemes for flat rayleigh fading. *IEEE Trans. Commun.*, pages 1525–1533, February/March/April 1995.

[Sh91]   W-H. Sheen. *Performance Analysis of Sequence Estimation Techniques for Intersymbol Interference Channels*. PhD thesis, Georgia Institute of Technology, May 1991.

[ShSh90]   G. R. Shafer and P. P. Shenoy.   Probability propagation. *Ann. Math. Art. Intel.*, 2:327–352, 1990.

[Si89]   S. Simmons. Breadth-first trellis decoding with adaptive effort. *IEEE Trans. Commun.*, 38:3–12, 1989.

[SmBe97]   J. E. Smee and N. C. Beauliue. On the equivalence of the simultaneous and separate MMSE optimization of a DFE FFF and FBF. *IEEE Trans. Commun.*, 45(2):156–158, February 1997.

[St96]   G. L. Stüber. *Principles of Mobile Communication*. Kluwer Academic Press, 1996.

[Ta81]   R. M. Tanner. A recursive approach to low complexity codes. *IEEE Trans. Inform. Theory*, IT-27:533–547, September 1981.

[Th00]   P. Thiennviboon.   Data detection for complex systems using message-passing algorithms.   Technical Report CSI-00-05-02, Communication Sciences Institute, USC, Los Angeles, CA, May 2000. (Ph. D. Disertation proposal).

[Th00b]   P. Thiennviboon, September 2000. Private communication.

[ThCh00]   P. Thiennviboon and K. M. Chugg. A low-latency SISO via message passing on a binary tree. In *Proc. Allerton Conf.*, 2000.

[Ul87]   R. A. Ulichney. *Digital Halftoing*. The MIT Press, Cambridge, MA, 1987.

[Ul00]   R. A. Ulichney. A review of halftoning techniques. In *Color Imaging: Device-Independent Color, Color Hardcopy, and Graphic Arts V*, volume 3963 of *Proc. SPIE*. SPIE, January 2000.

[Un74]   G. Ungerboeck.   Adaptive maximum likelihood receiver for carrier-modulated data-transmission systems. *IEEE Trans. Commun.*, com-22:624–636, May 1974.

[Va68]   H. L. Van Trees. *Detection, Estimation, and Modulation Theory Part I*. John Wiley & Sons, 1968.

[VaFo91]   G. Vannucci and G. J. Foschini. The minimum distance for digital magnetic recording partial responses. *IEEE Trans. Inform. Theory*, 37(3):955–960, May 1991.

[VaWo98]   M. C. Valenti and B. D. Woerner. Refined channel estimation for coherent detection of turbo codes over flat-fading channels. *Electronics Letters*, 34(17):1033–1039, August 1998.

[Ve84]   S. Verdú. *Optimum Multi-user Signal Detection*. PhD thesis, U. Illinois, Urbana-Champaign, August 1984.

[Ve86]   S. Verdú. Minimum probability of error for asynchronous Gaussian multiple-access channels.   *IEEE Trans. Inform. Theory*, 32:85–96, January 1986.

[Ve87]   S. Verdú. Maximum likelihood sequence detection for intersymbol interference channels: A new upper bound on error probability. *IEEE Trans. Inform. Theory*, 33:62–68, January 1987.

[Ve98]   S. Verdú. *Multiuser Detection*. Cambridge University Press, Cambridge, UK, 1998.

[VePo84]   S. Verdú and H. V. Poor. Backward, forward, and backward-forward dynamic programming models under commutativity conditions. In *23rd IEEE Conf. Decision Contr.*, pages 1081–1086, December 1984.

[Vi67]   A. J. Viterbi. Error bounds for convolutional codes and an asymptotically optimum decoding algorithm. *IEEE Trans. Inform. Theory*, 13:259–260, April 1967.

[Vi98]   A. J. Viterbi. Justification and implementation of the MAP decoder for convolutional codes. *IEEE J. Select. Areas Commun.*, 16:260–264, February 1998.

[ViMaPiRo00]   F. Viglione, G. Masera, G. Piccinini, M. Ruo Roch, and M. Zamboni. A 50 Mbit/s iterative turbo-decoder. In *Proceedings of the Design, Automation and Test in Europe Conference and Exhibition*, pages 176–80, March 2000.

[ViOm79]   A. J. Viterbi and J. K. Omura. *Principles of Digital Communication and Coding*. McGraw-Hill, 1965.

[ViTa95]   G. M. Vitetta and D. P. Taylor. Maximum likelihood decoding of uncoded and coded PSK signal sequences transmitted over Rayleigh flat-fading channels. *IEEE Trans. Commun.*, 43(11):2750–2758, November 1995.

[Vu97]   B. Vucetic. *Wireless Communication: TDMA versus CDMA*, chapter Iterative Decoding Algorithms, pages 99–120. Kluwer Academic Publishers, 1997. Eds.: S. G. Glisic and P. A. Leppänen.

[VuYu00]   B. Vucetic and J. Yuan. *Turbo Codes: Principles and Applications*. Kluwer Academic Publishers, 2000.

[WaJuBe70]   C. C. Watterson, J. R. Juroshek, and W. D. Bensema. Experimental confirmation of an HF channel model. *IEEE Trans. Commun.*, COM-18(6):792–803, December 1970.

[WaPo99]   X. Wang and H. V. Poor. Iterative (turbo) soft interference cancellation and decoding for coded CDMA. *IEEE Trans. Commun.*, 47:1046–1061, July 1999.

[We68]   C. L. Weber. *Elements of Detection and Signal Design*. Springer-Verlag, New York, 1968.

[We00]   Y. Weiss. Correctness of local probability propagation in graphical models with loops. *Neural Comput.*, 12:1–41, 2000.

[WeEs93]   N. H. E. Weste and K. Eshraghian. *Principles of CMOS VLSI Design*. Addison-Wesley, 2nd edition, 1993.

[Wi96]   N. Wiberg. *Codes and Decoding on General Graphs*. PhD thesis, Linköping University (Sweden), 1996.

[WiLoKo95]   N. Wiberg, H.-A. Loeliger, and R. Kötter. Codes and iterative decoding on general graphs. In *Proc. IEEE Symposium on Information Theory*, page 468, 1995.

[Wo97]   P. W. Wong. Entropy constrained halftoning using multiple tree coding. *IEEE Trans. Imag. Processing*, 6:1567–1579, November 1997.

[WoJa65]   J. M. Wozencraft and I. M. Jacobs. *Principles of Communication Engineering*. Waveland Press, 1990. (reprint of 1965 original from John Wiley and Sons).

[WuWo99]   Y. Wu and B. D. Woerner. The influence of quantization and fixed point arithmetic upon the BER performance of turbo codes. In *Proc. Vehicular Tech. Conf.*, volume 2, pages 1683–1687, 1999.

[YuPa95]   X. Yu and S. Pasupathy. Innovations-based MLSE for Rayleigh fading channels. *IEEE Trans. Commun.*, 43:1534–1544, Feb./Mar./April 1995.

[ZhFiGe97]   Y. Zhang, M. P. Fitz, and S. B. Gelfand. Soft output demodulation on frequency-selective Rayleigh fading channels using AR channel models. In *Proc. Globecom Conf.*, pages 327–331, Phoenix, Arizona, November 1997.

Index